ESCAPE FROM SHADOW PHYSICS

Adam Forrest Kay studied classics and physics at the University of Colorado, and did his graduate work in England and France. He is the recipient of many scholarships and academic distinctions. He has two PhDs, one in literature from the University of Cambridge and the other in mathematics from the University of Oxford. His maths dissertation discussed the possibility of three-dimensional hydrodynamic quantum analogues. Adam's current research interests centre around realist models of quantum mechanics, relativity theory, and partial differential equations, particularly variable coefficient wave equations.

ESCAPE FROM SHADOW PHYSICS

THE QUEST TO END THE DARK AGES OF QUANTUM THEORY

ADAM FORREST KAY

SCRIBE

Melbourne | London | Minneapolis

Scribe Publications
18–20 Edward St, Brunswick, Victoria 3056, Australia
2 John St, Clerkenwell, London, WC1N 2ES, United Kingdom
3754 Pleasant Ave, Suite 100, Minneapolis, Minnesota 55409, USA

Published by Scribe 2024

Printed and bound in Australia by Griffin Press

Scribe is committed to the sustainable use of natural resources and the
use of paper products made responsibly from those resources.

Scribe acknowledges Australia's First Nations peoples as the traditional
owners and custodians of this country, and we pay our respects to their
elders, past and present.

978 1 922310 42 2 (paperback)
978 1 761385 66 7 (ebook)

Catalogue records for this book are available from the
National Library of Australia.

scribepublications.com.au
scribepublications.co.uk
scribepublications.com

Veniet tempus quo posteri nostri tam aperta nos nescisse mirentur.

The time will come in which our descendants marvel at us not knowing such apparent things.

—Seneca the Younger[1]

To my father
and
the future discoverers

Contents

Chapter 1

Weird Science

When I was an undergraduate I found a little nook in the library. It was far away from everything, in a section of the stacks that few people used. I remember it as a warm, quiet, secret place. Tall windows let in sunbeams that swam with motes of bookish dust. The desks and walls were splashed with ballpoint-pen graffiti, and one of the scrawls made me laugh. It had a little arrow pointing at nothing, with the caption *Heisenberg may have been here.*

This is a nerd joke. It refers to the Heisenberg uncertainty principle of quantum mechanics. Heisenberg was a physicist who was in the smithy when the theory was being forged, and his principle says that if you know exactly how fast something is going, you have no idea where it is. Conversely, if you know exactly where something is, you have no idea how fast it is going. This means you can't ever know exactly where something is and where it will be the next moment. In the quantum theory this applies to very tiny specks of matter like electrons, but it's funny to pretend it applies to Heisenberg too.

Back then, I was vaguely aware that "quantum," as we called it, was attended by some question marks. I met another undergrad who was studying physics and my conversation starter was: Do you think quantum physics is right? (He didn't.) This is a bad question, for a few reasons, but in my defense I had not studied the subject. My awareness of it was based on the farrago of poetic exclamations by popular science writers who describe the theory as "crazy, compared with common sense,"[1] as well as the pseudoscience and nonscience about "quantum healing"[2] and how "all the physicists have become mystics."[3] In the years since, I have come to understand just how deep the rabbit hole goes.

If you investigate the subject, you soon come across statements by very knowledgeable people like Richard Feynman: "I think I can safely say that nobody understands quantum mechanics."[4] Or Peter Holland: "Quantum mechanics is the subject where we never know what we are talking about."[5] Or Jean Bricmont: "It's a fact that ordinary quantum mechanics does not make sense. Period."[6] Or John Bell and Freeman Dyson, who speak of the "fundamental obscurity in quantum mechanics"[7] and the "fog of misunderstanding that still surrounds the interpretation of quantum mechanics."[8]

The fog is one hundred years old. New attempts to dispel the fog always run into trouble. At least one famous descent into madness was blamed on such an attempt.[9] The usual outcome is simply a new interpretation of quantum mechanics, of which there are now something like twenty. Every interpretation has some supporters, and there is division even at the highest levels of professional practice. As Steven Weinberg wrote in 2017: "It is a bad sign that those physicists today who are most comfortable with quantum mechanics do not agree with one another about what it all means."[10]

The number of interpretations is not even the strangest thing. A striking feature of almost all of them is their extravagance. To see this, we can compress some of them into oversimplified formulas. According to these interpretations, quantum mechanics implies the following:

- The moon is not there when nobody looks at it.
- An entirely new universe is created each time something happens.
- The mind creates the physical world.
- The universe has more than one past.
- The future affects the past.
- Objects are not real, but their relationships are.

This list contains the two most popular views, a few minority ones, and at least one that was supported by famous physicists and mathematicians but is now mostly favored by people who are spiritual but not religious. As we can see, they share something rather glaring in common: each is an enormous modification of anything resembling common sense.

The question of what quantum theory means was discussed intensely for about twenty years, and physicists reached a majority consensus mere months after the mathematical theory was discovered. This consensus is still the most accepted view, and is formulated above by the denial of the moon's ongoing existence. That is an exaggeration of sorts, but it came from Einstein so it can be trusted. In case you are doubtful that the formula is accurate, consider this assertion of David Mermin, not a fringe figure by any means: "We now know that the moon is demonstrably not there when nobody looks."[11]

Scholars call the consensus view the Copenhagen-Göttingen school, after the two scientific centers where the majority of physicists subscribing to the consensus lived. In Copenhagen, Niels Bohr reigned over his very own institute; in Göttingen, Max Born had a chair in theoretical physics and, eventually, seventeen doctoral students. There was much communication between these two centers, and some of the most famous names in the field worked for a time at one or the other, or both.

Most people call it the Copenhagen interpretation, and I will discuss it extensively throughout this book. Let us here oversimplify and say that its main philosophical commitment, dramatized in the denial of the moon, is called *antirealism*. Antirealism rejects the following sentence: Objects (electrons, trees, people, stars) and their properties exist separately of any observation or substantiation. Outside of quantum physics, antirealism has never been a popular view. Nearly all people, including scientists, throughout history and today, are realists, because they believe the outside, objective world has a real existence that does not depend on humans. They disagree with Copenhagen implicitly. Consequently, much of the dissatisfaction with Copenhagen over the years has been centered on its antirealism.

Einstein in particular had an unshakable faith in the reality of objects, which he sometimes expressed dramatically. He said to the philosopher Hilary Putnam, on the one occasion that they met for tea at Einstein's house,

> Look, I don't believe that when I am not in my bedroom my
> bed spreads out all over the room, and whenever I open the
> door and come in it jumps into the corner.[12]

This was in 1955, the year Einstein died, and about five years after Einstein had been walking home, discussing quantum measurement
with Abraham Pais (another physicist who later became Einstein's biographer). Einstein suddenly stopped and asked Pais if he really believed the moon only exists when he was looking at it. This is where the
moon example comes from.

Einstein was the first to really push back against the Copenhagen
interpretation, which he compared to the "system of delusions of an
exceedingly intelligent paranoiac."[13] He called the position of its main
supporters, Heisenberg and Bohr, a "tranquilizing philosophy—or religion,"[14] and said that "this epistemology-soaked orgy should come
to an end."[15] Over several decades, Einstein produced many barbed
comments about the Copenhagen interpretation. But the criticism had
little effect on his contemporaries. Einstein's influence waned, and he
withdrew, as he admitted, into "deep solitude."[16]

Nevertheless, dissent against the standard interpretation has become
a tradition. Einstein's dissatisfaction continues to this day, carried forward by a minority of scientists and philosophers. Its main impulse is
contained in the adjective *classical*. This can mean a variety of things,
but all physics that is classical has a few distinguishing features. It
deals with real-world quantities like motion, pressure, density, weight,
and the like, and because of this it makes intuitive sense to human beings. It assumes that the quantities it needs to define in order to specify
and predict a situation are, in fact, definable. For example, size is a real
thing that is measurable, and it is presumed that both the Earth and
the sun had a definite size before life emerged. The promise of classical
physics at the turn of the twentieth century was that the universe is
intelligible. In other words: everything that happens can be understood
in terms of its underlying mechanism.

In the minds of the majority of physicists involved in the 1920s and
'30s, this idea of mechanism was vigorously attacked by the quantum
theory. Abandoning the ideals of physics situated in space and time
was, for Einstein and a few others, tantamount to abandoning science

itself. Dissent against the quantum orthodoxy has two main impulses. The first is an unwillingness, some might say a shameful inability, to let go of classical concepts as a meaningful basis for physical analysis. The second is a dissatisfaction with antirealism and the other claims of the Copenhagen interpretation.

Having the first impulse means you will also have the second one, since classical physics is realist. But many have the second impulse alone. They abandon classical intuitions and keep realism. Some of the new quantum interpretations—for example, the idea that new universes are continually spawning from each other, most of them differing in imperceptible ways—arose out of this kind of approach.

The denial of classical models is a common feature of almost all quantum interpretations. This is why quantum physics was described by its founders as a "revolution,"[17] and why all of the interpretations in the list above seem so strange. Whether or not Copenhagen is the right way of viewing things, there is a deep assumption that classical physics is not up to the task of describing reality. For example, after telling his students about the double-slit experiment, something we will explore in future chapters, Feynman calls it

> a phenomenon which is impossible, *absolutely* impossible, to explain in any classical way, and which has in it the heart of quantum mechanics. In reality, it contains the *only* mystery. We cannot make the mystery go away by "explaining" how it works.[18]

It's truly remarkable for any physicist to call any aspect of nature "impossible, *absolutely* impossible" to explain classically. Because this amounts to saying we must lower our standards of what constitutes understanding. We cannot make the mystery go away, or formulate things in terms of their causal mechanisms. Even more, Feynman makes explanation itself sound like a dirty or foolish thing to attempt when he refers to "explaining" in scare quotes.[19] This attitude might surprise us, but it is common. As John Bell observed, "The founding fathers of quantum mechanics rather prided themselves on giving up the idea of explanation."[20]

The attitude persists beyond the founding fathers and Feynman, who wrote those words in the 1960s. In their book on quantum mechanics, physicists Brian Cox and Jeff Forshaw say that it's a good mental exercise to try to explain the stripes that are observed in the double-slit experiment, but not for the reason you might think:

> It's a good exercise because it's futile, and a few hours of brain racking should convince you that a stripy pattern is inconceivable.[21]

One should not spend more than a few hours on this problem before concluding it is impossible to solve. Obviously this is not because the authors think that any problem can be solved in a few hours. The problem is assumed to be insoluble from the beginning. Trying to prove an impossibility is a waste of your time. If you must waste your time, don't waste more than three hours.

The physicist Jim Al-Khalili, in his popular account of quantum mechanics, says much the same thing:

> Physicists have been forced to admit that, in the case of the double-slit trick, there is no rational way out. We can explain what we see but not why.[22]

Beneath these quotes and many more like them is a debate, almost as old as the quantum theory itself, about "hidden variables." Some people believe that quantum physics is a kind of approximation and that there is more detail hiding at a deeper level than quantum theory can describe. In other words, that quantum theory is incomplete and we will one day discover the underlying mechanism. Others deny this, saying that the quantum theory cannot be expressed in the terms of any deeper language, because the theory provides the deepest possible language, and therefore is complete. If the deepest possible language makes statements that are "inconceivable" and for which there is "no rational way out," we have to suck it up. We have to accept that the universe is deeply extravagant and contrary to our quaint common sense.

As we discuss in Chapter 32, most people through the twentieth century believed quantum theory was complete. The philosophical implications of this view are still being worked out, but they are all deeply extravagant. Einstein, as you might have guessed, was certain that quantum theory was incomplete and provisional and that a deeper theory existed. His most cited paper was on this topic, "Can Quantum-Mechanical Description of Physical Reality Be Considered Complete?"

The attitude today is that quantum physics is the correct underpinning of classical physics. As just mentioned, one of the defining features of classical theories is that they do not need interpretations because they make intuitive sense to the human mind. Because quantum theory does not make sense, we now must explain our commonsense world using a world that makes no sense. In the quantum world, explanation removes no mystery, which is why it is mere "explanation."

Is this a problem? Many would say no, and that we are asking too much for the universe to be mathematically consistent *and* comprehensible. After all, why should God's thoughts make sense to the human mind? This idea is summarized in J. B. S. Haldane's maxim: "The universe is not only queerer than we imagine, it is queerer than we *can* imagine."[23] Unsurprisingly, this quote is sometimes attributed to Heisenberg.

And yet, it is likely that this fatalistic short selling of human knowledge will never be truly satisfying. Jim Baggott, talking about the quest to explain the mysteries of quantum mechanics, says "I honestly think that it goes against the grain of human nature not to *try*."[24] Aristotle would agree, as the first sentence of the *Metaphysics* makes clear: "Man by nature desires knowledge."[25] There will always be visionaries and explorers who strike out into uncharted seas, even if nobody ever returns from that voyage.

Whether the facts go against the grain of human nature or not, we must accept them. So it is important to ask: What is the fate of Einstein's program of dissent? Is the spirit of classical physics dead? Has humanity truly passed a threshold, into a realm where mystery cannot be removed by "explanation"? Are we forced into antirealism, or a multiverse, or something equally extravagant?

As I will show in this book, the answer to these questions is "No." Extravagance is not forced on us, at least not yet, because there is good reason to believe that classical physics may have reserves of unexpected richness that can give rise to quantum phenomena. New work has emerged that seems to strike decisively at the heart of the historic question of interpretations, and, as one would require of such a thing, to also point the way forward with unprecedented clarity.

This new work is called *hydrodynamic quantum analogues.* As this is something of a mouthful, I will often abbreviate it as HQA. They involve little drops of oil bouncing on vibrating baths of oil. The droplets can even move in straight lines across the surface of the bath, in which case they are called "walkers." To be clear, the systems of HQA are not quantum mechanics, but an analogy of many quantum phenomena, generated in fluids. This in itself is highly fascinating for two reasons.

The first reason is that analogies have driven many of the greatest discoveries in the history of science. Kepler said "I especially love analogies, my most faithful masters, acquainted with all the secrets of nature....One should make great use of them."[26] They were Thomas Young's faithful masters as well, as we will see in Chapter 4. As the mathematician George Pólya said: "Analogy has a share in every discovery, but in some it has the lion's share."[27] To somebody versed in the history of science, finding an analogy to quantum physics is good news, the kind of event that should be accompanied by a sunrise and flying cherubs blowing horns.

The second reason is that fluid mechanics is *classical.* As we have seen, since its creation, quantum mechanics has been considered the nonclassical theory par excellence, a "revolution" that swept away our understandable, classical (read primitive) view of the world. Classical and quantum have been treated as opposites. But what if this is wrong? The fact that we now have classical analogies to quantum phenomena raises the astonishing possibility that particles and atoms might be, in some sense yet unimagined, causally determined and fully comprehensible in terms of mechanism.

Probably the most encouraging feature of this discovery is that *it is not new.* The basic idea is that a particle interacts with a wave and its motion is guided by the wave. Called pilot-wave theory, it was imagined by Louis de Broglie, one of the most significant figures in the history of quantum mechanics.[28] His pilot-wave interpretation and the Copenhagen interpretation were presented in the very same year.

Pilot-wave theory has a fascinating history which I will explore in Chapters 25 and 41. It is not taught in classrooms today. In 1927, it was criticized briefly and abandoned, and then rediscovered in a modified form in 1952 by David Bohm. The criticisms from twenty-five years earlier were all shown to be wrong, but rather than being welcomed, the theory was again dismissed. In the words of Oppenheimer: "If we cannot disprove Bohm, we must agree to ignore him."[29]

People did ignore Bohm, or they paid attention to him and made his life hard. But pilot waves were not so easily forgotten after that, and the interpretation has maintained a small number of supporters. John Bell, whom we will meet in Chapter 42, became famous for his treatment of these problems, and he was shocked that Bohm's work was ignored and that nobody had told him about it. Everyone said quantum theory was complete and that no deeper theory was possible, and yet Bohm had written down such a theory! As Bell said, "I saw the impossible done."[30] This was a figurative statement; what Bell actually saw were equations. But now, in bouncing droplets, pilot waves have been rediscovered *again*, and now you can literally see the impossible done with your own two eyes.

It may not be an exaggeration to say that HQA has started a revolution. Experiments are now being done in over a dozen laboratories worldwide, and a growing number of scientists and philosophers believe that this subtle hydrodynamic system is saying something important about quantum foundations. HQA has suggested classical reinterpretations of a number of traditionally bewildering quantum notions, many of which we will explore in this book. These include wave-particle duality, wave function collapse, superposition of states, statistical projection effects, single-particle diffraction and interference, nonlocality, uncertainty, and entanglement. Collections of droplets also have remarkable features, many of them resembling behaviors we see in matter. These include quantized static and dynamic bound states, crystal vibrations, and spin-spin correlations.

This work points in a direction that is probably the least extravagant possible explanation for quantum phenomena. Thus, rather than writing about weird ideas like retro-causality, multiple universes, or psychic wave-function collapse (all of which has already been done multiple times anyway), I can now write about *classical physics*, which everyone can understand. The emerging science of HQA is the opposite of weird, yet it is full of surprises. HQA systems give rise to physical and

philosophical issues of considerable subtlety. As we examine these issues, we find ourselves traveling old paths with fresh eyes, and new
paths with surprising vistas. For me, this is more exciting, and more
likely to lead to progress, than the fun of wild metaphysical speculation.

If all this quantum weirdness actually had a logical, rational explanation that we could understand, and that made sense even in the
framework of classical physics, wouldn't that be the biggest surprise of
all? This is what Einstein thought would happen. It would be incredible
poetic justice if, long after the battle was judged in favor of the antire-
alists who gossiped about Einstein going senile, it turned out Einstein
was right all along.

Only time will tell, and it is reasonable to be skeptical now. The
bouncing droplet system could be a disanalogy, and the work built upon
it a bizarre school of red herrings. But, for many reasons that we will
explore, this is unlikely. The droplets are almost toys, an experimental
system that classrooms can set up for fifty dollars to inspire children.
But it seems clear that they are providing new and striking commentary on deep questions.

Though deep, these questions can be understood in essence by the
layperson. In my opinion, the best way to grasp them is to take a historical view. It is always worth going back to the beginnings of things.
Beginnings are surprising and extremely rich, and in the beginning,
problems often present themselves in their simplest form. In this book
I try to grasp the simple essence of things, and because of this I will
usually consider problems as they first arose. Sometimes, it is worth
following the development of the problem as well, what Nietzsche called
the genealogical method. When we are able to trace the whole evolution
of a question, we can say we understand it fully.

And so, to really understand the significance of these new experimental systems, I will try to impart a more expansive sense of particle-wave
duality than is usually done. This duality, also called a "dilemma," is
almost synonymous with quantum physics, and it is much older than
most people think. To see the first flash of the problem, we must go
back more than two millennia, to the city of Alexandria and the very
beginning of science.

Part I

Parable of Waves and Particles

The central philosophical puzzle posed by the quantum theory is what attitude should one adopt to the apparently mysterious dual nature exhibited in the microphysics of matter and radiation.

—M. L. G. Redhead[1]

Chapter 2

A Visual Fire

The personal life of Euclid (he flourished circa 300 BC) is lost to history. We know that he lived in Alexandria under the rule of Ptolemy I.[1] Euclid was the author of the *Elements*, the standard textbook in mathematics and geometry for over two millennia, and the longest-read book in the history of science. Anybody who has even skimmed one of the thirteen books of the *Elements* will not soon forget the experience. As Byrne said, this work is "by common consent, the basis of mathematical science all over the civilized globe."[2] The name Euclid evokes clarity, the precision of logic, and conclusive demonstration that transcends time. He showed the world how mathematics is done.

Euclid also wrote a much less famous essay, called the *Optics*, in which he studies visual perception. This is the first known text on mathematical optics. The path that led ultimately to quantum physics starts here.

In the *Optics*, Euclid followed the same deductive scheme as the *Elements*, stating his assumed truths at the beginning and exploring the consequences. These assumptions are called axioms, from the Greek word for "worthy," and so this type of inquiry is said to use the axiomatic method. In such a system, if the assumptions cleanly explain a lot of things, then you are probably onto something. Euclid was. With seven simple assumptions he was able to explain why things far away seem smaller, why the parallel lines of a hallway or road seem to

13

converge as they recede, how to measure depths and lengths, why the wheels of chariots sometimes appear distorted, and much else besides.

As is often the case with first attempts at science, Euclid was not studying exactly the right thing. He was actually studying vision, how we see things and why they look the way they do. In so doing he uncovered the behavior of light, but this was not his primary goal. At that time, the theory of vision was mixed up with the theory of light. Euclid followed Plato, who believed that sight was more like touch: the eyes themselves shed an invisible, extremely fast substance. Plato called this a "visual fire" in the *Timaeus*, believing that visual rays extended from the eye and, touching the external light from the sun or a lamp, melded with it.[3] This resulted in a visual perception, just as the vibration going up a cane can be felt in the hand.[4]

This theory of vision is today called *extramission*, Latin for "sending out." The opposite idea, that something comes into the eyes, is the correct theory. Called intromission, it also had its supporters in the ancient world. Euclid believed in the extramission theory of vision. What had been to Plato an immaterial and visual fire became, in Euclid's hands, sixty years later and six hundred miles away, a mathematical theory with assumptions, deductions, and predictions. For our purposes, the most interesting part of Euclid's theory is his seventh axiom: *Let it be assumed that things within several angles appear to be more clear.*[5]

What he means by several angles is several lines. Thus, the more lines you can draw from the eye to a thing, the more clearly that thing will be perceived. Therefore the eye emits a finite number of visual rays. Otherwise, the same number of lines (infinity) would hit every object, and some other explanation for the varying clarity of vision would be needed. Euclid's very first statement, after the axioms, is that *Nothing that is seen is seen at once in its entirety.* This is because the visual rays diverge from each other as they travel out, and so "they could not fall in a continuous line" upon any line in the distance. Again, this means there is a discrete set of rays, some finite number of them. To Euclid, the reason we seem to see things all at once is that "the rays of vision shift rapidly."[6]

The discreteness of the rays also explains why close things appear clearer than distant ones: more visual rays can fall on close things. And it also explains why small things, at a certain distance, can no longer be seen: two visual rays that are right next to each other diverge as they travel outward, and the small thing lies between them.[7] In a discussion of Euclid's visual rays, Arthur Zajonc introduced a striking example: Imagine you are searching for a needle you dropped. All of a sudden, you spot it! You were looking straight at it before but did not see it because the rays shooting from your eyes were not hitting the needle.[8] None of these, or any other explanations that Euclid gives about disappearance and clarity, would make any sense unless the visual rays were discrete. Maybe there are thousands or even millions of them, but the number is finite and there are spaces between the rays.

The visual rays that Euclid imagined were drawn into the eighteenth century. This engraving is from *Oculus artificialis teledioptricus sive telescopium* by Johann Zahn, 1702, p. 210.

Some four hundred years after the death of Euclid, in the same city of Alexandria, the next decisive step in the science of optics was taken by a man named Claudius Ptolemy. He was named after the pharaoh (it was certainly a very popular name), but today, Ptolemy the thinker is more famous than Ptolemy the Savior, and more famous than the fourteen other Ptolemies in the Alexandrian dynasty.[9] We know Claudius Ptolemy as the champion of the geocentric worldview placing the Earth at the center of the universe, orbited by the moon, sun, planets, and stars. These ideas were canonized in his *Almagest*, a book that, along with the

Elements, had the furthest reach of any text in ancient science.[10] It was accepted as truth for more than twelve hundred years.

Ptolemy also worked on the problem of visual perception. We are lucky to know what he said, because his book *Optics* was almost destroyed by time. It survived in only a single Latin text, created around 1160 AD.[11] One of the most striking features of this text is "the formulation of theory based on experimental results, frequently supported by the construction of special apparatus."[12] In other words, proper science.

Like Euclid, Ptolemy assumed the extramission theory, but he went much further than Euclid in providing an account of how light reflected and bent when passing through different media. Most relevant for our purposes, however, is that Ptolemy differed from Euclid on a critical point. He agreed that the eye emitted what he called a "visual flux," but he denied that it was formed by discrete rays:

> On the contrary, it must be understood that, as far as visual sensation is concerned, the nature of visual radiation is perforce continuous rather than discrete. . . . For, according to that claim [of discreteness], every large object ought to appear fragmented [like a mosaic], rather than continuous, and small objects lying the same distance from the eye ought to appear and disappear by turns as they are moved to the sides.[13]

For Ptolemy, then, the eyes emitted a continuous substance that reached outward in a cone, and the rays of Euclid were little more than convenient mathematical fictions, helpful for tracing how the visual fire moved.

Ancient Alexandria was the cradle of science, and we have gone back to the beginning and retraced the very first dusty steps of the theory of light for one simple reason: this disagreement between Euclid and Ptolemy is much more serious than it appears.

One of the great advantages of Euclid's theory is that the direction of the rays is unimportant. His many diagrams present the rays as completed lines, and their speed was considered to be so great (possibly infinite, otherwise how do you see the stars the moment you open your eyes?) that no effects of their travel were measurable. The rays shoot out, but they might as well shoot in, because the geometry is the same. This is why Euclid and Ptolemy were able to make so much progress, while simultaneously founding their theory on the wrong idea. The mistake does not affect anything much. We may reverse the direction of the rays and the theory is unchanged.

Let us therefore modernize Euclid and Ptolemy, imagining that they were intromissionists all along. This allows us to sidestep the confusion of false ideas. Now that the eyes are not sending out a visual fire, what kind of visual fire is coming into them? The answer, obviously, is light. When they talked about the visual fire propagating in straight lines, reflecting at equal angles and refracting through water and glass, Euclid and Ptolemy were actually studying the behavior of light, without even knowing it. However, the dispute still exists.

Euclid thought the visual rays were discrete, like infinitely long, rapidly shifting needles shooting out of the eyes. So for a modernized Euclid, light is a bunch of discrete rays converging on the eye. He might say that the things that we see emit an endless shower of fast sparks that bounce off things in all directions and converge upon us. The rays are the paths of the sparks; they grow longer as the spark advances. On one end of a ray is the spark's origin, and on the other, the spark's current location.

Ptolemy thought there were no rays, and that a continuous substance, a flux, shot out and filled the visual cone completely. Our modernized Ptolemy does not believe in the sparks. He would say that the objects we look at emanate a continuous flame, much like a vibrating rock in a pool sends ripples out in all directions. Because this continuous visual fire fills all of space, Ptolemy might liken it to flame without gaps. This flame, somehow emanated by all visible objects, converges on us and enters our eye.

Is light discrete or continuous? Is it a stream of tiny sparks, or an unbroken flame? This is the particle-wave debate in its most basic form.

Chapter 3

Baroque Animadversors

Entering the Great Gate of Trinity College, Cambridge, walking down the path of stones between the sacrosanct lawns, taking the first right to pass beneath the clock tower and through the archway of a Victorian porch, you come into the college antechapel. Here can be seen six imposing statues. The one with pride of place has an inscription: *Qui genus humanum ingenio superavit*: Who surpassed humankind in genius.[1] Standing above this (arguably correct) hyperbole is a likeness of Isaac Newton. He is leaning slightly backward, with short hair and a distant look. In his delicate hands he holds something small, trifling. It has a triangular cross-section. What is it? A little case? On closer examination, one sees it is a child's toy. Newton is holding a prism.

This prism was the beginning of Newton's fame, and it reminds us that "the man who has been to all subsequent generations the archetype of preeminent scientific creativity"[2] came to the world's attention by fine observation and reasoning, after curiosity led him to convert a plaything into a sophisticated optical instrument.

Newton's very first paper, entitled "A Theory Concerning Light and Colours," was published in early 1672 in the newly established *Philosophical Transactions of the Royal Society*. It tells of an experiment he did with a prism six years earlier.[3] He darkened a room and, with a single quarter-inch hole in his "window-shuts," placed a prism in the path of the light ray. He was surprised that, on his wall twenty-two feet away, he did not see the circular image of the hole, with a simple glow of light, as predicted by the current theory of refraction. Rather, he saw a long spread of different colors, with semi-spherical caps on the end. These caps were the projection of the circular hole. Bizarrely, the prism had pulled and stretched the circle into a long band, with the very colors of the rainbow in the middle.

Newton realized that the unexpected length of the image on his wall, and the fact that it was separated into a rainbow-like spectrum of colors, occurred because white light consists of a mixture of colors, and that each one refracts at a specific angle.[4] And a truly profound implication followed immediately: color is an inherent quality of light itself. As Newton wrote (and this one sentence became the summary of the entire theory), "Colours are not *Qualifications of Light*, derived from Refractions, or Reflections of natural Bodies (as 'tis generally believed), but *Original* and *connate properties*, which in divers Rays are divers."[5]

In this paper Newton also reveals, only slightly, what he thinks light is. Early on in his investigations, Newton suspected that the diverse refraction of colors was caused by light moving in curved lines, and "I remembered that I had often seen a Tennis ball, struck with an oblique Racket, describe such a curve line." In the same paragraph, he says, "If the Rays of light should possibly be globular bodies." He discards both of these ideas early on, as though they are incidental. Indeed, little can be made of these two brief mentions, and Newton's contemporaries did not try. But with the benefit of scholarship we know that these discarded ideas were in fact an integral part of Newton's thinking, and that the mention of little globes is revealing of his process. Newton used that image as a guide to help him conceptualize light. His notebook from 1661 to 1665 is full of mentions of light rays composed of globules, which he thought of mechanically, "traveling with finite velocities and interacting in accordance with the known laws of impact."[6] He would, for example, speculate about balls of light of different sizes hitting different bodies and losing different degrees of their speed, which he computed.[7] It is clear that already, at the age of twenty-nine, Newton was inclined more toward Euclid's theory of discreteness, imagining the sparks of light as "multitudes of unimaginable small and swift corpuscles of various sizes, springing from shining bodies."[8]

At the time, Latin was the language of academia, and inevitably some Latin terms snuck into English discourse. Among these were several derived from *animadvertare*, which breaks down into "adversarial mind." An *animadversor* was a critic or objector to a scientific claim, and the criticisms themselves were *animadversions*. These Baroque chaps also used *animadverting*, a jewel of a verb that everyone should deploy at the nearest cocktail party.

Within only a couple of days of Newton's paper being read "to so much applause," an animadversor emerged. This was Robert Hooke. Almost ten years Newton's senior, Hooke was professor of geometry at Gresham College London, the Royal Society's curator of experiments, a skilled architectural surveyor, artist, and mapmaker, and that's just a fingernail clipping of what he got up to. Hooke studied springs in the attempt to make a decent pocket watch, leading to a force law named after him, and his invention of the spring balance.[9] Hooke discovered the cell (and coined the term), was the first person to image a microorganism, and from his observations of fossils he advocated for biological evolution almost two hundred years before *The Origin of Species*. He had a theory about the Earth, which, like many of his other theories, was "startlingly on target."[10] Hooke had a lot of good ideas.

One of Hooke's good ideas was that light is a wave. He had already expressed this seven years before Newton's publication, and among his animadversions, Hooke brought up his wave theory no fewer than three times, saying, for example, that "all the experiments and observations" he had made in his life, and even Newton's new experiments, seem to him to *prove* that light is a wave motion, "propagated through a homogenous, uniform, and transparent medium." This medium was called the ether, and was relied upon by generations of scientists to account for many mysterious physical effects. Ether itself was not in dispute—Newton believed in it as much as Hooke—but Newton thought particles of light traveled through the ether like tennis balls through air, or even like stones skipping on a pond, and Hooke thought that the ether supported the light the way air supports sound or a pond supports ripples.

Hooke's clearest statement of his view on light took the form of a statement of fact, not opinion:

> The motion of light in an uniform medium, in which it is
> generated, is propagated by simple and uniform pulses or
> waves, which are at right angles with the line of direction.[11]

This is an extremely prescient claim, about 150 years ahead of its time. Along with this gem, Hooke raised some specific problems with Newton's theory. If light is a body and the different colors go along with different types of bodies, how exactly is white light created? Newton says white light is a combination of all the colors. But it is hard to imagine bodies combining to change their qualities, precisely because bodies are solid. As Hooke puts it, "All the coloured bodies in the world compounded together should not make a white body."[12]

Newton was most displeased by Hooke's various criticisms, and his reply was "caustic."[13] To him, Hooke's idea "it self seems impossible," for how can light be a wave if it travels in straight lines? One expects waves to travel outward in all directions. The waves of any fluid travel with "a continual and very extravagant spreading and bending every way into the quiescent Medium, where they are terminated by it."[14] If light was a wave, then the hole Newton made in his window-shuts should not generate a shaft of light as it actually did, but rather just a low-level diffuse glow through the entire room. Newton made this same point a few years later, saying that if light were a vibration, it "ought always to verge copiously in crooked lines…and to comply readily with any crooked pores or passages, as sounds do."[15]

This argument became classical, and is probably the strongest one against Hooke's ideas. About 150 years later, it was repeated and simplified by the French philosopher d'Ortous de Mairan, into what has been called the "night argument": if light was a wave, then shadows ought always to be destroyed, and there would never be night because the sunlight would just curve around the dark side of any planet, the way sounds can be heard when huge boulders or even houses are in the way. If light is a body, however, it should travel in straight lines just like any other piece of matter, creating shadows and darkness. Furthermore, Newton's point about the termination of waves is also troubling. Sound waves in water travel farther than waves in air, but they both dissipate before long.[16] Waves in the earth propagate even farther still, but even they die out eventually, thank goodness. But the light from the stars reaches us from extreme distances. What kind of wave can travel that far?

Hooke reproduced Newton's experiment and confirmed the result, but denied the conclusions, saying that "these experiments were not cogent to prove, that light consists of different substances or divers powders, as it were."[17] Newton, likewise, would continue to pursue the idea of the corporeity of light, despite the animadversions of Hooke. Neither man could disprove the other, and so both battled on.

The Royal Society would debate these issues for years to come. Newton's ideas, inspired partly by tennis, would come to be called the *emission theory*, after the idea that luminous bodies like candles were emitting countless tiny pieces of matter in every direction. Hooke's ideas, inspired partly by water, became known as the *undulation theory*, from the Latin *undus*, or wave. The particle-wave dilemma, implicit and latent for fifteen hundred years, had now come to life in a battle between two Englishmen.

We know how this battle played out: the spirit of Euclid triumphed. A poetic expression of the outcome can be found in Westminster Abbey. This place, central to British cultural identity, is where William the Conqueror was crowned in 1066, and where the world said goodbye to Princess Diana. On the floor is a stone commemorating the death of *Robert Hooke 1703*. It was installed in 2005, a mere 302 years after his death.

Newton also has a memorial in Westminster Abbey, but unlike Hooke, Newton's bones rest in this glorious building. His tomb is carved with cherubs, comets, planets, pyramids, and weighty books. The sarcophagus and sculpture are enclosed in carved and painted wood. The sheer grandeur, dare one say ostentation, of this monument expresses the esteem in which the British held Newton during and after his lifetime. Even Alexander Pope was reduced to gushing in his epitaph for Newton, which was not allowed on the tomb despite being of matching aesthetic sentiment:

Nature and Nature's laws lay hid in night;
God said Let Newton be! And all was light.[18]

Newton believed that light was a stream of "multitudes of unimaginable small and swift corpuscles," and Newton was sent by God, as the epitaph says. The epitaph rhymes, and things that rhyme sound true.[19] Newton had, as Clairaut said, *gloire*. Glory. He was so right about everything, so universally acclaimed, so why would anyone doubt emission theory? Plus, look at his tomb in Westminster Abbey. He was buried there a week after he died in 1727. Hooke wasn't smart enough to be buried in the abbey; he only got a crummy little floor tile, hundreds of years later.

What I am describing, of course, are animal instincts and the irresistible deference to status that seems burned into human nerves. It has nothing to do with real science, as anyone can see. But real science is done by real animals, it is an animal institution, and as such it is subject to many unscientific forces. The spirit of Euclid triumphed not because of any arguments that Newton made, per se, but because it got a free ride into history on Newton's long coattails. As Kuhn put it, "The next great step in optics, the development of an adequate wave theory, was retarded by the grip of Newton's corpuscular hypotheses upon the scientific mind."[20]

While most scientific work on light for the next century assumed Newton's emission theory, there were some exceptions. The great mathematician Leonhard Euler attempted to construct a wave theory of light in his *New Theory of Light and Colors*, in 1746, but this was not successful in the end. Euler did get the nature of colors right, attributing them to different lengths of the waves, and in this he was following a speculation that Newton made in toying with Hooke's ideas. Euler's work also influenced others, including the man who was to raise the first real challenge to Newton in over a century. This was another Englishman, a gifted prodigy and polymath named Thomas Young.

Chapter 4

The Doctor and the Engineer

It is a commonplace when talking about Young to mention that he could read by the age of two, had read the Bible twice a few years after that, and upon growing up was fluent in over a dozen languages. His linguistic acumen was instrumental in using the Rosetta stone to decipher hieroglyphics. He was a man of "unrivalled acquirements," who if asked a most difficult scientific question would answer "in a quick, flippant, decisive way, as if he was speaking of the most easy."[1] And yet, by trade, Young was not a scientist but a doctor. He lived in London. In his spare time he followed a wide-ranging and voracious curiosity. He has been called "a jack of all trades and a master of them all,"[2] "the last man who knew everything," and "the man who knew too much."[3] In other words, Young would have been an awesome guest at a dinner party.

Just for fun, here are the titles of some of his papers:

"An Essay on Music"
"Hydraulic Investigations"
"On the Functions of the Heart and Arteries"
"A Numerical Table of Elective Attractions"
"Remarks on the Astronomical Measurements of the Ancients"

"An Algebraical Expression for the Value of Lives"
"On the Habits of Spiders"

It so happened that, working as a doctor, Young got interested in how the throat produces sound. From this it was a short step to becoming interested in sound itself. Nobody doubted that sound was carried through the air in all directions, just as ripples traveled outward in a pond. So Young began thinking about waves. On January 16, 1800, over seven decades after Newton's death, a letter on these investigations was read before the Royal Society. Young discussed the speed of sound, how one can see sound using smoke, how sound decayed, the harmonics of pipes, and other things as well. The tenth section was entitled "Of the Analogy Between Light and Sound," and it begins like this:

> Ever since the publication of Sir Isaac Newton's incomparable writings, his doctrines of the emanation of particles of light...have been almost universally admitted in this country, and but little opposed in others.[4]

Young then added his voice to those who, through history, supported the undulation hypothesis. He also noted a fresh objection to the emission theory: if light corpuscles are so small, capable of being generated by even the slight friction of two pebbles, a burning candle, the white heat of a wind furnace, the putrefaction of fish, turning on your iPhone, biting into a LifeSavers candy, and so on, then why do these tiny bodies always move with the same speed? Presumably they should move like other physical bodies, with a wide variety of speeds depending on the motion that sent them flying. On the other hand, undulations in a wave medium, like sound, always move with the same speed, because that speed depends on properties of the medium, not on the state of motion of the emitter. Whether you are standing on a cliff or falling toward the base of it, your yodels will always travel at about 767 miles per hour.

This slim section in a letter otherwise devoted to sound waves had little effect on anybody. However, two years later the same author was invited to give the Bakerian Lecture before the Royal Society. These lectures, first begun in 1776, are an institution that continue to this

day. Such an invitation revealed the esteem afforded the good doctor.
He did not disappoint.

Young chose to elaborate and magnify upon his previous speculations, and made a conceptual exploration of the undulation theory. In
doing this, he took a stand directly against Newton, and was soon
violently attacked for it in the pages of a young yet influential cultural
journal, the *Edinburgh Review.* The author of this attack, Young deduced later, was a cofounder of the journal, an amateur scientist and
bully who went on to become a lawyer and then a member of parliament. His name was Henry Brougham.

The *Edinburgh Review*'s motto was *Judux damnatur cum nocens
absolitur,* which is a saying by the pithy Publilius Syrus meaning "The
judge is condemned when the guilty is absolved."[5] Clearly, the *Edinburgh Review* set itself up as a very smart gatekeeper and protector of
all true culture.

Brougham's review is not pleasant to read, unless you are enter-
tained by public executions. The tone of unbridled contempt is revealed
by a few choice sentences:

> We demand, if the world of science, which Newton once il-
> luminated, is to be as changeable in its modes, as the world
> of taste, which is directed by the nod of a silly woman, or a
> pampered fop? Has the Royal Society degraded its publica-
> tions into bulletins of new and fashionable theories for the
> ladies, who attend the Royal Institution? *Proh pudor!*[6]

Pudor is "shame," the root of the word *pudenda* (that for which one
should be ashamed). So *proh pudor* is for shame!

Young ignored Brougham's poisonous attack, and instead focused on
where his ideas were leading. He realized his ideas could explain a most
curious feature of light that had been known for hundreds of years,
called *diffraction.* It was first studied by an Italian Jesuit named Francesco Maria Grimaldi in the book *A Physicomathematical Thesis on
Light, Colors, the Rainbow, and Other Related Topics* published in
1665, two years after his death.[7] The first line of this book is: "Of light
we can speak only darkly,"[8] certainly a true statement at the time.
Grimaldi understood that light did not always travel in straight lines,

the way both Euclid and Ptolemy and everyone since had thought. His book contains sixty propositions, the very first of which is

> Light is propagated or spread not only in lines, in refraction, and in reflection, but also in another specific fourth mode: in diffraction.

This mode of diffraction occurs when light encounters obstacles. Going through a small pinprick hole, for example, does not result in a tiny point of light on the other side, but a diffuse circle. The smaller the pinprick, the larger the circle becomes. The light spreads in a cone from the pinprick. Again, if you shine light directly at a knife blade, there will be alternating bands of shadow and light just beyond the edge of the blade's shadow.

Newton was made aware of Grimaldi's work shortly after his own publication on the prism, and he devised an explanation that involved light corpuscles wiggling back and forth when encountering obstacles. The particles underwent "fits of easy reflection and transmission" that altered their course and gave rise to the observed interference fringes. His ideas were accepted, unsurprisingly, and became known as the theory of fits.

It was this behavior of diffraction that Young realized the wave theory could explain. He was interested in the possibility that light waves, when you combine them properly, might cancel each other out as water waves do. Water waves do this all the time because of something called *interference*.

This notion of interference is crucial, and easy to understand with waves. The idea is that when two disturbances in water meet, they simply add up. Because a disturbance in water can be higher or lower than the flat surface of the water at rest, the result of the interference can be high + high = higher, low + low = lower, or, the important one: high + low = zero. This region of zero disturbance, when it comes to light, is a shadow. If light was a wave, the shadows of diffraction could potentially be explained by interference.

This line of thought led Young to devise the famous double-slit experiment, in which a wave is passed through a screen with two small holes in it, and then measured on a second screen on the other side. The

waves generated from each hole spread out spherically, just as Hooke
said, and then they interfere with each other. In 2002 readers of the
Science Times voted Young's double-slit experiment to be among the
most beautiful in the history of physics. The winner was actually just a
modification of Young's original idea, which we will discuss in Chapter
19. So this London doctor, working by himself, devised arguably the
most beautiful experiment in the world.

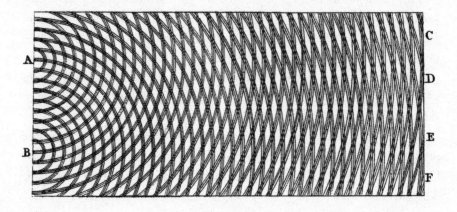

Young reported his latest results in another Bakerian Lecture, in
1803. Then came more attacks, from Brougham again, in the *Edinburgh Review*. This time, Brougham's sarcasm became abuse. He spoke
of "more fancies, more blunders...all from the fertile, yet fruitless,
brain of the same eternal Dr Young." The effect of this invective upon
public opinion was surprisingly large, and it seriously damaged Young's
reputation as a scientist.

Young composed a "masterly reply" to this new attack, and published it as a pamphlet that would have devastated Brougham if it had
sold more than a single copy. In it, Young's superiority is plainly on display. He called the articles "the repeated anathemas of the self-erected
Inquisition of the North." He observed that "they have much less the
appearance of the impartial discussion of a long disputed question in
natural philosophy, than of the buffoonery of a theatrical entertain-
ment, or of the jests of a pert advocate, endeavoring to place in a ridiculous light the evidence of his adversary."[9] He said that "to answer such
an attack in similar language would be degrading; to attempt to oppose
it by argument would be futile,"[10] because Brougham "misapprehends

and misrepresents completely the whole subject,"[11] and "his business is to censure others, and not to inform himself,"[12] and "it is unfortunate that he either has not patience enough to read, or intellect enough to understand, the very papers that he is criticising."[13]

And yet, Young had taken these attacks to heart. He ended his reply by saying, "With this my pursuit of general science will terminate: henceforwards I have resolved to confine my studies and my pen to medical subjects only."[14] If you ever needed evidence that science is a delicate human affair, look no further than this heartbreaking comment. It is as though a bird, having had its wings clipped for executing an original dive, vowed never to fly again. Naturally, birds must fly, and Young got over the sting of shame. He went on to exercise his brilliant mind for decades. But he did begin publishing anonymously more and more, and he did not pursue studies in the interference of light. As such, Young never formalized his ideas. In his hands they remained analogies, glowing signposts. The next step was left to another.

When he was a child, Augustin Fresnel was considered by adults to be dim, because his verbal skills were far behind those of his friends. His friends, however, called him "the genius." This was because Fresnel, when he was nine years old, did precise experiments to determine the best designs for toy guns and bows. These experiments were so successful that the children, upon following the genius's designs, made not toys but real weapons. The parents were obliged to call a meeting and expressly forbid their use.

Fresnel began his career at the age of sixteen. He was an engineer who built roads throughout many areas of France. Very exacting by temperament, for the most part a calm, quiet, and astute man, he could apparently explode upon being given a sloppy form by a subordinate. Because he was so good at his job, he advanced easily. But because most people were less detail-oriented than him, Fresnel did not like managing others.

Around 1810, seven years after Young's last ill-fated publication, Fresnel developed an interest in optics. He was twenty-eight years old, and came to the subject fresh, with neither the skills nor prejudice of training. For a long time he thought about light, in isolation. To him, the constant speed of light was a big clue that light was a wave.

Fresnel compiled a short essay of his thoughts, and passed the manuscript to his uncle, Jean François Léonor Mérimée, who was a professor at the École Polytechnique. The uncle submitted it to Ampère, one of the leading physicists of the day and the man after whom the amp, a measure of electric current, is named. Ampère did not respond, and it may be some consolation to modern professors who are always receiving the attempts of amateurs in their email that this is not a new phenomenon. Later, Mérimée had dinner with both Ampère and François Arago, another member of the Academy of Sciences, where he mentioned his nephew's attempts. Arago agreed to look at Fresnel's essay.

Through Newton, the spirit of Euclid reigned in Paris at the time. All of the fashionable scientific work assumed that light was a stream of tiny particles. However, Arago happened to be a close personal friend of Thomas Young's, who remained the sole champion of the wave theory. The two of them visited each other in London and Paris; they wrote letters back and forth in English and French. So it was natural for Arago to take this engineer of roads under his wing, and at least tell him about Young's work.

When Arago met briefly with Fresnel in Paris in 1815, Arago used a French saying that no researcher wants to hear: *vouz enfonçez les portes ouvertes*: you are breaking down open doors. In the fashion of a good supervisor, he suggested that Fresnel familiarize himself with the literature, and passed along a reading list. Fresnel had no interest in languages and could not read English. Moreover, he was on the wrong side of Napoleonic politics at the moment and, dismissed from his position, was on the way to the small town of Mathieu where his mother lived. He would not have access to any specialized books there.

Nevertheless, Fresnel continued with his investigations. In a darkened room of his mother's house, he focused sunlight through a drop of honey.[15] He placed small objects in the way of this light and then a lens beyond them. The lens focused the light to a delicate micrometer that he had designed himself. In this way, Fresnel was able to measure shadows with a precision that was probably greater than anyone before him.

What he saw, clearly for the first time, were images of shadows that we now call Fresnel diffraction. He placed, for example, a thin wire in the way of the light and observed fringes of alternating light and shadow, just as Grimaldi, Newton, and Young had done. But he also noticed something else. When he covered one side of the wire, so that the light was essentially just hitting the edge of a screen, the fringes were very different.

Fresnel reported his results to Arago, who helped sift what was new from what was already known. And Fresnel indeed was making new discoveries. With Arago's support, he got more time off from his post in the Corps des Ponts, and published his research in a journal that Arago edited. In the same journal, and just a month earlier, the scientist Biot had been explaining the same facts using Newton's theory of fits. Young stood alone in England, and Fresnel stood alone in France. The battle between the spirits of Euclid and Ptolemy was joined again, but it was hardly a fair fight. Biot, like Ampère, was a professor of physics in the Collège de France and a member of the French Academy of Sciences. Fresnel was a nobody, in a dark room in his mom's house.

For Fresnel, the natural way to explain diffraction is with Young's idea of waves overlapping and interfering with each other. When light can bend around both sides of the wire, the effect is very different than if it can only bend around one side of a screen. In this respect, Fresnel was essentially walking in the same field that Young had explored years earlier. The difference is what he did with Young's brilliant idea. Always fastidious, always practical, he began looking for a mathematical principle that explained the diffraction patterns.

There was one classic of optics that Fresnel could read, the *Traité de la Lumière* or *Treatise on Light*, by Christiaan Huygens. This Dutch scientist discovered the rings of Saturn, tutored Leibniz in mathematics, and took Hooke's side in the dispute with Newton. The main idea of his treatise was that light was a wave, and its shape could be computed by thinking of each point on a wave as its own little emitter of waves. Then, you could evolve the current shape forward in time by drawing a lot of circles and tracing their furthest envelope. Fresnel took this idea and added something fundamental: he said that the waves were not just any shape, but oscillating back and forth, unendingly. This is the famous Huygens-Fresnel principle, called by Born and Wolf "the basic postulate of the wave theory of light."[16]

The physical idea behind the principle is so clear that it can be expressed in an extremely simple equation. Computing the results was another matter, and mathematical tricks were required to sum up so many waves all at once. Though it makes some approximations, Fresnel's formula is probably the primary tool of optical science to this day. It predicts that light will move in straight lines, reflect from mirrors at equal angles, refract at specific angles through water and glass, and diffract upon encountering obstacles, just as water waves do. It explains everything.

The way Fresnel's ideas became known was rather dramatic. In 1817, the Academy of Sciences in Paris announced an essay competition on the subject of diffraction. By this time Fresnel was working in Paris, close to Arago and the center of French science. He took some time off from his job and wrote his ideas in their most comprehensive form. In 1818, he submitted his *Mémoire sur la Diffraction de la Lumière*: *Paper on the Diffraction of Light*. Poisson, who sat on the review committee, used the theory to deduce the existence of a bright spot in the center of the shadow of a small opaque disk and concluded that the theory was absurd. Luckily for Fresnel, Arago was the head of the committee. He did the experiment and found the prediction confirmed.

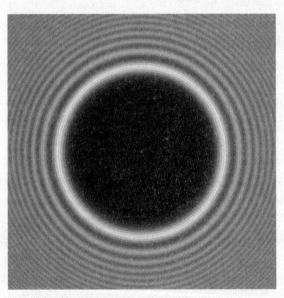

A simulation of a laser of wavelength 633 nanometers illuminating a four millimeter diameter disk. One meter behind the disk, this shadow would be observed. Credit: Thomas Reisinger.

The existence of this tiny pinprick of light was a striking victory for the wave theory, snatched directly from the particle theorists when they felt most sure of themselves. Fresnel won the grand prize, and began lengthy correspondences with career scientists, including Poisson, Laplace, and Young himself. His careful investigations drove the optics research program in both Europe and Great Britain for the next sixty years. The outsider had become, to use Arago's phrase, "one of the greatest savants of his day."[17]

The wave theory had been a joke only a few years earlier, but from this moment, the particle theory was on the defensive. Its believers now had to fight. It was not a rout, and the debate would remain lively for many years. But at this stage, Ptolemy's idea of continuity seemed more flexible, accounted for more facts, than Euclid's idea of discreteness. Sometimes. Other observations seemed better explained by particles. One continuing problem, raised even by Young when he was informed of Fresnel's ideas, was Newton's "night argument," the observation that waves bend around obstacles and so sunlight should bend around the Earth, destroying night.

Fresnel answered this objection. There is, as it turns out, a big difference between waves of sound and of light. For example, the pitch G♭5 (in the upper range of a piano) is produced by something moving back and forth about 740 times per second. The color red, on the other hand, is produced by something moving back and forth about 740 *ten trillion* times per second. Fresnel answered Newton by pointing out that waves bend around objects only when the wavelength was comparable to the size of the object. We make sound by moving large things, like strings and vocal cords, back and forth. That does not happen very fast, so sound waves have very large wavelengths, and will bend around large objects like boulders and houses. Light, on the other hand, is produced by vibrations ten trillion times faster, and has a wavelength that much smaller. On any human scale, therefore, light will appear to move in straight lines and cast sharp shadows. To observe the bending of diffraction, you must zoom in to very small scales.

This explanation, and others that Fresnel was able to make, brought an increasing number of scientists over to the wave theory. Finally, Arago devised an *experimentum crucis* to decide once and for all between particles and waves. Because Newton predicted that light particles should speed up as they entered a denser medium, and Huygens predicted they would slow down, it only remained to send light through some water or glass and measure its speed. This is a hard experiment to do but it was finally done in the 1850s, by Fizeau and Foucault. The result was unambiguous. Light slowed down. The emission theory was dead. Euclid and Newton and Biot and all the rest of them were wrong. Ptolemy, Hooke, Huygens, Euler, Young, and Fresnel were right. Light is a wave.

Fresnel did not live to see this moment of victory. He had struggled with tuberculosis most of his life, and was often haunted by thoughts of death. In 1827, at the age of thirty-nine, the man known by his childhood friends as "the genius" finally succumbed to his ill health. On his deathbed, he received the Rumford Medal from the Royal Society of London. Thomas Young had a hand to play in this, and in the letter to Fresnel informing him of the award, dated one month before Fresnel's death, Young wrote:

> I trust you will no longer have to complain of the neglect which your experiments have for a time undergone in this country. I should also claim some right to participate in the compliment which is tacitly paid to myself in common with you by this adjudication, but...I can only feel it a sort of anticipation of *posthumous* fame, which I have never particularly coveted.[18]

Chapter 5

Seemingly Monstrous Assumptions

Forty-four million years ago, during the Eocene epoch, the area near the Baltic Sea was covered with an enormous forest of pine trees. For time unimaginable they swayed in the wind, glimmered in the sun, stood dark and dripping under storms. The resin from this ancient forest collected on the forest floor and hardened with the millennia into copal, and then amber. This forest generated the largest known amber deposits in the world, and to this day amber washes up on the shores of the Baltic Sea.

Most of the amber in ancient Greece originated from this far more ancient forest. The Greek name for amber was *elektron*. They also used this word to mean a pale gold produced when gold was mixed with a small amount of silver, and also for shining things like sunbeams. The Greeks noticed something interesting about amber: after being rubbed with a cloth, little things, such as a piece of straw, would fly through the air and stick to it. Amber is electrical, probably the first electrical material known. But to the Greeks, electricity was ambery, and this is why electrons are named after it.

Magnets have a similar poetry to them. The most amusing story of their discovery, cited by Pliny the Elder, tells of a shepherd named Magnes, who was tending his flocks on a hillside in the Greek region of Magnesia. Walking innocently along, he suddenly felt the iron nails in his shoes adhere to the stones of the mountain. The story is very old, and is held by some to date from around 900 BC.

In the 1800s, amber and magnets made a startling appearance in the story of light. The details have occupied entire books, but the essence can be stated in a single paragraph.

The Londoner Michael Faraday was thirty-five when Fresnel died. Like Fresnel, Faraday was largely self-taught and yet wound up among the greatest savants of his day. His experiments into electricity and magnetism expanded minds by showing how certain objects, such as a piece of amber or a magnetic stone, threw their forces beyond their bodies, through space. The carrier of these forces came to be known as a "field." Next, the Scot James Clerk Maxwell organized and systematized Faraday's work in a series of papers that Einstein called "the most profound and fruitful [change] that has come to physics since Newton."[1] Maxwell regulated all of the observed facts of electricity and magnetism into a set of mathematical relationships. In the process, he revealed that they were two aspects of the same thing—electromagnetism. Maxwell also discovered that the electromagnetic field could support wave propagation. Most amazingly, the speed of the wave, easily calculated from the known constants of the theory, was exactly the speed of light. This was in 1865, fifteen years after light had been proved definitively to be a wave, and it remains not only the crowning achievement of nineteenth-century physics, but a landmark in the history of our species.

The wave theory had been vindicated by its elegant explanation of diffraction, by Huygens and Fresnel's mathematical model which could explain all the observed properties of light and also by Fresnel's defeat of Newton's night argument. It had been vindicated by the experiments of Fizeau and Foucault, showing that light slowed down in water and glass, according to Huygens's and not Newton's prediction. With Maxwell, however, the wave theory was vindicated in an entirely new and uniquely powerful way—by the first unification ever accomplished in physics. Nothing like this had happened before and it raised the tantalizing possibility that there was one single theory, and that the human mind was capable of formulating it. In terms of light, at least, there was no doubt any longer. Newton and Euclid were defeated, and each passing year was more dirt on their proverbial graves.

The nineteenth century brought more than theoretical break-throughs. The use of electricity had, in a very short time, transformed the cultural and economic landscape of Europe, and the wider world. The maddening speed and technocracy of our life today began here, with a great bull market in electromagnetic innovation. A short handful of decades generated light bulbs, electrical motors, the telegraph, the telephone (billed as an improved telegraph), the radio (billed as a wire-less telegraph), the electric dynamo, and much more. It was the age of Morse, of Edison and Siemens, of Tesla and Marconi. Two and a half millennia after Magnes walked the windblown hills, humanity could say for the first time that it was beginning to understand nature at a deep level.

So rapid was the rate of innovation, so complete were Maxwell's equations in their success, that the optimism of the age became hubristic. In 1894 Albert Michelson said in a speech at the University of Chicago that

> While it is never safe to affirm that the future of Physical Science has no marvels in store even more astonishing than those of the past, it seems probable that most of the grand underlying principles have been firmly established and that further advances are to be sought chiefly in the rigorous application of these principles.... An eminent physicist has remarked that the future truths of Physical Science are to be looked for in the sixth place of decimals.[2]

This comment has become rightly famous for how completely wrong it was. Michelson, though cautiously, dared to speculate that physical theory had reached its end and was now complete. It was not the first time (Michelson was, after all, quoting someone else, probably Lord Kelvin). Moreover, the sentiment seems to have been rather widely shared. In the exact same year that Michelson gave his talk, Robert Millikan was a graduate student at Columbia. He recalled of his room-mates: "I was ragged continuously by all of them for sticking to a *finished*, yes, a *dead subject*, like physics."[3]

It was in this atmosphere—an ever-expanding vista of electrical engineering victories, a satisfying sense of finality and completion in physics, the complete triumph of the wave theory—that an unknown patent clerk named Albert Einstein suggested that light was made of particles after all.

After graduating with a teaching diploma in 1900, Einstein tried for two fruitless years to secure an academic position. For a while, he sought employment offering private lessons in math and physics (the first lesson was free). Necessity finally forced his hand and, in the year 1902, he obtained a position as "patent examiner, third class," with the help of a friend's father. Shortly after, he married Mileva, the only girl of the five students in his mathematics classes. A year after that, they had a son, Hans Albert.

While he examined patents for gravel sorters and electric typewriters, Einstein worked to finish his PhD thesis. He finally accomplished this shortly after his twenty-sixth birthday. The year was 1905, and it holds a special place in the history of science. In this year, in the space of about seven months, Einstein published four papers in *Annalen der Physik*, the leading physics journal of the day. Every one of them was a significant advance. Collectively, they shook the world, and are now fundamental to modern science. It was an outpouring of scientific creativity that rivaled Newton's miraculous year in 1666.[4]

One of Einstein's papers was titled "Concerning a Heuristic Point of View Toward the Emission and Transformation of Light." It begins with the observation that "a profound formal distinction exists" between matter and light. According to the theory of the day, a particle of matter is localized to one small region of space, whereas a light wave is continually spread over an endlessly growing volume of space, precisely like Hooke's wave in water.

This is the duality of discrete and continuous. It lies at the heart of the dispute between Euclid and Ptolemy. Einstein admitted that the wave theory of light "will probably never be replaced by another theory," but nevertheless, he warned, this theory only described *time averages* of processes. Thus, it is conceivable that something else is going on at a deeper level, much faster than we can see. Only after some period of time (which is practically instant to us lumbering giants) does the deeper behavior become identical with the wave theory. Einstein's heuristic point of view was that

> when a light ray is spreading from a point, the energy is not distributed continuously over ever-increasing spaces, but consists of a finite number of energy quanta that are localized in points in space, move without dividing, and can be absorbed or generated only as a whole.[5]

Almost a century after Young, Einstein was standing up in exactly the way Young had done, but in the opposite direction. Einstein knew that, given the climate of the day, not to mention the experimental fact of interference, supporting Newton's emission theory was bordering on scientific heresy. And yet he was compelled to argue for it. The most likely outcome after publishing a paper like this (other than being attacked in the pages of a literary journal) would have been deafening silence from the scientific community. Einstein, however, could not be ignored after 1905. His other papers of that year, particularly the special theory of relativity, which was immediately hailed as a work of genius, had established his name.

For several years Einstein's reputation grew, largely based on relativity theory. He continued, throughout this time, to promote Euclid and Newton's ideas, not called the emission or particle theory, but now the "light quantum hypothesis." In a talk in Salzburg in 1909, he claimed that Newton's particle theory "seems to contain more truth," despite the fact that some of its support rests on a "seemingly monstrous assumption."[6] He admitted that diffraction made everyone believe in the wave theory of light, and he was keenly aware of its usefulness. His concluding sentence was: "The two structural properties... (the undulatory structure and the quantum structure) displayed by radiation... should not be considered as mutually incompatible."[7]

Max Planck, the sober Prussian and professor of physics at the University of Berlin, was in the audience at Salzburg. He had, by that time, realized that his own work of a decade earlier had something to say about discreteness. But Einstein's proposal was a step too far for him. In the discussion following the talk, Planck expressed a very reasonable opinion.

> I, too, emphasize the necessity of introducing certain quanta....
> The question is now where to look for these quanta. Accord-
> ing to the latest considerations of Mr. Einstein, it would be
> necessary to...give up Maxwell's equations. This seems to
> me a step which in my opinion is not yet necessary.[8]

This opinion was nearly universal. The early supporters of relativity theory, for example Planck, Laue, Wien, Sommerfeld, and Lorentz, all rejected Einstein's light quantum hypothesis. Maxwell's equations worked (they still work, unchanged, to this day). But it was the interference of light, predicted by Young and so beautifully captured in mathematics by Fresnel, that was the coup de grâce that made almost all physicists confident in rejecting this bizarre revival of Newton and Euclid. There were two notable exceptions who did support Einstein: Stark in Germany, and Ehrenfest in Russia. And so it was that in a few short years, the particle theory of light went from being totally dead to actively debated.

Impossibly, the battle had been joined again.

Chapter 6

Lifting a Corner of the Great Veil

And now we come to Louis Victor Pierre Raymond de Broglie, the physicist who plays a most important role in this book. He was born in August 1892 in Dieppe, France. His family were aristocrats, and for hundreds of years influenced the political and military life of France. Louis's father was a duke (the title eventually fell to Louis himself in 1960); his mother was the granddaughter of one of Napoleon's marshals; his older sister by seven years, Pauline, was a countess and writer of belles lettres; and his older brother by seventeen years, Maurice, was a naval officer.

Actually, Maurice was more than an officer. Shortly after 1906, the year his father died and the title duc de Broglie fell to him, Maurice took a leave of absence and, in his Paris town house on the rue Chateaubriand, set up a little laboratory. Work on experimental physics was more pleasing to him, and he soon withdrew from the navy entirely. His family was dismayed. "Tinkering with strange pieces of apparatus, even if at the beginning it was practiced only as a hobby in his own house with his own mechanic, instead of becoming a general, a statesman, or at least an admiral, did not seem right for a Duc de Broglie."[1] Maurice's grandfather put the same point rather more pointedly, saying that science was "an old lady content with the attractions of old men."[2]

41

Yet this tinkering was the seed of the greatest fame to ever accrue to the famous house, for the younger brother Louis was strongly influenced by his older brother, and it is doubtful whether, without this laboratory, Maurice and Louis would have become physicists at all.

Louis was the youngest of four living siblings, the baby of the family. He was described by his sister in her memoirs as

> a charming child, slender, svelte, with a small laughing face, eyes shining with mischief, curled like a poodle. Admitted to the great table, he wore in the evenings a costume of blue velvet, with breeches, black stockings and shoes with buckles, which made him look like a little Prince from a fairy tale. His gaiety filled the house....He had a prodigious memory and knew by heart entire scenes from the classical theatre that he recited with inexhaustible verve.[3]

When the young aristocrat neared eighteen and went to university, he took a degree in history. And yet, immediately after completing his studies, Louis abandoned them. Louis spent a year studying law, abandoned that too, and finally took up a mathematics course that would prepare a path into the sciences.

Around this time, in common with many reflective and imaginative people of his age, Louis had a crisis. The voluble child was gone, and the young man stood on the verge of a transformation.

His reflections became continually deeper, and an "intellectual impulse, firmly rooted" carried him toward the great problems of theoretical physics. He passed his examinations for a license of science, doing "brilliantly." His enthusiasm returned, and with it, a sense of certainty. He had navigated the straits of adolescence and found his calling.

Then came the great war. Louis had begun his mandatory service only a year earlier, and so despite his efforts to avoid the military, he was dragged in not as an officer, but a simple private. Assigned to the engineering corps, Louis was sent to a fort outside Paris where he was "bored to death."[4] A friend of his brother's recommended a transfer, and Louis began working as an electrician and radio specialist. He was stationed underground, beneath the Eiffel Tower. At the top a wireless transmitter was installed.

He spent six of his most productive years, the majority of his twenties, in a uniform. Some have called this a "waste," and Louis bitterly resented the disruption of his deep reflections.[5] The war finally ended, and he was decommissioned in August 1919. A mere three months later, the name of Albert Einstein became world famous as his prediction for the bending of starlight was confirmed by the Eddington experiments. This amazing result required expeditions to the West African island of Príncipe and the Brazilian town of Sobral to gaze at the sun during an eclipse. Newspapers across the world reported on the event; the *New York Times* said that the world of science was "more or less agog."

Louis wanted to pursue physics, but had been dragged into a war that shattered his inspiration. Now he was free, and some of the most interesting and profound things were happening in his chosen field. There was no time to delay.

Now twenty-seven years old, Louis began spending time in his brother's private laboratory, following Maurice's work closely and interacting with his young collaborators. In this setting, Louis began doing serious research for the first time. He presented his work, no doubt guided by Maurice, at the weekly sessions of the Paris Academy of Sciences. This was the same institution that had granted Fresnel the grand prize for his essay on the wave theory of light a century earlier, and it provided Louis with his first publications in its proceedings.

The two brothers had long talks about what Louis called "beautiful experiments." The focus of the lab was on the interaction of light and matter. They studied the way atoms emitted different colors of light, how high-energy light (called X-rays) was absorbed by matter, and how electrons were knocked off of charged metal when light shone on it. This last phenomenon, called the photoelectric effect, was the main reason Einstein argued for light quanta in 1905.

Working as they did on the interaction of light and matter, both de Broglies were keenly aware of the particle-wave problem. Louis observed the photoelectric effect closely in the laboratory, and rather than concoct excuses the way most others had done, he believed the physical picture that Einstein had imagined. And yet, light was clearly undulatory as well, as everyone knew from Fresnel's explanation of diffraction and the success of Maxwell's equations. In going back and forth between these two viewpoints, eventually Louis began thinking of light as some kind of combination of them both.

This in itself was a remarkable position, and as we have seen, it was first taken by Einstein. Rather than thinking of Euclid and Ptolemy, Newton and Hooke, Brougham and Young, Biot and Fresnel, as the conscripts of two armies battling to the death, Louis began to wonder how both sides could be revealing parts of a deeper truth. The very fact that the history of optics had for so long hesitated between these two possibilities led de Broglie to conclude that "these two representations are thus without doubt less in opposition than we have supposed."[6] And this line of thought led him to the threshold of a profound discovery.

If light behaved as both a wave and a particle, what about straightforward particles like electrons? Might they have wavelike properties?

In 1923, Louis wrote a note on this problem called *Ondes et quanta*: *Waves and quanta*. It was presented to the Academy of Sciences in September 1923 by Jean Perrin, a professor at the Sorbonne.[7] This note and several more that followed in the space of weeks were "the result of a very long reflection"[8] on two different streams of thought, both originated by Einstein in 1905: the energy relation of special relativity, $E = mc^2$, and the essential formula of Einstein's light quantum postulate, $E = hf$. Combining these resulted in the beautiful equation

$$hf = mc^2.$$

Because f is the frequency of a wave, and m is the mass of a particle, and both of them are proportional to the energy, then the mass of a particle must be associated with the frequency of a wave. Here in these few symbols we have Euclid and Ptolemy, not arguing, but dancing.

De Broglie's proposal has been hailed universally as one of the most profound breakthroughs in quantum theory. Popper called it "without doubt one of the boldest, deepest, and most far-reaching ideas in the whole development."[9] Abragam said, "It is hard to overestimate the extraordinary daring and the far-reaching consequences of this simple hypothesis."[10] Schrödinger spoke of de Broglie's "notions so interesting and so precious."[11] However, as will not surprise anyone who has read this far, when de Broglie first advanced the matter-wave hypothesis, few understood or believed a word of it.

The next year, he converted the notes into a PhD thesis. His examination committee consisted of three very distinguished Sorbonne scientists, including Perrin, who had presented the first note to the academy. It was probable that "none of them had the mental equipment to appreciate the brilliance and daring of de Broglie's work."[12] They called in Paul Langevin and leaned on his expertise in quanta and relativity to evaluate the thesis. But even Langevin referred to Louis as "the little brother," and considered the thesis to be "far fetched."[13] Langevin himself was not sure what to make of the thing and so he had de Broglie send a copy to Berlin, addressed to one Albert Einstein.

Einstein immediately recognized what de Broglie had achieved. He wrote back to Langevin, saying that de Broglie's work was sound and that "he has lifted one corner of the great veil."[14] Einstein's authority carried the day and de Broglie obtained his PhD. But it was a close call. The examiners did not believe it was correct, though they did respect Louis. Afterward, when Maurice de Broglie asked Perrin for his opinion on the thesis, Perrin said, "All I can tell you is that your brother is very intelligent." The recollection of Van de Graaff, then a student who happened to be in the audience during the public examination, surely says it all: "Never has so much gone over the heads of so many."[15]

This old issue rests at the heart of the quantum mystery. De Broglie did not show why particles and waves are so closely involved with each other, or even how they could be reconciled. He showed that the problem was much worse than anyone ever suspected. The problem is not just with light, it is with matter too. The wave-particle duality is everywhere.

Part II

Interlude

Chapter 7

Nature Is Talking

Historically, the particle-wave duality was considered to lie at the heart of the quantum difficulties. Redhead wrote in 1977: "The central philosophical puzzle posed by the quantum theory is what attitude should one adopt to the apparently mysterious dual nature exhibited in the microphysics of matter and radiation."[1] Einstein in 1949 called this "the riddle of the double nature of all corpuscles (corpuscular and undulatory character)."[2] Louis de Broglie wrote that "the great drama of contemporary microphysics has been, as you know, the discovery of the duality of waves and particles."[3]

We now understand the historical debate about particles and waves in respectable detail. My ulterior motive for going into such depth was to really grasp the significance of the bouncing droplets and hydrodynamic quantum analogues. The droplets of HQA are not just another beautiful fluid mechanics experiment. They have historical significance to science, and this needs to be stressed. Now that it has been stressed, we can focus increasingly on the specifics. So let us take a detailed look at the droplets for the first time.

The analogy was first discovered in Paris, completely by chance, in the lab of Yves Couder. He had long been one of the leading lights in fluid mechanics, known for "conceiving extremely elegant and deceptively simple experiments."[4] He worked for most of his career at the École Normale Supérieure. In 2005, the year he discovered bouncing droplets with Emmanuel Fort, Couder was sixty-four years old, a smiling, soft-spoken man whose white hair still had a touch of dark in it.

In the Paris university system, there is a course called Experimental Work. It so happened that in 2004, Yves and Emmanuel were teaching it together. In this class, everybody starts from scratch, and the students have to build an experiment. It is usually from an article, or some project that the professors have. Couder and Fort had a problem on drop non-coalescence. The idea was to understand what would stop a droplet from being absorbed into its parent fluid (this absorption is what coalescence means). They would set up the experiment with a droplet on a syringe, and press the droplet into the oil bath. Or try something different, with a rotating bath, and notice that coalescence was harder to obtain. They were just exploring.

Left to right, Emmanuel Fort and Yves Couder in the lab. Credit: Julien Moukhtar.

Strangely, they got very different results if they were holding the syringe, rather than clamping it. Eventually, as Fort said later, they saw that "our hands were in fact vibrating, and this was making some movement that was sufficient enough to prevent coalescence. So here we said, Oh, maybe we don't need to have the bath rotating. Maybe we just need to have shaking. We put the whole syringe on the shaker. It did work, perfectly. It was not coalescing. Then we thought at some point we should freeze the droplets and make the whole bath vibrate."[5]

This was the origin of the beautiful discovery: if you fix a bath of oil to a vibrating plate and put a droplet of oil on the bath, the droplet can bounce indefinitely. As Couder said, on a TV segment devoted to the discovery: "For several days if you wish."[6]

The droplet bounces because the air separating the oil bath needs to be pushed to a critical thinness before the droplet can coalesce with the bath. Before this can happen, the force of that air on the underside of the droplet overcomes the droplet's weight, and it leaps upward and away from the oil. The air layer never gets thin enough, and so the droplet never coalesces into the bath. It can bounce there forever.

This is cool, but not too surprising. On café tables standing under melting icicles, during misty drizzles over parking lots, or even in the shower, tiny shining droplets can be observed sliding across wet surfaces before blinking out of existence as they coalesce. Indeed, it has been known since 1978 that droplets can bounce for extended times on the surface of their parent liquid.[7]

But this system is different. Each bounce creates a wave around the droplet, just as when a raindrop hits a puddle. But quite unexpectedly, rather than shooting out and disappearing, these waves set up what are called standing waves, vibrating up and down. The standing waves last for a long time because they are fed with the energy from the vibrating plate. As a result, there is a vibrating wave field surrounding the droplet. Every time the droplet comes down, it will hit its own standing wave and interact with it. And this is tantamount to being influenced by its own past.

Because the droplet is affected by its own past, it is said that the dynamical system has *memory*. The philosophical issues this raises are surprisingly subtle, and lead immediately into issues of causality. The idea of a system with memory is not exactly new; it was introduced by Volterra in 1830, and in the literature such systems are called non-Markovian, or hereditary. The idea of system memory is new to quantum physics. When dynamics are driven by the state at the current time *and* a number of past times, very unusual things can happen. Systems with memory can be challenging to think about.

Couder said, "In any physics experiment, you only see what you are prepared to see. Of course it was very obvious that there was a memory, but it took us some time to realize... because you have to adapt to this new idea."[8] And as Fort described it,

> The particle-wave duality of the system lies in this strange
> wave memory, but if we look at memory in physics, usually
> it's a simple feedback or something like that. Of course, if
> you look outside of physics, everything is memory. In biol-
> ogy, economics, sociology, geology, history, it's all a matter
> of memory. We were even cited by some people studying the
> way the brain is functioning. So this is fascinating. Memory
> in physics in this way is new.[9]

Here we have a particle and a wave, in a single experiment that can be put together for as little as fifty dollars. Both occur together, each depends on the other, and they involve a concept that nobody really thought to apply in this way before: wave memory.

This memory can be controlled by the vibration of the bath, and this is where the surprises start to happen. When you increase the vibrational amplitude, the memory also increases, and everything suddenly changes. The droplet is no longer bouncing there, generating waves. It starts moving in a straight line. What happens is this: the droplet no longer hits directly in the center of the wave field it created, but ever so slightly to the side. This tiny discrepancy is enough to push the droplet off in one direction. This gets repeated until the droplet picks up speed in a single direction. As a result, it turns into what is called a walker.

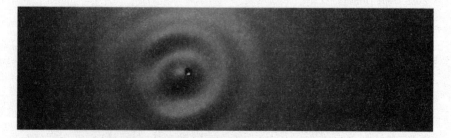

A walking droplet gets pushed around by the very standing waves it created in the past, in turn creating new standing waves centered in other places. A standing wave generated earlier can cause an obstruction *ahead* of the droplet. And yet, droplets generally move in straight lines unless they are interacting with other walkers or boundaries. This is surprising, given that they are moving on a wave field that, to them, looks like a round hillock.

Adding more than one droplet, an unbelievable complexity and richness immediately emerged. Small droplets might walk, but larger ones do not. The small ones would be drawn into the orbit of the larger ones and then repulsed by them when they got too close. Two droplets of the same size could come into stable orbits with each other. They could walk side by side, going in the same straight line, but wobbling together as though dancing. By controlling the droplet size and placing them at specific distances, Couder and Fort could produce stable lattice configurations of bouncing droplets, where each individual droplet contributed to a collective wave in the bath surface that influenced them all.

When they first discovered this system, neither professor was working on quantum physics. Couder was working on philotaxy (he was studying the arrangement of leaves on a stem), and Fort had a problem in optics. However, as they worked on the droplets, watching them walk around on the fluid bath, it did not take them long to think about quantum mechanics, and to start asking themselves, "Just *how* similar is this system to quantum mechanics?"

This began the research program of hydrodynamic quantum analogues. The first analogy to quantum mechanics, that of the particle and the wave, has led to more and more. There are now so many analogies that I cannot mention them all in this book. As we will see, however, some of them are truly dazzling.

As Fort put it: "This experiment is *very* beautiful. Nature is talking. Sometimes, people ask us: How did you discover this? In the sense of, how did you build this? No, of course we did not build it. Nature is talking. It is just that we are listening."[10]

In 2006, Yves Couder was invited to MIT to give the Simons Lectures, considered to be among the most prestigious lecture series in mathematics. It was only a handful of months after he had published the first droplet paper. His last talk, on the third day, was called "On the Dynamics and Self Organization of a Wave-Particle Association at a Macroscopic Scale."

Two of the mathematicians in the department, Peter Shor and John Bush, were sitting next to each other in the audience. Shor is known, among other things, for Shor's algorithm, a factorization result in quantum computing. Bush is a fluid dynamicist, specializing in surface tension. His work got global attention during the COVID-19 pandemic when he and his colleague Martin Bazant actually computed safety times for limiting indoor airborne transmission of pathogen-bearing aerosol droplets.

During the talk, Shor leaned over and said: "This looks like pilot-wave theory."

Bush responded quizzically: "What's that?"

In those days, Bush was examining interfacial phenomena of all sorts, including how insects like water striders walked on water. He had an aesthetic eye, and many of his photographs wound up in magazines or the Gallery of Fluid Motion.

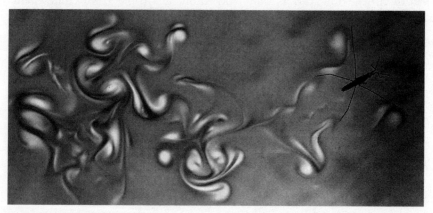

A water strider creates vortices as it moves. Credit: John W. M. Bush, MIT.

Couder's talk was the first time the droplets were presented in America, but Bush had, in fact, known about the work before it was published. He traveled frequently to Paris and always visited Couder. He had seen the first droplets bouncing in the lab at Normale Sup.

As he recalls, "Couder was a great hero of mine. He had excellent taste in problems. Always did. He solved that beautiful pine cone problem.

That's how I met him. He came to MIT and gave this talk. You look into the heart of sunflowers, and you see spirals. You count the ones going clockwise and counter-clockwise, and they're always adjacent numbers in the Fibonacci sequence. Why is that?

"After that talk, Yves came to my office. It was a memorable meeting of the minds. We were interested in precisely the same things. And he also had a strong aesthetic sense. We were friends for a long time before he initiated the bouncing droplets."[11]

Bush did not immediately see the significance of the droplets. He had intentionally left quantum physics early on because he was not learning to solve problems he had actually thought about. "My entire graduating class abandoned modern physics," he said. "They went into geophysical-environmental fluid mechanics, biophysics, or other variations of continuum mechanics. Those who did go into particle physics disappeared—God help them. Without input from experiment, it's very easy to get lost."

But Bush soon realized that the droplets were potentially commenting on deep mysteries. The turning point came when he attended a conference in Tuscany. It was held in a sixteenth-century monastery owned by Mike Towler, a physicist who spent almost twenty years at the Cavendish Laboratory. Towler himself had taught a course on pilot-wave theory at Cambridge, and had a personal interest in it.

Bush was watching someone change a quantum equation into a fluid equation, a procedure known as the Madelung transformation, when suddenly the light bulb came on. This transformation gives rise to a term called the *quantum potential*. But the mathematical form of this potential looks like a curvature term, as arises in interfacial fluid dynamics, owing to surface tension. In quantum mechanics, Bush realized, Planck's constant relates the particle energy to its frequency. But in the droplet system, the surface tension plays that same role. It prescribes the energy in the wave field.

After returning from this conference to his Boston flat, Bush returned to modern physics with new interest. As he described it, "I went into manic mode and was insistent on figuring this out. That's when I did all the Madelung stuff and figured out the connection with Klein-Gordon and de Broglie's mechanics. It was really an electrifying time in my life. I'd be up till four in the morning, and there was nobody to talk to, except for a friend in Saudi Arabia. So I'd call him. It was super fun."[12]

This is how MIT became one of the first universities to host a lab on bouncing droplets. At that time, the two other research groups were in Paris and Liège. Bush began encouraging his friends in applied math to work on aspects of the problem, in order to forge connections between quantum phenomena and their hydrodynamic counterparts. As they worked, the analogies increased in number and strength, making others sit up and take note. As Bush put it:

> A beautiful thing about HQA is that we can work on quantum foundations from the safe haven of fluid mechanics. Studying walkers is a perfectly good thing to do as a fluid mechanic and mathematician. And yet, in doing so, we are redefining the lines between the classical and quantum. So far as I know, it's the one place where you can work on foundations without getting sacked, labeled as a lunatic, or mired in philosophy.[13]

Today, the droplets are studied in over a dozen labs and two dozen universities across the world. In Europe and the US, as well as the UK, Brazil, China, and, most recently, Africa. The goal of researchers in HQA is to really understand the droplets in all their rich complexity. As Bush says, "This must be done at the mathematical level. Otherwise, one cannot expect a physicist's reaction to go beyond *How curious. It looks something like quantum mechanics!*"[14] By recruiting so many people, and providing them with such clear guidance, Bush became the driving force behind the growth of the field.

I myself am one of those people, and without Bush this book would not exist. I first heard about the droplets in 2015, from an article in *Wired* magazine. I thought and read about them, off and on. Four years later, I was doing a DPhil at Oxford (their fancy term for a PhD), while simultaneously living in Los Angeles and driving two or even three hours a day for a commute to a full-time coding job. Eventually, something clicked in my mind, and I folded in some of my ideas about the droplet system into the DPhil. John agreed to be my external examiner, and in the summer of 2019 I met him in person for the first time at my thesis defense. This day began, by long twists and turns, my arrival as a postdoc at MIT.

Unfortunately, I did not arrive in the usual sense. A global plague kept me isolated in the mountains of Colorado, doing video calls with my colleagues. I started my appointment in winter, but it wasn't until the next summer that I landed at Boston Logan and met John again for the first time in two years. I also met my colleagues in person for the first time. They are a varied bunch, equally likely to be into pure mathematics, fluid mechanics, and quantum mechanics. The conversations are always interesting, and you can always find someone to check your math by scribbling on the back of a piece of acrylic.

A typical setup. *Left*: The computers used to monitor and analyze the experiment. *Right*: The experimental rig itself, with lights, cameras, and droplet action. Credit: Adam Kay.

John Bush with graduate students Hugo Bitard (left) and Kyle Davis. They don't actually dress like this in the lab. Credit: Adam Kay.

Chapter 8

Parable of Phlogiston

What is fire? Why do some things burn and others don't?

In 1635, a boy named Johann Becher was born in Speyer, an ancient town founded by the Romans on the bank of the Rhine. At the age of nineteen he wrote his first book on philosophy, and three years later he was a professor of medicine at the University of Mainz.

The continent of Europe had just emerged from thirty years of slaughter with pikes, axes, swords, and glaives. Trials of witches had only recently begun to decline. Galileo had died fifteen years before, and it was heresy to say the Earth went around the sun.

Becher answered the question of fire in his book *Physica Subterranea, Underground Physics*. He built upon the four elements—earth, air, water, and fire—an idea going back to ancient Greece that held fire was a substance just like air. Becher thought that things combusted and burned because they had something fiery inside of them.

Becher's student was named Stahl. He developed Becher's ideas into a full system, and he used it to explain many things. Stahl named this fiery something *phlogiston*. It was absorbed by the air, but it is a substance, and so air quickly becomes saturated with it. The technical term was *phlogistinated*. This explained why a fire goes out if you starve it of new air. The old air can't absorb any more phlogiston.

The phlogiston theory was passed from teacher to student. Philosophers read Stahl and built upon his system. It gained strength. In the 1700s and 1800s, phlogiston was the explanation for many things.

Plants absorb phlogiston out of the air, so they are full of the stuff, which is why they burn so easily. What is the flame that licks the wood? It is made of phlogiston mixed with water. What is the wood itself? It is made of phlogiston mixed with ash. When you release all the phlogiston from it through burning, you are left with ash alone.

Phlogiston theory, developed continually for two hundred years, became a jewel of science.

Of course, there were problems. Phlogiston was considered lighter than air, but burning some metals made them heavier in a way that could not be explained by some quantity of escaped phlogiston. Other things didn't fit. But the supporters of the phlogiston were not dissuaded, and worked to understand phlogiston more deeply.

In Paris, in 1743, a boy named Antoine Lavoisier was born to a family of aristocrats. At the age of twenty-nine, Lavoisier became interested in combustion, and through that the study of chemistry. He began reading, doing experiments. And he proposed a set of new ideas that was to change the world.

Lavoisier asked: Why should we try to explain everything with earth, air, water, and fire? He wrote: "The fondness for reducing all the bodies in nature to three or four elements, proceeds from a prejudice which has descended to us from the Greek Philosophers."[1]

Lavoisier went against the idea of phlogiston, and the scientists of the day argued against Lavoisier. Yet, with Lavoisier's ideas, the problems began to disappear. They withered like a thirsty plant, they faded like mist in the sun. Instead of dephlogisticated air, or empyreal air, or vital air, or base of vital air, there was a single word: *oxygen*. Sharp maker. Lavoisier named it. Instead of inflammable air, or the base of inflammable air, there was *hydrogen*. Water maker. He named that too.

Because fake problems are hard to solve, the way out is to drop them. After Lavoisier, there was more simplicity. There was also more complexity because the new ideas led to new questions and new problems. An efflorescence of problems.

The new chemistry was fertile, productive. More and more, explanations fell into place. Working at this time was like coming out of a dark forest into a bright field full of new flowers, stones, and animals that had never before been imagined. There were more questions than before, and more answers. There were more things to learn than before, and more things to say. Workers needed the idea of oxygen more and more, and the idea of phlogiston less and less, and then not at all.

As Priestley put it in 1796, there had been "few, if any, revolutions in science so great, so sudden, and so general" as this new idea of the Antiphlogistians. In but a brief time they demolished "the doctrine of Stahl, which was at one time thought to have been the greatest discovery that had ever been made in the science" of chemistry.[2]

From the greatest discovery that had ever been made, to a footnote in history. Little-known, false, and irrelevant.

Out of the death of phlogiston, modern chemistry was born.

Chapter 9

Something That Has Not Been Seen

Early in 1926, in Switzerland, the best school for technical subjects was the Eidgenössische Technische Hochschule, or ETH. This institute was founded by the federal government in 1854. One of the most illustrious physicists at the ETH in those days was Peter Debye. Only a few years earlier he had explained the scattering of high-energy light, by starting from Einstein's hypothesis of "needle radiation" (that is, the light quantum hypothesis, aka Euclid and Newton's particle theory). For a time, in Europe this was known as the Debye effect, but because the American Arthur Compton had obtained the same result both experimentally and theoretically, and slightly earlier, Debye himself graciously called it the Compton effect, and this is the name we use today.

Physics was also taught a six-minute walk down the Rämistrasse, at the University of Zurich, by "a smaller and rather less illustrious faculty than that at the ETH."[1] The head of theoretical physics at the University of Zurich for the last five years was a thirty-nine-year-old Austrian named Erwin Schrödinger. He had moved back home after having held professorships in both Jena, Germany, and Breslau, Poland.

Physicists like talking about physics, and in those years there was a regular colloquium held, alternately, at the ETH and the University of Zurich. It was modestly attended; on a good day it had about twenty people. At one of these colloquia Felix Bloch, a young student at the time, witnessed an interesting exchange. Fifty years later, he told the story, admitting he could not "render the exact words I heard," but also promising that "in content, I shall report the truth and only the truth."[2]

They were meeting at the University of Zurich. At one point Debye turned and said,

> Schrödinger, you are not working right now on very important problems anyway. Why don't you tell us some time about that thesis of de Broglie, which seems to have attracted some attention.

So at one of their next meetings, Schrödinger did this. He gave a "beautifully clear account" of de Broglie's ideas, explaining how de Broglie linked waves and particles, and deriving some of the consequences of this idea for atomic physics.

> When he had finished, Debye casually remarked that he thought this way of talking was rather childish. As a student of Sommerfeld he had learned that, to deal properly with waves, one had to have a wave equation. It sounded quite trivial and did not seem to make a great impression, but Schrödinger evidently thought a bit more about the idea afterwards.

> Just a few weeks later he gave another talk in the colloquium which he started by saying: "My colleague Debye suggested that one should have a wave equation; well, I have found one!"

> ...I was still too green to really appreciate the significance of this talk, but from the general reaction of the audience I realized that something rather important had happened.[3]

We know a little more than Bloch did about what happened during those intervening weeks. It was nearly Christmas, 1925. Schrödinger went into the Swiss Alps, to a town called Arosa. He brought de Broglie's thesis with him, as well as two pearls, and a woman, not his wife, whose name remains unknown. The pearls he placed in his ears while he was working, to minimize distraction. No doubt the mystery woman helped him maximize distraction. He spent a little over two weeks there, returning on January 9, 1926, with a discovery that "has been universally celebrated as one of the most important achievements of the twentieth century."[4]

And yet, from the very beginning, people were asking questions about Schrödinger's equation. In the same year that Schrödinger presented his discovery for the first time, there was a physics conference followed by a boat trip to a restaurant on Lake Zurich. There, presumably with the aid of alcohol, the students of the ETH and University of Zurich composed a verse which in loose translation runs:

Erwin with his *psi* can do
Calculations quite a few.
But something that has not been seen:
What does this *psi* really mean?[5]

This *psi* was the name for the Greek letter ψ, the conventional symbol for a solution to Schrödinger's equation. Its meaning was a puzzle not only for undergraduates, but also the professors, including Schrödinger himself. It immediately became the subject of a considerable debate that has only grown more extensive and vociferous during the intervening century.

We have now reached a fork in the road. The usual procedure in books of this kind is to start talking about wave functions interfering and collapsing, particles being in many places at once until they're not anymore, and other mysteries that sound very imposing and leave us scratching our heads. The road less traveled, dare I say never traveled at all, is to take a big step back, and a lot of time out, and develop real understanding. We want to grasp the essential point that sets the whole debate in sharp relief, and generates that golden moment of *eureka*.

There is a famous movie from the 1980s called *The Karate Kid*. It is about a young man named Daniel, newly arrived in Los Angeles, and his relationship with Mr. Miyagi, an old karate master who serves as the handyman in his building. In classic style, a gang of thugs is involved, and Daniel must learn karate from the master to defend himself. In the process he becomes, as Mr. Miyagi calls him, Daniel-san (Mr. Daniel).

At the very beginning of the karate lessons, Mr. Miyagi makes him wash and then wax his cars. "Wax on, wax off," he says, and directs Daniel-san to use specific circular arm motions, and to breathe. The next day, it's "sand the floor," and Daniel-san has to sand the entirety of a huge deck. The next day, it's "paint the fence," and Daniel-san again spends all day moving a paintbrush, again with specific direction, up and down. The next day, Miyagi goes fishing while Daniel-san paints his entire house. Understandably after the fourth day of this, Daniel-san is fed up. He quits and walks away, but Mr. Miyagi calls him back, reviews all the arm motions, and then throws a series of punches that Daniel-san blocks without even knowing how he did it.

For the next several chapters, we will be waxing cars and painting fences. It might not make sense at first, but I promise you we will be talking about quantum mechanics the whole time. The history we are about to explore is shockingly relevant to the current situation. My goal is to impart Daniel-san-level defenses in your intuition, so that nobody will ever be able to push you around with outlandish and extravagant ideas.

Part III

The Hidden Nature
of Heat

Indeed the matter of heat or caloric is sometimes talked of
with the same confidence as water, or any common ponder-
able bodies....The truly philosophical inquirer into nature
will not consider it as a disgrace, that he is unable to explain
every thing; he will wait, and labour with hope, tempered by
humility, for the progress of discovery—and he will feel that
truth is more promoted by the minute and accurate examination of a few objects, than by any premature attempts at
grand and universal theories.

—Sir Humphry Davy[1]

Chapter 10

The Heat Equation

Jean-Baptiste Joseph, Baron Fourier, was born in 1768 in the region of Burgundy. He did not inherit his title, or enjoy any privileges by reason of birth. It was rather the opposite. His father, a tailor, had fifteen children from two marriages. Joseph was late to the party—he had only three younger siblings. When he was nine years old the unthinkable happened: both of his parents died within months of each other.

This was the inauspicious beginning of "one of the most gifted scholars of modern science,"[1] who was favored by Napoleon (where the barony came from in 1809), and ultimately became the permanent secretary of the French Academy of Sciences, a member of the French Academy and the Academy of Medicine, and a foreign member of the Royal Swedish Academy of Sciences. Fourier's name is engraved on the Eiffel Tower, along with seventy-one other French luminaries including Fresnel and Arago. (De Broglie did not make the cut, presumably because he was three years old at the time of construction.)

Fourier had an eventful life. In 1794, at the age of twenty-six, he was imprisoned because of his involvement in the French Revolution and yet escaped the guillotine. The same year, he was nominated to study at the newly formed École Normale in Paris. His teachers were some of the leaders of European science, people like Lagrange and Laplace. Fourier began teaching his own students in 1795, was imprisoned and released again, and a few years later joined Napoleon's Egyptian campaign as a scientific advisor. This experience led him to becoming a noted Egyptologist, and while he was there he helped found and then became the secretary of the Cairo Institute. Fourier came back to France two years after Napoleon, and was nominated by the soon-to-be emperor to an administrative position in Grenoble. He was unhappy about leaving academia, but refusing Napoleon's wishes was not really an option.

It was in Grenoble, during what has been called "years of exile" working to drain swamps and build highways, that Fourier began his researches into heat. A few years later, in 1807, Fourier presented a short summary of his results, titled *On the Propagation of Heat in Solid Bodies*, to the Paris Institute, in which he derived what is today called the heat equation. This itself is impressive. Even more impressive, he solved the equation with a bold mathematical idea, today called Fourier analysis.

As will surprise nobody who has read this far, such a crucial advance in science was not received favorably.

> In fact, far from receiving the universal acceptance and acclaim it can now be seen to have deserved, the memoir gave rise to a lively, many sided, and at times acrimonious controversy.... [Fourier's responses to criticism] are written with such exemplary clarity—from a logical as opposed to calligraphic point of view—that their inability to persuade Laplace and Lagrange...provides a good index of the originality of Fourier's views.[2]

In 1811, Fourier arranged for there to be a prize competition on the subject of heat propagation, set by the Paris Institute. Fourier's work was granted the prize (there was only one other submission), but again the report of the judges was not entirely favorable. They held that Fourier's work "still leaves something to be desired on the score of generality or even rigour,"[3] and did not publish the essay. The work was only published, by Fourier himself with the help of others, in 1822, just three years after Fresnel won the prize for his essay on the diffraction of light.

The work itself was epochal. Fourier's book is 639 pages of dense analysis of specific physical situations, composed over almost two decades. The effect it had on the future of mathematics can hardly be overstated. The Irish mathematician Hamilton looked back a century and said this:

Poisson...does not seem to me to have nearly so logical a mind as Cauchy, great as his talents and clearness are; and both are in my judgment very far inferior to Fourier, whom I place at the head of the French School of Mathematical Philosophy, even above Lagrange and Laplace, though I rank their talents above those of Cauchy and Poisson.[4]

There is one idea we need in order to understand Fourier's heat equation. Imagine a suburban neighborhood where everyone is intensely aware of their neighbors' wealth. They are not trying to outdo each other, though; they are trying to *not stand out*. If everyone around them is wealthier, they will stop at nothing until they are also wealthier. But if everyone around them is poorer, they will spend money as fast as they need to. We can call this *Keeping up (and down) with the Joneses*. If some neighbors are wealthier and some poorer, the rule is simple: take the average wealth and go for that.

If we wanted to make this idea perfectly precise, we would talk about something called the laplacian, named after Pierre-Simon, marquis de Laplace, the successor to Newton whom we will meet several times in this book. But we don't need to understand the laplacian in its full glory. In one spatial dimension (which I will restrict everything to for simplicity), the laplacian is particularly simple: it is just curvature.

Imagine you are walking along in a park, and the path is perfectly flat. A moment later the path bends up in a perfect ramp. These are two different things. If you put a ball on the flat part, it will not move. If you put it on the ramp, it will not stay still. In either of these cases, though, the path does not curve, and so it has zero curvature. Now imagine that beyond the ramp, the path curves smoothly up and down. In this case the ball can *both* move and stay still, depending on where you put it to begin with. If you put the ball exactly at the peak or trough of the path, the ball will not move. Anywhere else, it will roll downhill toward a trough.

The third type of path that curves up and down in elevation has a property called *curvature*. You can tell immediately by looking at a surface if it has curvature, but how will we define this mathematically?

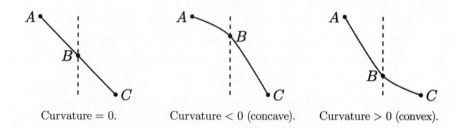

Curvature = 0.   Curvature < 0 (concave).   Curvature > 0 (convex).

Another way of expressing this is to ask yourself: Do the points surrounding B tend to be below or above B? Or is there no tendency? If there is no tendency, then the same number of points are above and below B, and the curvature is zero so B is on a straight line. If the points tend to be below, then the shape is concave at B and the curvature is negative. If they are mostly above, then the shape is convex and the curvature is positive.

In our above example of the path we would have a variable modeling elevation, and how steep the path was would tell you the rate of change of elevation. A flat path has zero elevation change per step you take. A straight ramp has a constant elevation change per step. What is the rate of change of the rate of change in both of these cases? It is zero. But if the path curves at all, the elevation change per step has to change, so we have nonzero curvature.

The elevation of a garden path is given by a number that changes as you go left or right along the path. The number need not be modeling elevation. It could be modeling pressure in air and give the exact same *shape*. Where the elevation of the path was high, now the pressure in the air would be high. Or, we can choose for the number to be the temperature reading of an iron bar along its length. It is in this way that we can say that heat has curvature.

The evolution law that Fourier's heat equation expresses says the following. If the heat is convex, then heat change will be positive in time: that region is getting hotter. If the heat is concave, which is negative curvature, then the heat change will be negative in time: that region is getting colder. If the heat is neither convex nor concave, then it has zero curvature, which means it is a straight line. In this case, the heat change in time is zero; there is nowhere for the heat to go, so the heat does not change at all.

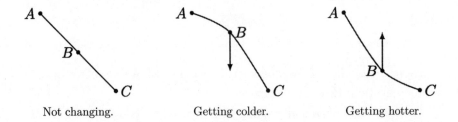

| Not changing. | Getting colder. | Getting hotter. |

We may now state the rule that Fourier discovered, which works for all time. Written first in words, and then in mathematics, it goes like this:

How fast heat changes in time = How curved heat is in space

$$h_t = h_{xx}$$

This is the heat equation. It says that when we fast-forward time (written t), just a little bit, how the heat h changes actually depends on the spatial curvature of the heat. The spatial curvature is written h_{xx}. This is the rate of change of the rate of change of the heat, as we move left and right in the iron bar. Whether the temperature at a point gets bigger or smaller, and how fast this happens, depends entirely on how heat is currently distributed. As time advances, the pockets of hot will get cooler, and the pockets of cool will get hotter. This is exactly what we expect from our commonsense experience of how heat works. It *diffuses* through things, and evens itself out. How fast this happens depends on how quickly the temperature jumps to another value.

The remarkable thing is how simple this rule is, and how universal. It works. It captures something essential about heat flow. Of course there are many subtleties I am leaving out; for example, heat flows at different speeds through wood and rubber and iron. But Fourier's equation can easily handle all of that. Fourier discovered the fundamental rule underlying something that is so obvious and pervasive that only children and scientists stop to wonder about it.

But this is just half of what Fourier did. Not only did he discover the correct differential equation, he *solved* it, too. This latter accomplishment is what laid the foundation of his fame. Mathematicians value new techniques very highly, because the more tools they have, the more problems become solvable. Fourier's solution to the heat equation has become fundamental to many fields. It is practically a magic trick.

To get a handle on the magic trick, let me ask a seemingly uncon-nected question: How do we hear?

The mechanism basically works like this: Air is composed of count-less little particles, all moving around with their own speeds. When a disturbance, like the snap of the fingers, jostles these molecules, their average behavior is changed en masse. They push against nearby par-ticles, which push on nearby particles, outward in an expanding shell. This forms a wave in the air entirely analogous to Hooke's and Huygens's picture of waves on a pond spreading outward from the disturbance of a thrown rock.

Then, the waves enter our ears, which are specially designed to col-lect the disturbances and funnel them into the ear canal and up against the eardrum. It picks up these transmitted disturbances and vibrates, sending a copy of the disturbance deeper, into the inner ear. Here is a remarkable structure of bone that contains the cochlea, a tube that spirals into itself like a shell (*cochlis* in Greek means spiral, and also snail shell).

Inside the cochlea there is a delicate array of hair cells of differing length, width, and rigidity. When the vibrations pass through the inner ear, the hair cells vibrate in sympathy. The vibration of these hair cells gets converted into electrical signals that travel (quite a bit slower than the speed of sound) up into the brain. The mechanical disturbance of air particles jostling has been transduced into electrical disturbances in brain cells. Every sound that we hear has a "fingerprint" given by how strongly each of these hairs is vibrating.

But what determines whether a hair vibrates or not? The answer is amazing, and it depends on frequency. The hairs respond to *pure tones*—in other words, pure cosine waves. If we played one of these waves on a synthesizer and went up the scale with it, we would see one region after another in the cochlea vibrating. Starting with the low tones, the big, loose hairs toward the base of the spiral would vibrate. One tone higher on the synthesizer, the hairs a little up the spiral would vibrate. And so on until we reached very high pitches, which vibrate the short, stiff hairs that are deep inside the innermost spiral.

Now, imagine that instead of playing one note at a time, we played a chord. Say an octave, which is the two closest pitches that are the same note. Two regions of the cochlea would vibrate. Just from this fact we can deduce something else quite remarkable. It must be the case that all sounds—the snap of a finger, the bark of a dog, a loved one calling your name—are complicated chords. It means that every sound you can hear can be broken down into a collection of pure tones, of varying strengths and durations. Imagine you had a very sophisticated synthesizer, with far more than eighty-eight keys. With the right program, it could reproduce any sound you can hear. Not only the sound of a violin, but also birdcalls, explosions, and the Gettysburg Address.

Because our hair cells vibrate in response to vibrations of air, it follows that the shape of the vibrations of air, caused by things moving back and forth in the air and jostling the air particles, must also have a fingerprint, given by a collection of pure tones. But if we can do this for the shape of vibrations in air, we can also do it for any shape at all: the pressure in a stone column, the light in an optical fiber, and the heat in an iron bar.

When we think in these terms, we have two things to keep track of. The first is the function of interest, and the second is the fingerprint. This fingerprint, called the *spectrum*, tells us which waves are present and how strong they are. The two things to keep track of exist in different domains. For a sound traveling through space, the signal washes over the eardrum as time passes, and so this is called the time domain. The domain of the spectrum is called the frequency domain, and it does not have anything to do with time.

Let's take an example. Imagine silence. In this case, both the time domain (the sound wave) and the frequency domain (the fingerprint in your cochlea) are still:

Now, imagine that someone plays a single note on an idealized player piano. It is a perfect cosine wave, maybe with a wavelength of 4. We

will give it an amplitude (or loudness) of 1, which tells us how strongly
this tone is present and how vigorously our hair cells will be vibrating.

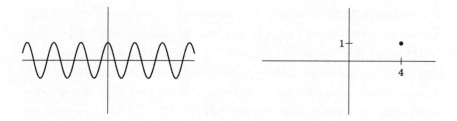

Then, after the room has fallen silent again, they play another note,
an octave lower, twice as loud. Then we get:

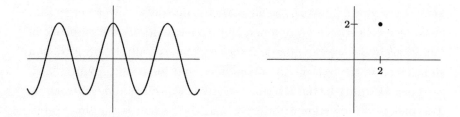

If we play both of these notes together, we will get a chord, in this
case an octave. We know that the frequency domain will have two dots,
and we know where they are. But the time domain, what does it look
like? What is the rule for combining two waves? Luckily for us, the rule
is about as simple as you can imagine. You *superpose* the two waves on
top of each other. That is, you add them together. Where both waves
have peaks, you will get even higher peaks. Where both have troughs,
you will get lower troughs. This is called *constructive interference*.
And when one has a peak and the other a trough, you will get something closer to quiet, which is *destructive interference*. In our case, the
time and the frequency domains will look like this:

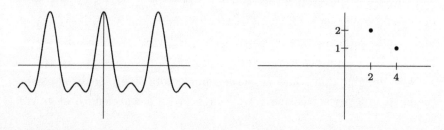

We can add up as many waves as we want, and Fourier's great idea was that, if we can find just the right wavelengths and amplitudes, we will be able to recreate any shape whatsoever. It was this claim that aroused skepticism, if not outrage, among Lagrange and the rest of the French Academy. It would be many years before the theory was developed in all rigor.

Now, rather than thinking about the time domain, with a pressure wave impinging on your ear as time passes, let's think of a space domain, with temperature changing as we go left and right along an iron bar. The frequency domain is the same as before.

Fourier was led in this direction in his researches on heat because these simple waves are solutions to his equation. If you had heat set up in a long pipe, and the temperature looked like one of these waves, then Fourier's heat equation says that the wave pattern will die down, its amplitude shrinks, as time passes. How fast this happens depends on the wavelength. The solution is particularly simple. However, if we can get *any* shape of temperature by adding together enough waves, then we can describe the behavior of any temperature in an iron bar as the sum of many simple evolutions.

This trick of breaking apart the complexity of heat evolution into a collection of many simpler evolutions is Fourier's great contribution to humanity. The idea is not limited to heat, as we saw in our analysis of sound. Nature discovered this trick long before Fourier, and today it is used absolutely everywhere. It is not an exaggeration to say that Fourier analysis makes the modern world possible. It is used to solve many different types of equations beyond the heat equation; it is used in radar and sonar, in software packages that interpret and modify images and sound, in data transfer, and, indeed, anywhere a signal needs to be processed digitally. The Fast Fourier Transform, a version of Fourier analysis, has been called "the most important algorithm of all time,"[5] and this is probably accurate.

Fourier's results were about heat, but they spread far and wide into every corner of science. They have echoed and amplified for hundreds of years, and are why every mathematician, engineer, physicist, and even coder today knows Fourier's name.

Chapter 11

Parable of Caloric

What is heat? In Fourier's day, the dominant theory went like this: Heat is caused by an invisible fiery fluid that cannot be created or destroyed; is self-repulsive, extremely light, or even weightless; and is so subtle that it permeates the finest pores of matter. This fluid was named *caloric*, by Lavoisier at the French Academy in 1787. The term comes from the Latin word *calor*, heat.

There was an alternative on the scene, called the vibrational, or kinematic, or dynamical theory. This idea held that heat was a modification or state of matter, caused by "insensible motions." In other words, a hot iron bar is seething in ways we cannot perceive with our eyes. A cold iron bar might also be moving, but differently. Not seething, per se. Maybe it is slowly writhing.

Already in 1620 in the second book of his *Novum Organum*, the Right Honourable Francis Bacon proposed that heat was a modification of matter and not a thing itself. After carefully analyzing all known facts on the subject, and several ways that heat is produced, Bacon concluded that "heat itself, its essence and quiddity, is motion and nothing else."[1] Among his most interesting examples was

> a practice which blacksmiths have sometimes recourse for
> kindling a fire; they take a rod of soft iron, half an inch or
> less in thickness, and, laying the end of it upon their anvil,

76

turn and strike that end very quickly on its different sides
with a hammer; it very soon becomes red hot.[2]

This practice is the origin of the lovely aphorism: *Do not wait to
strike till the iron is hot; but make it hot by striking.*[3]

Newton, Hooke, Boyle, and Locke were all of the view that heat is
motion. However, the theory was hard to develop, and none of these
men could describe the motions better than Boyle's "various, vehement,
and intestine commotion of the parts among themselves."[4] In other
words, seething.

The core idea of the vibrational hypothesis surely comes from the
observation that rubbing things together, like your hands, makes them
warm. Bacon believed that *new* heat was created upon striking an iron
rod, quite to the contrary of the calorists who believed that the amount
of caloric never changed but instead was released and made evident
when you hammered iron. Another example: in 1722 John Locke wrote
that heat is a form of motion, and mentioned the fact that "the axle-
trees of carts and coaches are often made hot, and sometimes to a
degree that it sets them on fire, by rubbing of the naves of the wheels
upon them." The logical conclusion is that, as he presciently claimed,
"the utmost degree of cold [absolute zero] is the cessation of that motion
of the insensible particles."[5]

Benjamin Thompson was another keen supporter of the view that
heat is motion. Born in the colonies, he took Britain's side in the American Revolutionary War and went on to become a fellow of the Royal
Society, a knight of the British Empire, and a count of the Holy Roman
Empire. Count Rumford, as he came to be known, wrote seventy-two
papers on a wide variety of subjects, and in many of them he reiterated
his belief that the dynamical theory of heat was "basically the sounder
one....His primary object was to attack the caloric theory from as
many different points of view as possible."[6]

In 1789, Rumford published his most influential attack, titled "An
Inquiry Concerning the Source of Heat Which Is Excited by Friction."
He said that it came about out of his "habit of keeping the eyes open
to everything that is going on in the ordinary course of the business of
life." The business, in this case, was the boring of cannons for the arse-
nal in Munich, where Rumford was superintendent.

The cannon was of brass, about five feet long when finished. It came from the foundry rough, with an extra two feet at the front. This extra part, called by the Germans *verlorener kopf* (the lost or doomed head), was meant to be cut off. It was always included so as to prevent the real head of the cannon from being cast with too many pores. Boring proceeded by placing the cannon horizontally on a stand, and attaching it to an apparatus that would allow the cannon to rotate on its axis. Two horses did the hard work by walking in circles (one could have been employed, but Rumford got two so he could "render the work lighter").

For his experiment, Rumford maintained a bit of the doomed head and carved it into a cylinder, which he then hollowed out. He had now, at the front of this cannon, a sideways vessel. Into this vessel he placed a fixed, blunt bore. The whole point was to generate friction between the bore and the bottom of the vessel. He set the horses moving, and at once there leapt out a terrible screeching sound of the metal parts grinding together. Rumford said the noise was "very grating to the ear, and sometimes insupportable."[7] Nevertheless, he did no fewer than four experiments this way, including one where he fixed a wooden box around the metal vessel, made the box watertight with "collars of oiled leather," and filled it with cold water. The horses then turned the cannon, and Rumford reported that the water became increasingly hot until, after two hours and twenty minutes, it "ACTUALLY BOILED!"

> It would be difficult to describe the surprise and astonishment expressed in the countenances of the bystanders on seeing so large a quantity of cold water heated, and actually made to boil, without any fire. Though there was, in fact, nothing that could justly be considered as surprising in this event, yet I acknowledged fairly that it afforded me a degree of childish pleasure, which, were I ambitious of the reputation of a *grave philosopher*, I ought most certainly rather to hide than to discover.[8]

This experiment shows that the heat in the cannon seemed to be inexhaustible. Material things are not inexhaustible, so how could heat be a material thing? Rumford concluded that the heat generated came from motion. The motion of the horses translated motion to the cannon, which, in rubbing against the bore, imparted some kind of vibratory motion to the brass itself. This in turn imparted motion to the water. Rumford boiled a little less than a cubic foot of water, but with enough horses, cannons, and time, he could boil away a lake with the same method.

Another supporter of the vibrational theory was the British chemist Humphry Davy, who later in life became the patron of Faraday. When Davy was seventeen years old, he brought a friend to a river to show him that ice could be melted by rubbing it together. Four years later, in 1799, Davy published a more rigorous account of his childhood investigations. Much later, this idea was held by Joule to be even more convincing than Rumford's experiment with the cannon.

As a final champion of the minority theory that heat is motion, there is someone we have already met, the great Thomas Young. In precisely the same years that he was thinking about the wave theory of light, originating the double-slit experiment, and dealing with Brougham's wrath, he also criticized caloric and supported the vibrational theory. In his capacity as professor of natural philosophy at the Royal Institution from 1801 to 1803, Young taught courses in which he claimed that all the main phenomena of heat could be explained by the vibrational theory just as well as with caloric, "and in the case of friction, very considerably better."[9]

Research into the hidden nature of heat began with confusion. The entire process of formulating thermodynamics took hundreds of years and was called "tragicomical" by Truesdell, and its history "accursed by misunderstanding, irrelevance, retreat, and failure."[10] Benighted, the scientists of this age groped in the dark. Their books have memorable titles like *History of Cold*, *Animal Heat*, and *Pyrometrie*. The truth, and the true extent of the complexity, became apparent only slowly. One important step was made by the Scottish doctor and chemist Joseph Black, who spent most of his life in Glasgow and was among the first to use a thermometer rigorously.

Black discovered something that is today called *latent heat*, the idea that heat may change within a body without any evident temperature change. His reasoning was memorable. For ice to melt *must* require a great deal of heat entering the ice over a period of time, because if the ice suddenly all melted the moment a single temperature was reached, all the snow in the mountains would melt at once, and the "torrents and inundations would be incomparably more dreadful [than they already are]. They would tear up and sweep away everything."[11] On the other hand, it takes a continual increase of heat to boil water and then eventually boil it all away. If all the water boiled away the moment the boiling temperature were reached, there "should be an explosion of the whole water with a violence equal to that of gunpowder."[12]

Black also discovered what is today called *specific heat*, a phenomenon that seemed to set the dynamical theory in doubt. Before Black, it was thought that the amount of heat required to increase the temperature of matter a fixed amount was proportional to the amount of matter in that body. For example, the same volumes of quicksilver (i.e., mercury) and water have radically different weights, since quicksilver is about 13.5 times denser than water. A gallon of that stuff would be 113 pounds and very hard to carry. It just seems common sense that to raise the temperature of a fixed volume of mercury one degree, we should need about 13.5 times the heat required to raise the same volume of water by one degree.

Black discovered "very soon after I began to think on the subject (anno 1760)"[13] that this was completely wrong. Collecting a variety of identical volumes of various types of matter, all with the same original temperature, he found that the amount of heat required to raise each sample by a fixed temperature was all over the place. It depended entirely on the material. Completely contrary to expectation, mercury actually required *less* heat than water did. Black said that "no general principle or reason could yet be assigned" to explain this.[14]

This observation creates a problem for the dynamical theory. If heat is motion, then it should be *harder* to heat mercury than water, because mercury weighs so much more. Saying heat is motion seems absurd because it implies that it can be easier to move a heavy stone than a light one. As Black commented: "I do not see how this objection can be evaded,"[15] and he was inclined to the caloric theory because if heat was supposed to be a "tremor," he could not "form a conception of this internal tremor which has any tendency to explain even the more simple effects of heat."[16]

What was this thing that flowed into and out of quicksilver and water, but had different effects on both of them? It was a *quantity* of something. Black imagined an actual physical substance, a type of matter that could move in and out of other matter, like water moving in and out of a sponge. He wrote: "This idea of the nature of heat is the most probable of any that I know."[17] This was not Black's idea; it went back at least to Boerhaave, a contemporary of Newton's a century before. But in Black's day, this idea was supported by the majority of European scientists. Most people supposed heat to actually be a *thing*,

> the particles of a subtle, highly elastic, and penetrating fluid matter, which is contained in the pores of hot bodies, or interposed among their particles—a matter that they imagine to be diffused through the whole universe, pervading with ease the densest bodies.[18]

Here "highly elastic" means that the particles of this invisible heat ocean are self-repulsive; pressing two of them together is just like pressing two ends of a spring together. This makes qualitative sense. We know that heat tends to spread out with time. Hot things get colder and cold things get hotter. The immediate explanation, according to the calorists, is this self-repulsion of the heat particles.

Building on Black's discovery that different types of matter have different affinities for heat, Cleghorn wrote a dissertation in 1779 called *On Fire* that gave a systematic description of a fluid called "fire," which had the essential properties soon associated with caloric. His main idea was that the heat particles are actually attracted to matter particles with a strength depending on the matter in question. This explained why quicksilver required less heat than water to be raised the same temperature: the particles of mercury had a higher affinity for the particles of heat. This interplay of attraction and repulsion created rich possibilities, developed by many of the century's most celebrated scientists.

For example, Lavoisier, the brilliant chemist who explained combustion by banishing phlogiston, supported caloric. His authoritative *Elementary Treatise of Chemistry, Presented in a New Order According to Modern Discoveries* was published in 1789, the very same year that

Rumford published his work about heating cannons.[19] The first chapter of the first volume begins with the properties of caloric. Lavoisier even included caloric as a chemical element in that work, alongside hydrogen, oxygen, nitrogen, zinc, and sulfur. This is why the caloric theory was sometimes called at this time the *chemical* theory of heat.

In short, there were two theories of heat going into the first decade of the 1800s. Nobody knew what heat really was and doubt was frequently expressed by both sides. The textbooks of the time acknowledged this, though they invariably gave precedence to caloric because, in the eighteenth and nineteenth centuries, the caloric theory was simply more successful in accounting for the observations.

Many problems circled around the phenomena of heat during this time, and the caloric theory rose to answer them in a way the kinetic theory could not. A Frenchman named Gabriel Jars had noticed something in a Hungarian mine in 1758: If air compressed by a column of water over forty meters high was suddenly allowed to escape, "compact ice was deposited on any object placed in the way of the issuing air."[20] The kinetic theory would predict the opposite, right? If faster air means a higher temperature, compressed air should *melt* ice. This phenomenon, today called *adiabatic cooling*, could be explained by the caloric theory of Cleghorn, who wrote in his thesis that the rarefication of air causes cold

> because, with the air ... now being rarified, the repulsion between the fire particles was diminished and so fire flowed out of the thermometer.[21]

This is also connected to another famous problem of the day, a discrepancy in the speed of sound as it was observed, and as it was computed using the wave equation. This problem was attacked by Newton, Euler, Lambert, Lagrange, both Bernoullis, and d'Alembert. None of them could resolve the difficulty. Laplace and Biot finally solved it by examining temperature changes within the density fluctuations of air itself, and dropping the critical false assumption that was causing the problem. This was all based on the mathematics of caloric theory.

Another very important theory based on caloric was given by the Englishman John Dalton. He is mostly remembered for the hypothesis that chemistry should rightly concern itself with atoms and their inter-actions. He believed atoms could be numbered and ordered, and that they had different sizes and weights. This idea is what finally delivered chemistry from alchemy and magic.

Through the first decade of the 1800s, the question of atoms and their sizes and weights was very much on Dalton's mind. He imagined that each different atom of matter had a specific attraction for caloric, and that atoms collected an atmosphere of caloric around themselves. Those with a strong attraction would have a larger atmosphere, and thus a larger size. It was the caloric, Dalton reasoned, that determined the stability and mixing properties of different gases.

He drew the caloric as rays emanating from the atoms, and termi-nated those rays in boxes of different sizes, depending on the size of the atmosphere. Dalton presented these ideas in his book *A New System of Chemical Philosophy*, published in 1810, a landmark work dedicated to Humphry Davy that begins, "The most probable opinion concerning the nature of caloric." It included many images like the ones below.

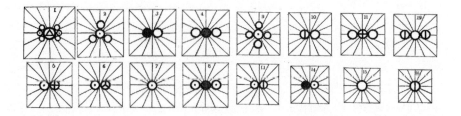

In short, the caloric theory was able to not only account for challeng-ing new experimental results, but also solve old problems. As Dalton's work showed, it provided a fertile ground for the development of new ideas and hypotheses. The vast majority of scientists supported it, as the third edition of the *Encyclopedia Britannica* showed when it said in 1797 that "[heat] is almost universally believed to be the effect of a fluid."[22] As late as 1811, Avogadro, a name known to every chemist, produced one of the most complex caloric theories. Roller summarized the theoretical situation nicely:

> Indeed, as a conceptual scheme serving to correlate most of the facts then known about heat, and as a convenient way to think about these facts or even to predict new ones, [caloric theory] was proving itself to be far more useful than the opposing and then relatively undeveloped doctrine that heat is a form of motion.[23]

Chapter 12

Positivism

Influenced by Laplace and Biot before him, Fourier had originally tried to address heat as a many-body problem, specifying how individual particles of matter might be moving. Then, around 1804, for "some reason that is not quite clear, Fourier...made a bold departure from convention, which eventually led to his masterpiece."[1] He considered the heated body to be continuous and undifferentiated. There was no knowledge of the body; what it might be made of Fourier did not care. He zoomed out and out, until there was nothing left to talk about except for the flow of heat in time.

This strategy appeals to those who think that looking for the "true nature" is dubious at best. There are several reasons this might be so. The problem of the true nature of a phenomenon might be too complex right now for anyone to solve, so simplifying and zooming out is the only way to make progress. Or, a more extreme position is to doubt that we can come to any knowledge of "true nature" in the first place. Fourier seems to have taken the second position.

The first sentence in his 1822 *Essay on the Propagation of Heat* is rightfully famous for its strong stance on the nature of science:

> Primary causes are not at all known to us; but are subject to simple and constant laws, which we may discover by observation, and the study of which is the object of natural philosophy.

By natural philosophy, Fourier means what we would today call science, especially physics. By primary causes, he means the hidden nature, the understory, the deepest possible explanation. The idea that physics does not deal with primary causes harmonizes with a viewpoint that is today called *positivism*. The term comes from the work of Auguste Comte, probably the first philosopher of science we would recognize today. Comte had a formidable mind, and composed his books the way Mozart composed music: in his head. Comte's "marvelous memory and power of mental concentration enabled him to think out an entire volume in all its parts."[2] He lectured without notes, and in preparing his lectures for publication, he simply sat down and copied them out from memory. He once wrote 343 octavo pages in 28 days.

The basic idea Comte presented was that there are three types of knowledge. Each type has its corresponding mode of cognition. He called these modes the theological, metaphysical, and positive. But these terms are not the easiest to use. John Stuart Mill, one of the finest writers England produced and a close friend of Comte's in the later years, wrote that "instead of Metaphysical, [we should prefer to speak of them as] the Abstractional or Ontological."[3] In short, metaphysics seems to be used by Comte to mean most forms of philosophizing. The term "positive" can probably without too much loss be translated in a modern sense as "scientific."

This distinction between theology, philosophy, and science is itself nothing very new, but Comte's ideas got juicy and weird when he claimed that scientific understanding was the best, most mature, and final one. Below the positivistic mode of gaining knowledge, inferior and antecedent to it, was the metaphysical, and below that again, the most primitive and worst, was the theological.

To the theological mode of thinking, all things happen as they do because divine intelligences and sentient wills impel them. Knowledge then consists of understanding the details, the wills and laws, of deities. Someone who sees the plan of God in every occurrence is, according to Comte, a primitive and childish thinker who has not grown beyond this first stage of inquiry. Even highly developed cultures can still be theological. In Kepler's time, for example, it was believed that the planets orbited the Earth fixed in translucent crystalline spheres, and that literal angels drove the planets along in their courses.

Next comes the metaphysical stage. No longer satisfied to explain things using the wills of supernatural beings, the metaphysical thinker replaces supernatural intervention with imagined essences. The essences are abstract "explanations" for phenomena which take the facts into account but are primarily concerned with the causes behind the facts. These are concepts of the absolute, the metaphysician's bread and butter. Knowledge, according to this mode, consists of finding the correct hidden essences and fitting them together to account for all the facts.[4] Comte believed metaphysics was a transitional state only. Its whole significance lies in how it lifts humanity out of theology, and prepares the path for the final, best, coolest, and most useful positive stage.

The final positive stage means mathematical empiricism: a careful quantitative study of phenomena themselves and their patterning, until the laws underlying the phenomena are expressed clearly and definitively. Neither gods nor essences are required for the positivistic mode of thought, which recognizes that primary causes and absolute notions are forever unknowable. This mode *renounces* (note well the word!) the search to know "the hidden causes of phenomena."[5] For Comte, this mode alone generates any knowledge worthy of the name: knowledge of how things happen—and with it, technological control over the world. This real knowledge must always begin with the renunciation of false hopes.

Comte believed that all fields of human endeavor, be they physics, biology, or politics, went through these stages of thought in order, and that human beings themselves went through these stages as they matured:

> Now, each of us is aware, if he looks back upon his own history that he was a theologian in his childhood, a metaphysician in his youth, and a natural philosopher in his manhood.[6]

As a child, you may have imaginary friends. You may believe in Santa Claus and the Easter Bunny. When you come to your adolescence, you begin to form many grand theories about the world. You debate with your friends about the true nature of time and the proletariat. At last you grow up, and betake yourself to real problems: a careful quantitative study of a single real phenomenon. Comte called this progression a "law." It came to be known as his Law of Three Stages. Each stage was associated with a mode of thought: theology with fiction and imagination, metaphysics with abstraction and discussion, and positivism with reality and demonstration.

Any philosopher who writes thousands of pages over multiple decades unfolding a single system will inevitably be blinded by their own ideas, and this is what happened to Comte. He thought that "positivism," rightly understood, condensed into a single term the "highest attributes of human wisdom." He believed the word would "soon become generally accepted."[7] He went on, in the 1840s and '50s, to write books of increasing zeal and ambition. He became interested in politics and literally tried to start a world religion, called the Religion of Humanity, in which "theorists and practitioners" become "servants of humanity."[8] As one scholar determined, "In place of the calm detachment supposed to characterize a scientist there is all the zeal of a preaching monk who has, moreover, gone a bit fanatic about details as well as objectives."[9]

In *The Catechism of Positive Religion*, Comte set up a "complete system of belief and ritual, with liturgy and sacraments, priesthood and pontiff, all organized around the public veneration of Humanity."[10] The details of worship were largely inspired by Catholicism, leading Thomas Huxley to describe the religion as "Catholicism minus Christ." Its followers—yes, the religion gained followers—believed it was rather "Catholicism plus science." Comte invented a new calendar, year zero being 1789, the year the Bastille was stormed, and he named the months after men, only men, like Moses, Homer, Aristotle, Caesar, Shakespeare, Gutenberg, and Dante. Every function of the brain was classified (there were only eighteen so it was relatively easy). In his grandiosity Comte requested that the Pantheon, one of France's most important buildings, be given over to the new religion.[11]

The Religion of Humanity is full of weird twists and turns and has influence in surprising places. Among the ideals Comte promoted was living for others, *vivre pour altrui*, which is the origin of the word "altruism." The idea of a secular religion found support in England as the *Origin of Species* gained influence, and George Eliot's poem "O May I Join the Choir Invisible!" is inspired by Comte's idea of humanity as the Great Supreme Being. Support was not limited to England, and the positivists made themselves felt in many places including France, the United States, and, above all, Latin America. Positivism was influential in Argentina, Chile, Cuba, and Mexico, where it was seen as a progressive political force in direct opposition to Catholicism. The most affected country in the world was probably Brazil. The motto of the religion was "Love for Principle and Order for Basis; Progress for End," which can be seen today on the flag of Brazil in the words *Ordem e Progresso*. Júlio de Castilhos, a member of the assembly that drafted the Brazilian constitution in 1891, was the leader of the positivists of Porto Alegre.

Today there are eight temples of positivism. Four in England, three in Brazil, and one in Paris. The temple in Rio de Janeiro is derelict. A storm collapsed part of the roof in 2009, and thieves stole "the original design of the Brazilian flag and other Positivist treasures."[12] In 2016, the *New York Times* published an article on Brazil's roots in positivism that begins with a description of the temple:

> Neighbors from surrounding apartment buildings toss empty beer bottles through a gaping hole in the roof of the once-majestic church. Pigeons roam the cavernous nave, their excrement piling up on the floor.[13]

The Religion of Humanity is all but forgotten today. Even as I wrote this section in a café, a group of Brazilian athletes from Goiânia sat down next to me. I struck up a conversation with them on this topic, and they were astonished to learn about the origin of the motto on their flag. They knew the Glória neighborhood in Rio, where the temple is, but had never heard of the temple itself.

Finally, Comte was also a deeply strange human being. Like many remarkable minds, he suffered from mental instability, perhaps what we would today diagnose as bipolar disorder. He broke off his first lecture series after going temporarily insane in 1826. A year later, so imbalanced by jealousy of his own wife, he jumped off the Pont des Arts in an attempted suicide. In his late forties he fell intensely in love with Clotilde de Vaux, a younger woman of ill health who would not share his bed. Of course, this is normal; the unusual thing was that he treated the armchair where she sat when she came to visit as an altar, and knelt before the empty chair twice daily in prayer. When Clotilde sent him a letter, he read it on his knees in front of this "altar." She was beatified as Saint Clotilde, and the new followers of the Religion of Humanity worshiped Comte as high priest and Clotilde as high priestess. There is a painting of her, in the mode of the Virgin Mary, in the temple of positivism in Rio.

Its richness and complexity mean that positivism does not lend itself to simplistic analysis. Happily, there is no need to study its many faults. We may invoke Nietzsche, who said, "I mistrust all Systematizers and avoid them—the will to a System is a lack of integrity."[14] As Comte was one of the most systematizing thinkers in European history, we will allow him to disappear into the rosy clouds of his own utopian dream. There is only one fault that is truly interesting for us, anyway. This crucial error, which has echoed through two centuries of scientific practice, concerns Comte's conception of the relation between the metaphysical and positive.

Positivism grew from the intellectual soil that Fourier nourished. He attended Comte's lectures on positivism which began on January 4, 1829, and eventually the two men became friends. Comte called Fourier "my illustrious friend" and "the immortal Fourier," and his work on heat Comte declared to be "the last great conquest of the human spirit." Comte even dedicated his greatest work, the *Course of Positive Philosophy*, to Fourier.

Positivistic ideas were in the air in France during the 1820s and '30s, and had many sympathizers, but Fourier was

> an intellectual leader of the younger French physicists who staged a silent revolt against Laplacian physics, an important part of which was a positivistic indifference or hostility toward Laplacian hypothesizing about microphysical forces and structures.[15]

There were changes brewing, many of them in reaction to the dominant figure of Laplace, whose approach to science had overshadowed all others for decades. In rebelling against Laplace's scientific practice,

Fourier was the head of the spear. His heat equation was everything that Comte admired.

Early in the *Course*, Comte searches for the "best illustration" of a correct application of positive philosophy, and lands on Newton's universal law of gravitation. In practically the same breath, looking for one more outstanding example, Comte praises the heat equation:

> I will choose the beautiful series of researches by M. Fourier on the theory of heat.... In this work, which is so eminently positive in its philosophical character, the most important and most precise laws of thermology find themselves unveiled, without the author inquiring one single time into the hidden nature of heat. Without him mentioning, except to indicate emptiness, the so agitated controversy between the partisans of the material caloric and those who made heat consist in vibrations.[16]

Comte admired Fourier for restricting his scope to the flow of heat alone, which he called "restraining himself in the research to an entirely positive order." Doing otherwise, and attempting to approach the hidden nature of heat, would be to "throw himself into inaccessible problems." They are inaccessible not because we are not ready for them, but because the human mind can never know them. As Fourier said in his opening sentence, primary causes are not at all known to us. Attempting to approach primary causes with hypotheses just ends in "interminable discussions," because this is an attempt to make pronouncements on "questions necessarily insoluble for our intelligence."[17]

For Comte, causes are metaphysical, not scientific. They belong in the realm of philosophy, and interminable discussions. He considered the cause of weight and attraction, and the hidden nature of heat, to be unknowable. He was quite wrong about both of these pronouncements, and it is interesting that he was wrong nearly every time he tried to place a boundary on knowledge by identifying insoluble problems. For example, planetary astronomers would be amused to learn that their entire field was declared impossible by the armchair musings of this nineteenth-century philosopher. Concerning the possibility of measuring the temperature of the planets in our solar system, Comte said,

> I regard this order of facts as for ever excluded from our
> recognition. We can never learn their internal constitution,
> nor, in regard to some of them, how heat is absorbed by their
> atmosphere... [and] we can never know anything of their
> chemical or mineralogical structure; and, much less, that of
> organized beings living on their surface.[18]

Similarly, he thought we could never learn anything about the chemical composition of stars, a claim that has been called "Comte's blunder."[19] When it came to biology, he blundered again. Thinking about
inorganic and organic matter, Comte said that "there is no need to ask
if the two classes of bodies are, or are not, of the same nature—that
is an insoluble question."[20] Embryology was also considered impossible
by Comte, who pronounced research into the physiological generation
and development of animals to be an "insoluble inquiry."[21] I could probably continue but will limit myself to a final example: Comte also said
some things about the particle-wave duality that aged like fine milk.
He thought both the particle and the wave picture were "chimerical
hypotheses," and that supporting one or the other hypothesis on the
nature of light was a "vicious method of philosophizing," and that these
hypotheses "only obscure scientific realities." All they can do, according
to Comte, is bring us out of philosophy and into science. Once they have
served their purpose, "their action can henceforth be only injurious."[22]

This is an impressive list of errors. As we shall see, once Comte
was replaced by other positivists, they carried on the tradition of being
spectacularly wrong.

It would seem that, for Comte, real knowledge looks like an algebra.
The heat equation is a perfect example because it is simply a trio of
empty symbols h, t, and x mapping to heat, time, and space, and in-
terconnecting in such a way as to capture the laws of sequence regulat-
ing the phenomenon of heat. This gives us knowledge of process, and
technological power and control. For a positivist, that is all we are entitled to. The origin of the laws themselves is a mystery and there is no
desire to ask, What are heat, time, and space, really? Such a question
is considered senseless, chimerical, obscurantist, vicious, insoluble, and
primitive. The moment we try to look for the nature of what "really
is," says Comte, we have devolved into metaphysics and interminable

discussions. We have entered the realm of abstract thought where we are once again treating words as things in a vain search for ultimate causes.

It is no wonder that Comte admired Fourier so ardently. Fourier's book on heat was long, dense, and mathematical. It has chapter headings like "Measure of the Movement of Heat at a Given Point in a Solid Mass," and "Equations of Uniform Movement of Heat in a Solid Prism of Infinite Length," and "Development of an Arbitrary Function in Trigonometric Series." The entire thing is calculations arising from the heat equation. Fourier does not mention the word "caloric" even once. This, for Comte, was a very beautiful thing. This was true thinking and true knowledge, infinitely superior to the babble of philosophers and their supposed invisible causes. Comte took much nourishment from Fourier and praised him often, for example writing that

> Fourier put away, in his study of heat, all fantastic notions about imaginary fluids, and brought his subject up to such a point of positivity as to place it next to the study of gravity.[23]

Or that

> the labours of Fourier...must soon establish a thoroughly scientific method; and this result cannot but be aided by the fact that the two great modern hypotheses about the nature of heat are in direct collision.[24]

Or that

> the genius of Fourier released us from the necessity of applying the doctrine of hypotheses, as previously laid down, to the case of thermology.[25]

In summary, the heat equation was developed in conscious disdain of the debate about the "real nature" of heat. Fourier believed primary causes were completely unknown. His treatise on the heat equation was subsequently praised as a model of positivism by Comte, who thought that talking about the real nature of heat was unworthy of science, and one step above praying to the fire god.

Leave this vain debate about what heat "really is" behind, Comte would say. These sorts of hypotheses are metaphysical notions of what really is, and as such they are not scientific at all. They are retrograde, a hangover from the attempt to shake off a world full of gods throwing thunderbolts and making us fall in love. Be free! Don't dispute! Compute! Stop talking about primary causes we will never know! Shut up, and calculate!

If Comte had his way, everyone would have forgotten about all "empty" and "completely insoluble" debates. They would have renounced hard questions as "metaphysical," and applied themselves to "real" problems, which positivism alone could identify. The theory they developed on this path could never have refuted the path itself for, as Bernstein pointed out: "It would have been perfectly consistent to take these laws [of thermodynamics] at their face value without searching for a deeper meaning."[26]

Today, we know that this would have been a complete disaster, freezing progress indefinitely until people rebelled against Comte's sentiment that the true nature of heat was unknowable. The debate between the caloric and vibrational theories was not empty; it was not interminable. It was completely necessary. Yes, the debate went back and forth. Yes, it was hard to decide in Comte's day between them. But we know today that the caloric theory is completely, utterly, fundamentally wrong. Heat is motion.

That there were two hypotheses about the true nature of heat and one was right is a total crushing defeat for Comte. If the relationship between metaphysical and positive thinking is not one of hierarchical supremacy, if hypothesizing about hidden natures is actually useful, then Comte's law of the three stages is wrong. But as Mill said, this so-called law "forms the backbone of his philosophy, and, unless it be true, he has accomplished little."[27] Those who support positivism today walk inside a church no less dilapidated than the temple in Rio, and, through a surfeit of piety or simple blindness, insist that the place of worship is very beautiful. History has shown that Comte's system is unsound, and we should not pretend out of politeness or the urge to appear sophisticated that anything here is worth saving, other than a few handsome fragments from the shattered stained-glass windows.

Chapter 13

Intellectual Dynamite

The vibrational theory was hard to develop because it required an entirely different branch of knowledge to blossom first. Because heat arises as an average of inconceivably many rapid motions, the theory of heat is actually a statistical theory. To understand heat, we must first understand probability.

Like every branch of knowledge, probability theory was created by people thinking about problems. The very first manuscript on this subject, *Liber de ludo aleae*, Book on the Game of Chance, was written by Girolamo Cardano, in 1564.[1] However, because his manuscript on gambling was not published until 1663, a hundred years after it was written, the origin of probability theory is usually attributed to a famed correspondence of 1654, between Blaise Pascal and Pierre de Fermat.

Blaise Pascal was one of those highly sensitive people who seem too delicate for this world. He had poor health from the earliest age, and was homeschooled by his father, Étienne. The father, thinking them too demanding, withheld mathematical subjects from his son. This is rather humorous in its irony, because the young Blaise was so brilliant mathematically that after finally beginning the subject and studying for two years, he joined his father as a member of the Paris Academy at the age of fourteen. This salon was founded by Mersenne, a smart guy who supported Galileo before it was cool, and had among its members some of the most learned men in Europe, Fermat and Descartes included.

When Pascal was sixteen years old, he published his investigations into conic sections, which are the shapes formed when a plane intersects with a cone. The manuscript was so complete and advanced (he proved over four hundred interconnected facts) that Descartes was astonished that such a young man had written it.

Seventeen years later, only a few months after cofounding probability theory with Fermat, the horses pulling Pascal's carriage spooked and bolted over the side of a bridge. Rather than taking the carriage with them, the traces broke, and it was only the horses that plunged to a violent death. Pascal interpreted this miracle as the grace of God, and that evening he had an intense mystical experience. He wrote an account which started, "From about half-past ten in the evening until about half-past twelve: FIRE," and carried the parchment with him for the rest of his life, sewn into his coat.

He abandoned science from that day and devoted himself to theology and philosophy. Eight years later, after developing brain and stomach cancer, Pascal died in 1662. He was thirty-nine years old. Even though he has a scientific unit named after him (the pascal, a measure of pressure), even though he cofounded probability theory, constructed a famous mathematical triangle, designed and built the very first working calculator, and almost discovered the fundamental theorem of calculus seven years before Newton and fourteen years before Leibniz,[2] Pascal is most remembered today as a deep religious thinker and a gorgeous prose stylist. This latter compliment, bestowed on a mathematician in a country with a high bar for gorgeous prose, tells us all we need to know about how special Pascal was.

Pierre de Fermat was a very different kind of man. Solid and responsible, he married Louise de Long, his mother's cousin, and with her had eight children, five of which survived to adulthood: Clément-Samuel, Jean, Claire, Catherine, and Louise. For the majority of his life, Fermat was a lawyer and municipal councilor of Toulouse. Today he would be of interest only to the dustiest of Renaissance scholars, if it were not for his mathematical abilities. Fermat was one of the most productive mathematicians of his time.

Fermat knew six languages. In an age which was rediscovering the wisdom of the classical authors, he was frequently consulted for his knowledge of ancient Greek. Some have even called Fermat the originator of calculus in its modern form because of foundational ideas which Fermat had developed in a treatise in 1636. The treatise establishes the scientific result for which he is best known. Today it is called Fermat's principle, and it expresses the amazing fact that the path that light takes from point A to point B always minimizes the amount of time required to make the journey. That is, every other path that light *could but does not* take from A to B would require more time.

Fermat's cofounding of probability theory with Pascal started because Antoine Gombaud, chevalier de Méré, a nobleman with a penchant for games of chance, became interested in the problem of points. This riddle had been proposed much earlier by Luca Pacioli, a Franciscan friar who invented double-entry bookkeeping and tutored his roommate Leonardo da Vinci in mathematics. The problem of points involved determining the fair distribution of the stakes when a game of dice was halted prior to completion. When de Méré turned to it, he discovered that a well-known gambling system predicted one thing, but his own calculations said something else.

De Méré proposed the problem to Pascal, who in turn interested Fermat, and they began corresponding. These two talented, educated, and interested men took energy from the collaboration, and produced between them what has been called "intellectual dynamite."[3] Their letters, sent through the summer of 1654 between Paris and Toulouse, mostly discussed technicalities arising from combinations of throws of dice, but they also contained small personal fragments. For example, Pascal, upon receiving a letter expressing an idea that already had occurred to him, wrote,

> I should like to open my heart to you henceforth if I may, so great is the pleasure I have had in our agreement. I plainly see that the truth is the same at Toulouse and at Paris.[4]

And Fermat, in a letter to Carcavi, who was acting as a messenger between them, said,

> I was overjoyed to have had the same thoughts as those of M. Pascal, for I greatly admire his genius and I believe him to be capable of solving any problem he attempts. The friendship he offers is so dear to me and so precious.[5]

So it was that two Frenchmen, one of them sick in bed, the other stealing time from family and municipal duties, established the basic concepts and rules of a theory that has become absolutely critical to any mature, modern approach to knowledge.

Now that we know when and how the bud first began, let us start thinking about the branch it eventually grew into. In the spirit of the founders, we will ground our entire discussion using dice. Let us think of a single six-sided die. If we roll it, the outcome will be one of six different things:

This exhausts the possibilities. When we throw the die, we will observe one, and only one, of the faces. In probability theory, this complete collection of things that can happen, this space of all possibilities, is called a *sample space*. The nomenclature comes from the term *sampling* for making an observation. So we sample the random process of a die by throwing it, the outcome we observe is a sample, and every sample is contained in the sample space.

Assuming the die is equally weighted, we are equally likely to observe any one of the elements in the sample space. How can we make this idea of "equally likely" more precise? How can we build a useful machine that allows us to calculate likelihood in general? This is how: We will attach likelihood to a number between 0 and 1. We agree that the number 1 represents something so likely as to be certain, and the number 0 represents something so unlikely as to be impossible. This number is called the probability. For example, what is the probability that, upon throwing the die, *some* die face appears? This is certain, so the probability is 1. Similarly, the probability that we will get anything other than a die face (such as a blank face, a seven, a rhinoceros, etc.) is 0. There is simply no way that can happen.

But what about the intermediate cases? And what kind of questions can we ask? And how can we be sure that the numbers are always between 0 and 1? The answer was formulated clearly by Laplace in 1814, in *A Philosophical Essay on Probabilities*. What we do is think in terms of events. An *event* is a thing that can happen (such as getting an even number, or a rhinoceros), and so events are the things we attach probabilities to. To do this, we go over the sample space and count the number of ways the event can happen, then divide that by the number of all things that can happen. This second number is just the size of the sample space, and so the probability of an event is that event's portion of all possible events. This provides the classical definition of a probability, and we will call it the frequency rule. In a formula:

$$\text{probability of event} = \frac{\text{number of ways the event can happen}}{\text{number of all things that can happen}}.$$

In order to think about events, and to attach a probability to them, we need to carve up the sample space in different ways. Each carve-up of this kind is called a *partition*. A partition is just like carving up a cake into different slices of different sizes. The main rule of a partition is that if we fit all the pieces back together again, we have the whole cake. When we partition a sample space, we create a number of subspaces, and each one is a different event. Different partitions therefore define different events, allowing us to ask different questions. Because the collection of all subspaces (events) in the partition must equal the sample space, if we sum the probabilities of all the events in the partition, we will always get the number 1.

For example, say we wanted to know the probability that we might roll an even number on a six-sided die. We make a partition of the sample space, and one subspace is the event "even," which contains the three ways a die can be even: ⚁ ⚃ ⚅. Because there are six elements in the sample space, the probability that we roll an even number is $3/6 = 1/2$. However, we know that the sum of all events must equal 1, so there must be another event in the partition. This event happens when we do *not* roll an even number. In other words, when we roll an odd number. Of course, the probability of rolling an odd number is $1/2$. The partition required to ask about even and odd is therefore:

When we put the two events back together, they equal the entire
sample space. This is like asking for the probability of rolling *either*
an even or an odd number. To ask about this, we would add the probabilities of the individual events: $1/2 + 1/2 = 1$. As expected, this event
is certain.

So far, we have three fundamental ideas. The sample space, a partition of the sample space into events, and the association of the events
with probabilities according to the frequency rule. With these ideas, we
now have a little more purchase on the commonsense idea of "equally
likely." If our event is just a *particular* die face turning up (say ⊡),
then there is only one way for that event to happen, and so the probability is $1/6 = 0.1666\ldots$

We can make a picture that expresses this idea by putting the specific events (the die faces) on the horizontal axis, and their respective
probability on the vertical axis. (The fact that we draw it as a picture
is convenient but irrelevant; we could write it as a table or an abstract
rule instead.) A picture like this is called a *probability distribution*, because it expresses how the probability of 1 is distributed among the different events, the way one birthday cake is distributed among different
children. The distribution of a fair die is also fair to the children—it is
just a straight line. This is called a *uniform distribution* because every
event is uniformly likely.

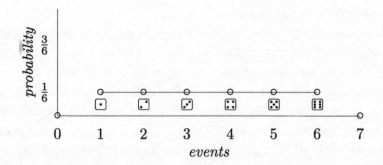

It is easy to get a nonuniform distribution, even from a uniform process. To see this, now roll two identical fair dice instead of one. What is
the sample space for this particular setup? If the first die is a ⊡, then
the second can be ⊡, ⊡, ⊡, ⊡, ⊡, or ⊞. There are six ways that we
can get a ⊡ for the first die. If the second die is a ⊡, it is the same, so

there are six ways that can happen. And so on all the way to 🎲 for the first die. This exhausts the possibilities when we roll two dice. Because there are six different ways for six different things to end up, the sample space has thirty-six elements in it.

Now, what will we do with these two dice? This is a subtler question than it might first appear. We could subtract one from the other, multiply them, or perform any number of other operations. The natural choice is to sum the numbers on the faces, and this is what we shall do. But it is important to realize that this is a choice.

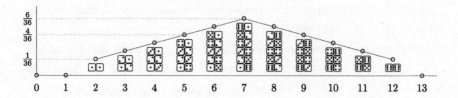

Summing two dice generates numbers between two and twelve, and we pick the most granular possible partition, so each number is a different event. We don't have a uniform distribution anymore. It is obvious that there are more ways to get the number seven (⚀⚅, ⚁⚄, ...) than the numbers two (⚀⚀) or twelve (⚅⚅). If we count all the ways, and make a distribution, we get a pyramid shape that starts small for the number two, gets larger and peaks at the number seven, then gets smaller back to the number twelve. The probability that you will roll a seven on two dice is $6/36 = 1/6$. The probability that you will roll a two or a twelve is $1/36$ because there is only one way each of those outcomes can happen, out of the thirty-six different possibilities.

If we roll three dice, the distribution changes again—it peaks more in the middle and falls away more gradually at the sides. As we add more and more dice, this shape-shifting continues. In fact, as we add more and more dice, something interesting happens. The distribution starts to resemble more and more the famous bell curve, also called the standard normal distribution, or a Gaussian function, after the German mathematician Carl Friedrich Gauss. The transformation is rapid, and is already well underway by the time we roll three dice. Here is a figure showing the distributions corresponding to one, two, and three dice.

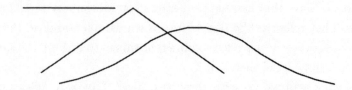

The more we look at the world, the more we find the bell curve. It describes test scores, errors in measurements, the sizes of things made by machines, blood pressure readouts, the weights of one-year-old babies, the antler sizes of white-tailed deer, the amount of light generated by light bulbs, the number of potato chips found in each bag, and on and on. Why is it so prevalent?

Whenever events vary because of a lot of independent factors, which can themselves vary, the probabilities of the events will be described by a bell curve. You can think of each factor as a die roll, and all of these rolls come together to determine the outcome. For example, there is a test in the United States called the SAT, which is taken by millions of college applicants every year. If you count the number of people who got each individual score, the results are very close to the normal distribution. Most people score close to average, with very few people doing exceptionally well or poorly. The test score of each person is here like the sum of multiple dice. The different "dice" being rolled for each person in this case are probably the student's inherent intelligence (whatever that is), the quality of teaching from the earliest days up to the present, the quality of test preparation, whether the student was intrinsically or extrinsically motivated, the support of parents, how much sleep the student had the night before, and so on.

From the simple basis of thinking about dice, we have explained something fundamental about nature. This is what makes science thrilling. The world can be understood, and this understanding is not reserved for ordained scientific priests. By simple and clear thinking, anyone can do it. The fact we have just described is today called the central limit theorem. In technical jargon, the theorem states that the average of almost any set of independent random variables rapidly converges to the normal distribution. This fact is fundamental to probability theory, and was first proved by Laplace in 1810.

We have met two of the most important distributions, the uniform and the normal. But there are an enormous number of different distributions—Wikipedia lists over one hundred. Despite their variety, every single one shares a common feature. In providing a mapping from events to probabilities, every distribution collects in one place the effects of the physical structure that creates the various likelihoods for the different events. Because it encodes patterns that emerge out of physical systems upon repetition, the distribution cannot itself be a physical system. It does not exist on the same level as the system. It is a theoretical construct which describes facts about the system that are impossible to observe directly.

This is so important that I will repeat it, instarred:

✳ ✳
✳   ✳
✳   *A distribution is not a physical thing.*   ✳
✳   *It is an encoding of the repetition*   ✳
✳   *patterns inherent within physical things.*   ✳
✳   *Physical things have repetition patterns*   ✳
✳   *because they have structure.*   ✳
✳   ✳
✳ ✳

Chapter 14

The Kinetic Theory of Gases

Eventually, after hundreds of years of confusion about the hidden nature of heat, scientists realized that there was something special about gases. It is here, not in solids or fluids, that we find the simplest nontrivial example. Gas structure has everything you need, and none of the bedeviling complications of heat that make the study so difficult. So let us take a closer look at something we all take for granted: the air.

Imagine you're having a picnic someplace dry. It's a nice day, about 77 degrees Fahrenheit (25 degrees Celsius), and you stare at a cubic centimeter of air right in front of your face. This is roughly the size of the tip of your finger, underneath the pinky nail. In this tiny portion of space there are 78 nitrogen molecules for every 21 oxygen molecules, and the total number of molecules is inconceivable, astronomical, it is so big you can't even fathom it, but I'll put it down here anyway: 253 with 17 zeros after it, or if you prefer meaningless words instead: 25.3 quintillion. A quintillion is a billion billion, so if there was a person with a quintillion dollars in this world, they would have a billion dollars for every one dollar a billionaire had. And since we're on this theme, consider this: each time you breathe, about 25 sextillion molecules flow into your lungs. This is far more than the number of stars in the milky way galaxy. In fact, if you breathe nine times, you have cycled more molecules through your lungs than there are stars in the entire universe.

We would like to talk about the properties of gas, but we can't even begin to talk about the individual molecules in a single cubic centimeter of air. If you tried to do it with a computer, your program would hang for the rest of your life at the first step, and that is assuming you had enough memory to specify the state of each molecule at one moment, which you don't. It would take 25.3 billion gigabytes of memory to represent this cubic centimeter of air at one moment, assuming each molecule only required one byte. The latest MacBook Pro comes with 512 gigabytes standard, so, yeah, just a little short.

With so many molecules packed into such a small space, we lose the ability to say anything specific about individuals. The only kind of systematic knowledge we can develop is statistical. But this trade-off is brilliant, because the statistical properties are the most important ones, and not only that, they are extremely robust. The robustness follows from something called the *law of large numbers*, which says that the statistical properties become more pronounced with more samples, and dominate in the long run. If you had one hundred dice and suspected a villain had weighted one of them, you could not tell if this was true by rolling all the dice once, or even a thousand times. But if you rolled them a *billion* times and looked at the distribution of results, you would certainly be able to tell whether something was just a little bit off.

But what are the general properties of these molecules? For a long time, most people thought that gas molecules adopted a lattice formation due to long-range repulsive forces between them, and then did not move unless disturbed. This was called the *static* theory of gases, and was successful for several reasons. The idea of long-range inter-particle forces was associated with the great Newton, and the origin of the forces was attributed to everyone's favorite self-repulsive material, caloric.

However, we know today that this is wrong. We should think that gases are formed of innumerable tiny particles, traveling at high speeds in straight lines. The first person to publish this idea was the Dutch-born mathematician Daniel Bernoulli. He came from a family of well-known mathematicians, and first learned the trade from his father, Johann, who also taught the great Euler.[1] So far, we have met some of the winners of the grand prize of the Paris Academy. For example, Fresnel for his memoir on diffraction. Bernoulli also won this prize—ten times. The majority of his triumphs were for nautical topics, drawing from his deep knowledge of hydrodynamics, a field he helped create. In fact, our modern term "hydrodynamics" comes from Bernoulli's *Hydrodynamica*, published in 1738. This is his most famous book, in no small part because chapter 10 contains an idea that was well ahead of its time, and deceptively simple.

Bernoulli's proposal is called the *kinetic* theory of gas structure, from the Greek word for motion, *kinesis*. It was not particularly influential for a hundred years after it came out, the static theory being predominant. But Bernoulli's idea was independently rediscovered in 1820 by John Herapath, an Englishman, and again in 1840 by John James Waterston, a Scot. They were too early. Prejudices, even in those like Davy, who supported the vibrational hypothesis, prevented their work from being influential. Waterston's paper was described by a referee as "nonsense, unfit even for reading before the [Royal] Society."[2] And yet, within a decade of this damning judgment, the kinetic theory supplanted the static one, and succeeded in "finally condemning that theory to an obscurity from which it has never since emerged."[3]

Let's remove the confusion of false hypotheses. The caloric does not exist, so we get rid of it. We also get rid of the static theory of gases, instead imagining that particles move rapidly all the time. The originators of the kinetic theory thought of a gas like a bunch of billiard balls, flying around in three dimensions. From time to time they hit the sides of the container or each other—*clack!*—and then recoiled according to the known laws of mechanics.

So now imagine what happens if all of the molecules in this square centimeter of air are going at random directions, but for some bizarre reason, they are all going the same speed. The moment they start colliding, this uniformity of speed will be destroyed. First of all, nitrogen weighs less than oxygen, so even if a nitrogen and oxygen molecule collide head-on at the same speed, after recoil they will have different speeds. Moreover, two molecules of the same type will hardly ever collide head-on. If they collide in a glancing angle, only part of the directional speed from one may be imparted to the other. What if a molecule catches up with another one and bumps it from behind? Or multiple molecules hit another one at the same time but from varying angles? There are a lot of possibilities.

Because of this, the randomness very quickly takes over everything. We now pretend we are able to sample the molecules in this tiny portion of space and measure their velocity. Each sampled molecule can have one component of its velocity going left-right, a second going up-down, and a third going forward-backward. We can focus on just the left-right component of velocity, since each of these components is the same because of spatial symmetry. If the molecule is going to the right, we say its speed is positive, and to the left will be negative speed. What are we likely and unlikely to see in this case? What will the probability distribution of velocities look like?

This problem was first attacked by James Clerk Maxwell. We met him, all too briefly, when we discussed the theory of light waves. In 1859, six years before Maxwell published his electromagnetic equations, he was working on the kinetic theory of gases. The first sentence of his breakthrough paper on this topic is

> So many of the properties of matter, especially when in the gaseous form, can be deduced from the hypothesis that their minute parts are in rapid motion.[4]

Maxwell computed the distribution of velocities for the first time. His work was expanded in 1868 by Ludwig Boltzmann, and for this reason the probability distribution of velocities is called the Maxwell-Boltzmann distribution. It is a bell curve, centered on zero:

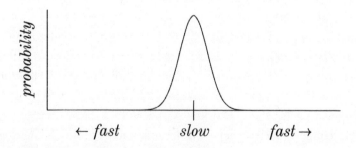

This is saying that most of the molecules will be going slowly to the left and the right. Some might not be moving left-right at all (though they could still be going up-down or forward-backward). A very few of these molecules will be going fast, either to the left or the right.

Now, it is useful to imagine what happens as the sun gradually warms the world. The temperature goes up, and this has an effect on your lemonade, on your skin, and on the cubic centimeter of air in front of your face. If we sample the molecules again, their speeds will still be distributed along a bell curve, but with a wider spread. Many more of them will have a higher speed, and in aggregate, they have sped up.

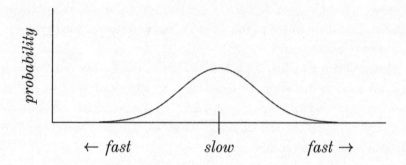

This is the essence of the answer. The hidden nature of heat is statistical. Heat is not a thing, but an emergent property of an inconceivably large number of molecules. Any molecule has a mass, which is a number specifying how much it weighs, and a speed, how fast it is going. The energy associated with this molecule's motion, formerly called *vis viva* and today called kinetic energy, is dependent on these two properties alone. The heat at some point in a cubic centimeter of air is the *average* of all the kinetic energies at that point. Today we know what heat is. It is mean molecular kinetic energy.

We can begin to understand now why heat behaves the way it does. The kinetic energy in a system will, over time and through an unimaginable number of collisions, spread evenly. If two neighboring regions of iron have different temperatures, then the probability distributions of speeds must be different. But because they are neighboring, they will interact with each other. The aggregates with a lower average speed will start going faster as they steal kinetic energy from the aggregates with a higher average speed, which, of course, start going slower. The system will ultimately seek a single probability distribution of speeds. If the distribution does not change from point to point in the iron bar, everywhere must have the same temperature. This is called thermal equilibrium, and no heat is flowing.

We have, in fact, just discovered why the heat equation works. We have solved Comte's "inaccessible problem," and uncovered the hidden nature, the primary cause, of why temperature varies in time as the spatial curvature. We can grasp the essential point by considering a single collision. If two identical particles are going the same speed and they collide head-on, they will simply change their direction. The kinetic energy of both will be unchanged. This is like having no difference in temperature between two neighboring points. If there is no temperature difference, then of course the curvature is zero, and there is no temperature change.

On the other hand, if one particle is still and another of equal mass approaches it swiftly in a head-on collision, *all* of the kinetic energy is transferred. The particle that was still is now going fast, and the fast one has come to a halt. This is like having a very high temperature difference between neighboring points. The cold (slow) region becomes fast (hot), and vice versa. A high spatial curvature in temperature means that there is a big difference, on average, in the kinetic energy of the countless molecules in two touching points. The Maxwell-Boltzmann bell curves will be very different. On average, collisions will have a big effect on both of them, and the probability distributions of speed will change shape quickly in time.

The most general idea here is *diffusion*, and it has its own history. In 1833, the Scottish chemist Thomas Graham first studied how fast different kinds of gases flow through tiny plugs in plaster of paris. He derived a law of relative diffusion speed that bears his name today, and his work inspired that of Adolf Fick, a German-born doctor who studied the diffusion of matter in general circumstances. Fick has two laws named after him, both involving the diffusion of mass undergoing numerous random particle collisions. As it turns out, his second law has an identical mathematical form to the heat equation.

The diffusion of dye in water is an accessible example. If we had a little dish of water and we dropped dye into it, Fick's second law says that the dye will diffuse through the water exactly as heat flows through an iron bar. Places of high concentration will lose dye, and places of low concentration will gain it. If we drew the concentration as a function, the shape of this function will melt. We already understand the law describing exactly how it melts: the way dye spreads in water is determined by how it is concentrated. The dye particles will, merely by the laws of statistics, seek the average concentration of the surrounding points. This is exactly what temperature does, too, as we know. It equalizes itself; it seeks the average of surrounding points as it melts through everything.

As physical systems, the kinetic energy of many iron atoms in a cubic crystal is completely different than the number of dye molecules in a dish of water. And yet, because they are both governed by the same differential equation, they are *necessarily* identical at a certain level of abstraction. In fact, wherever we have lots of random events that can be considered, roughly, as collisions, we may expect the heat equation to be involved in some way. The moment we think to look, we can find many other physical processes that diffuse like heat does. For example, the theory of asset prices, begun in 1900 by Bachelier, uses ideas of diffusion and is therefore linked to Fourier's heat equation. Prices, as it turns out, also diffuse. Other areas of science where diffusion is relevant include the spread of fluid through soil, the flow of blood through capillary veins, the theory of resistance in electrical circuits, the theory of compound errors, the theory of random walks, the spread of neutrons in graphite, and so on.[5]

This is exactly what we expect to happen as knowledge advances. A differential equation may be mysterious at first, but it must work for a reason. In time, this reason will be uncovered and lead into surprising places. As understanding of heat matured and combined with other investigations, it became thermodynamics, the science of "fire, ice, and the universe."[6] This resulted in much more than an understanding of heat. It yielded insights of a fundamental character, into energy, entropy, information, work, life, causality, and time. Thermodynamics is among humanity's greatest achievements. Einstein called it "the only physical theory of universal content, which I am convinced . . . will never be overthrown."[7]

Part IV
Interlude

Chapter 15

Parable of Atoms

The word *atom* comes from the Greek. *Tomos* is a cut, and the *A* in front is called an alpha privative; it negates what comes after. A-moral, A-political, A-tomic. The Greek idea of atoms was something uncuttable. Something you could not divide into anything smaller. An atom would be basic, a building block, the fundamental constituent of everything else around us that we see and interact with. Understanding atoms, then, is to understand reality at some core level, because atoms make up all the stuff.

Not everyone believed in atoms, even after the kinetic theory of gases took off. You could explain a lot of things using the atomic hypothesis, they said, but so what? That did not automatically mean atoms exist. You could explain a lot of things with caloric, too. And so the atomic hypothesis became a matter of hot dispute.

Ludwig Boltzmann believed in atoms enough to base his entire career on them. With dark hair, a thick beard, and a brooding mien, he cuts an imposing patriarchal figure in the many busts made of him. He relished food and drink and put on the pounds steadily, to the point that his wife Henriette called him "my sweet fat darling."[1] He had two sons, Ludwig Hugo and Arthur, and three daughters, Henriette, Ida, and Elsa. He was brilliant, and tenacious, and his scientific guiding star was well expressed at the beginning of his book *Principles of Mechanics*:

Bring forth what is true;
Make it clear to view;
And defend it till you're through.[2]

Boltzmann insisted on the existence of atoms. He wrote papers with titles like "On the Indispensability of Atomism in Natural Science." He said things like "Perhaps the atomic hypothesis will one day be displaced by some other but it is unlikely."[3] Boltzmann admired the power of Maxwell's reasoning, and he extended it in broad and deep channels.

However, many scientists and philosophers during Boltzmann's time did not believe in atoms. The most influential and well-known of them was Ernst Mach. He worked on the physics of shock waves, and his philosophy of science sent its own shocks throughout all of Europe.

Mach had very strong opinions about physics, which he asserted should obey something called the "economy principle." Make the most of the least. All we have access to is our perceptions and observations, so physics is exactly the theory of mathematically ordering what we can observe. Economy of thought meant that hypothesizing about the hidden natures of unseen objects was bad practice. Added baggage. A waste of time at best.

Is this ringing any sort of bell? This hostility to metaphysics led many to call Mach a positivist, and he was so influential that positivism is today more closely associated with Mach's name than Comte's.[4] While he drew hardly any inspiration from Comte, Mach did admire his French compeer. In 1902, for a memorial that can still be seen in the Place de la Sorbonne, Mach wrote:

> You have just erected a statue dedicated to the great French philosopher, Auguste Comte...to the man who has fought with energy against metaphysical superstitions; I would like to associate myself to the international homage to the great thinker....But he erected for himself a much more imposing monument through his life and his works, the influence of which is growing every day among us.[5]

Mach also admired Fourier, and for the same reasons that Comte did. Mach said that Fourier's theory of heat conduction was "an ideal physical theory" because it "is founded, not upon a hypothesis but upon an observable fact." Because of this, Fourier's theory "remains secure." But, he said, "a hypothesis like that of the kinetic theory of gases...must be prepared at any moment for contradiction by new facts."[6]

For Mach, atoms were just another metaphysical superstition. At a meeting of the Imperial Academy of Sciences in Vienna in 1897, a group of academics were debating atomism. Boltzmann was there, and in his words: "Mach suddenly said laconically: *I do not believe that atoms exist.* This utterance ran in my mind."[7]

Mach was much more influential than Boltzmann, at least when it mattered to both men. By 1883 he had already written the first volume of a historical treatise on physics that made him very famous, and had "a great influence on an entire generation of scientists."[8] Einstein said that everyone had imbibed Mach's ideas "so to speak, as mother's milk."[9] By 1901, he had ascended beyond scientific matters and become a member of the Austrian Parliament. After his death in 1916, Mach's influence and power only grew. Mach had such an influence on leftists that Lenin thought it worthwhile to write a book against his ideas. And Mach also inspired logical positivism, a radically anti-metaphysical movement that had a deep impact on modern analytical philosophy. It is associated with Wittgenstein and the Vienna Circle, which was originally called The Ernst Mach Society.

Boltzmann, on the other hand, was isolated and struggling. He wrote that, because of the influence of Mach's work in 1883, the attacks on the kinetic theory "began to increase." He was more than capable of defending himself, and participated in many debates about atomism, at one point ripping his opponents to shreds.[10] Boltzmann fought, admirably and continually and with all his might, for the truth. But he was outnumbered.

> I am conscious of being only an individual struggling weakly against the stream of time. But it still remains in my power to contribute in such a way that, when the theory of gases is again revived, not too much will have to be rediscovered.[11]

Boltzmann felt like he was drowning in a sea of falsehood, but he never doubted the atomic hypothesis. Note the tone of assurance above.

Boltzmann *knew* that atoms exist, which is why he also *knew* that the theory of gases, which must have seemed to him on the verge of being defeated, would eventually be revived.

There is only so much one man can do, and especially a man like Boltzmann, who was not particularly suited to be a scrapper. His health was failing him. He had asthma, he wore thick spectacles because his eyesight was going, he was overweight, and, above all, he had been emotionally unsettled his whole life. His brain was prone to debilitating swings between ecstasy and grief. He joked that this was due to his birth, which was in the night between Carnival and Lent.

At the end of 1906, when he was sixty-two years old, Boltzmann sought relief from his illness. He went on vacation with his wife and daughter Elsa, who was fifteen, to the village Duino. There the Duino Castle is built on top of a rock overlooking the Adriatic. Rilke knew the place; it served as inspiration for his *Duino Elegies*. The fifth of September was the end of the vacation. The family planned a return to Vienna, as Boltzmann was due to begin lectures the very next day. Henriette and Elsa went down to the sea for a swim. When they came back they found Ludwig Boltzmann, husband, father, teacher, had hanged himself.

This tragic end was just one year after Einstein proved the existence of atoms with his miraculous paper examining the random motion of pollen grains in water. If Boltzmann had held on just a little longer, he would have seen the tide turn. Today, we have the ability to see atoms, to manipulate them one by one, using them to spell words and even make stop-motion films.

The very beginning of Feynman's famous *Lectures on Physics* is this:

> If, in some cataclysm, all of scientific knowledge were to be destroyed, and only one sentence passed on to the next generations of creatures, what statement would contain the most information in the fewest words? I believe it is the *atomic hypothesis* (or the *atomic fact*, or whatever you wish to call it) that *all things are made of atoms—little particles that move around in perpetual motion, attracting each other when they are a little distance apart, but repelling upon being squeezed into one another.*[12]

Today, nobody in their right mind doubts the existence of atoms. Mach, like Comte, was fatally wrong about the role of hypothesizing in science. Out of positivistic convictions, Mach opposed what is arguably the most important idea anyone ever had. When evidence of atoms was presented to him, he still refused to believe that they were real, which is why Brush called him "the unrepentant sinner."[13]

Einstein said of Mach's philosophy that "it cannot give birth to anything living, it can only exterminate harmful vermin."[14] We should take this metaphor literally. Positivism is rat poison. Maybe you should keep some around in case of an infestation, but for god's sake, don't swallow it.

Chapter 16

Quantization

Consider a toaster. When it heats up, the coils start to glow. The same thing happens with a light bulb, lava, embers, or anything that is "red-hot," like a poker. Well, why do these things start glowing when they are hot? As it turns out, this is the wrong question, because they are always glowing. Everything is glowing with its own internal light. You are glowing right now. However, just as some sounds are beyond human hearing, we can only see this glow when things get hot enough.

Toaster coils glow red, but if they heated up more they would become orange, then yellow, then bluish-white. (These are roughly the different colors that stars can be.) Children, scientists, and other curious people might wonder: How much of each color do you get as you change the temperature? This is called the *blackbody problem*. It is easy to state, but it took almost half a century to solve.

The blackbody problem was first studied in 1859 by Gustav Kirchhoff, but it was not until 1893 that a rough answer was presented by Wilhelm Wien. Wien's ideas were clearly on the right track, but they did not match the observed colors for low frequencies of light. A different idea, the so-called Rayleigh-Jeans formula, fixes Wien's formula at the low frequencies, but at the high frequencies (above violet) it predicts that even everyday objects like a broken toaster would radiate enough energy to destroy the Earth. Luckily this prediction, called the "ultraviolet catastrophe," is fantastically wrong.

Max Planck, the Prussian physicist who eventually solved the black-body problem in 1900, described his breakthrough "simply as an act of desperation." He made a somewhat strange and arbitrary assumption in his mathematical derivation, one which Planck himself treated as a trick with no real meaning. As he said, it was "a purely formal assumption...actually I did not think much about it."[1]

The significance of Planck's idea was drawn out and clarified by Einstein, and today we consider these developments in 1900 and 1905 to be the very birth of quantum mechanics. Planck showed that matter can only absorb or emit light in tiny individual units; Einstein interpreted these units as particles of light. The amount of energy in each particle depends on the frequency (color) of the light. Recalling that red light has a frequency of 740 ten trillion cycles per second, Planck's idea means that the energy of red light inside a closed furnace has to be proportional to 740, or 740 + 740, or 740 + 740 + 740, and so on. The red light comes in chunks, indivisible bits you cannot cut into smaller parts. Just like atoms.

Physicists talked about this discreteness with the word *quantum*, Latin for "how much." We can only imagine how many times the Romans said *quantus est*? when asking for the price of something in a market. Planck's resolution of the blackbody problem used this discreteness. Einstein's resolution of a different problem (the photoelectric effect) also used this discreteness. But that was just the beginning. It turned out that many things could be explained by this idea of quantization.

Compton used the light quantum hypothesis to explain how light scattered from matter. The old problem Black discovered, of the differing specific heats of different materials (like mercury and water), had a classical theory that did not give the right answers at low temperatures. Einstein assumed that *molecular* energy was also quantized, and was able to remedy the classical defect. Then, in 1913, came a breakthrough so spectacular that we should take a few pages to really understand it.

The atomic hypothesis, as we saw, is likely the single most valuable idea ever conceived. But what are atoms supposed to be? What kind of thing is uncuttable, indivisible? Attempts to answer this question go all the way back to the beginnings of science, and by the twentieth century, there was one outstanding mystery about atoms that served as a clue.

It was 1666 when Newton darkened his room and shined a beam of light through a prism. Rather than a circle of white light, he saw the rainbow. The beauty, simplicity, and importance of this experiment are all evident, but even the simplest experiment can contain hidden treasures. In 1802, almost 140 years after Newton first did the experiment, another invaluable result came out of the simple combination of light and a prism.

The English chemist and physicist William Hyde Wollaston was comparing the ability of different materials to refract light. Among other things he looked at diamond, cold butter, the crystalline lens of an ox, the vitreous humor of an eye, arsenate of potash, the white of an egg, pitch, human cuticle, and horn (these last three refracted no light).

This paper of Wollaston's would likely be forgotten if not for an observation he made at the end, merely in passing. He had done Newton's famous experiment again, using a prism made of flint glass. He was very careful, and by looking closely he observed for the first time something that Newton himself could conceivably have seen. In Wollaston's own words:

> If a beam of day-light be admitted into a dark room by a crevice 1/20 of an inch broad...the beam is seen to be separated into the four following colours only, red, yellowish green, blue, and violet....There are also...distinct dark lines, *f* and *g*, either of which, in an imperfect experiment, might be mistaken for the boundary of these colours.[2]

Wollaston's original diagram.

Wollaston described the dark lines, but he did not even remark on how strange they were. Around 1820, however, the strangeness was clear. Joseph von Fraunhofer, a name known to all who study optics, had developed the diffraction grating, a tool which improved upon the prism and allowed him to separate sunlight out much more broadly than was previously done. He identified *thousands* of dark gaps in the continuous spectrum. As it turns out, the more you expand each color, the more gaps you find. Today we call these gaps *spectral absorption lines*. In the nineteenth century, they were quite a puzzle.

It had been known for a long time that throwing chemicals into fire could change the color of the flame. This experiment is popular even today in basic chemistry classes. The light from this flame could be put through a diffraction grating, where bright lines would show up in the rainbow. For example, dropping sodium bicarbonate powder (also known as baking soda) into a flame will turn the flame more yellow. Separating the light into a rainbow using a prism or diffraction grating will reveal two bright bands of yellow right next to each other. These are called the sodium D lines. The same thing can be done with gas. Passing a high-voltage charge through a tube filled with pure hydrogen gas results in a glowing red light. In fact, this is how neon lights work. A red neon light does not have neon gas in it at all—it has hydrogen. A yellow neon light actually has helium, and blue neon lights are full of mercury gas, so please don't break them.

Putting the light from these gases through a prism or diffraction grating reveals that there is more than just one color present. For example, electrified hydrogen gas emits light that breaks apart into four different colors, and it looks red to the naked eye because the red light is the strongest.

These are called the *spectral emission lines*, and it is natural to look for a connection between these and the absorption lines. In 1853, the Swedish physicist Anders Ångström speculated that the different colors emitted by electrified gas were the same colors this gas could absorb in different circumstances. Kelvin had a similar idea around the same time. Then in 1859, Gustav Kirchhoff and Robert Bunsen confirmed the suspicion. The dark lines Fraunhofer saw in the rainbow were essentially missing colors, and these exact same colors were observed in the emitted light of heated elements.

It was therefore discovered that a bright band means the atom is emitting light at that color, and a dark band means the atom is absorbing it. Every atom has its own emission *and* absorption spectrum, formed of exactly the same colors. Also, if two atoms are different, they will have a different spectrum. The spectral lines thus allow us to identify the presence of atoms. The dark Fraunhofer lines were like a bunch of fingerprints jumbled together saying *hydrogen, helium, calcium, iron, etc. were all here.* And so the mystery was solved: the gaps in the sun's rainbow were correctly attributed to the absorption of light by different elements in the sun's atmosphere.[3]

Research into hydrogen, the simplest, lightest, and most abundant element in the universe, found more than a hundred spectral lines. Because every line corresponds to a color, and every color is a frequency of vibration, the physicist Arnold Sommerfeld quipped that the hydrogen atom must be more complicated than a grand piano, which only plays eighty-eight frequencies. How the lines themselves were formed was still a complete and total mystery. Why did the colors emerge as they did? Atoms clearly had a rich deeper structure, but what was it?

This brings us to the promised breakthrough, made by the singular Niels Bohr. We will spend all of Chapter 33 talking about Bohr's personality. For now, just know that we are about to meet one of the most remarkable men of the twentieth century.

In the year 1911, Niels Bohr, twenty-six years old and freshly minted as a physics PhD, left his native Denmark and sailed for England. He had a postdoctoral year at the Cavendish, funded by the Carlsberg Foundation. His application, which says a lot about how much academia has changed in a century, went like this:

> I hereby permit myself to apply for 2500 kroner to spend one
> year at foreign universities.[4]

But the collaboration with his famous mentor (J. J. Thomson) did not go particularly well, and so Bohr wound up in Manchester, working with one of Thomson's many students, the New Zealand–born Ernest Rutherford. While Rutherford was generally suspicious of theoretical physicists, he said, "Bohr is different. He's a football player!"[5]

After the transfer, Bohr only had a few months left of his fellowship and nothing to show for it. But Rutherford's lab was a good place to be. Mere months earlier, two students, Geiger and Marsden, had discovered that particles scattered in a way "almost as incredible as if you fired a 15-inch shell at a piece of tissue paper and it came back and hit you."[6] This led to a new model of atoms, which everyone today knows, in which electrons orbit the nucleus like planets around a star.

But Rutherford's model had a problem: if electrons orbit a nucleus, they should radiate energy, and as they do so, spiral inward and collapse into the nucleus in a fraction of an instant. The newest theory of atoms on the scene in 1912 predicted, among other things, that matter is not stable and therefore that the world does not exist.

Bohr attacked this problem, and the answer that he published in 1913 laid the foundation of his fame. He simply *postulated* that electrons did not radiate energy when they orbited the nucleus. Not only that, he said that electrons were restricted to quantized values of angular momentum. They could only go around the nucleus in certain orbits. Electrons simply could not exist in any other orbits than the special stable ones, which again involved Planck's constant.

With his model, Bohr could compute all the spectral lines of hydrogen. A particular frequency of light was absorbed when an electron jumped to a higher energy orbit. It was emitted when an electron fell to a lower energy orbit. All of the facts of hydrogen's interaction with light were immediately, and perfectly, predictable. The problem of spectral lines was a door into atomic structure, and quantization was the key.

As Bohr put it, fifteen years after his first breakthrough:

> The essence of quantum theory is the quantum postulate: every atomic process has an essential discreteness—completely foreign to classical theories—characterized by Planck's quantum of action.[7]

So not only is light quantized, but molecular motion and even the energy of electrons in atoms are quantized. This discreteness is a fact of nature, but... *why*? The physicist John Archibald Wheeler, whom we will meet again, considered this question as recently as 1986 in a paper called "How Come the Quantum?" He wondered: "The necessity of the quantum in the construction of existence: out of what deeper requirement does it arise?"[8]

In modern theories, the existence of the quantum is assumed because it works. Some physicists might have proposals to answer Wheeler's question, but I would wager that none of them are classical and dynamical in nature. And yet, here's something amazing: the hydrodynamic pilot-wave system is quantized as a result of classical fluid mechanics.

Chapter 17

Hydrodynamic Quantization

With low vibrational forcing, the frequency of the droplet bouncing is initially the same frequency as that of the bath. However, when the forcing is increased, the droplet undergoes a transition called a period-doubling event. Then, the droplet's bouncing frequency slows down to one-half what it was before and resonates with the vibrating oil bath.

Resonance is a perfect matching between two frequencies. It's like pushing a child on a swing, at exactly the right moment each time. The swing will go higher and higher. If you are not pushing in resonance, eventually you will push forward when the swing is coming backward, an unpleasant experience for all concerned. Pretty much all physical objects have resonant frequencies, one common example being wine glasses. By sliding your finger around the rim of a wine glass you can eventually create resonance between your finger and the glass. This causes the glass to start vibrating at its resonant frequency, happily within the range of human hearing, and the glass will sing.

Without the resonance, the droplet would be creating waves at all different moments, like the parent incompetently pushing a child on the swing. If the droplet did this, many of the waves would interfere, destroying any regularity. So resonance is important. When the droplet locks into resonance, the bath comes alive with standing waves. These waves, moreover, are highly structured and simple, with a single wavelength.

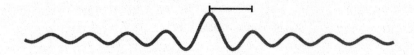

This is a side profile of the wave disturbance created by a single bounce, seen close up. (For the boffins reading this, the shape is a cylindrical Bessel function of the first kind and order zero.) The straight line measures out the distance between two consecutive peaks of the wave, a distance known as the wavelength. Because these waves are being generated in a vibrating bath that is just beneath the Faraday threshold, we call this the Faraday wavelength.

In HQA, the Faraday wavelength plays a role analogous to the de Broglie wavelength of matter in quantum mechanics. The wavelength is a constant feature of the wave, and so the droplet's dynamics are strongly influenced by it. The other important feature of the droplet system is the memory. As we described before, each wave created by a single bounce persists for some time, leaving behind historical traces for the droplet, called path memory. It is the combination of resonance and path memory that "is ultimately responsible for all of the system's emergent quantum features."[1]

In a 2010 paper, "Path-Memory Induced Quantization of Classical Orbits," Fort and coauthors in Couder's group set up the bouncing droplet system with a twist, literally. They put the whole system on a surface that could rotate at various speeds, and observed the trajectory of the droplet. What they found was that the droplet traveled in a circle, with a radius depending on the speed of rotation. As they increased the rotation, the radius got smaller.

There was a smooth dependence of radius on rotation; the system was continuous and had no quantization. However, when they increased the memory so that the droplet was influenced by waves going along the entire orbit, "the resulting closed orbits present a spontaneous quantization."[2]

The multiple waves that the droplet laid down over its whole orbit all add up to a completely different wave surface, radially corrugated in ripples going outward from the orbital center (not necessarily the center of the bath). The droplet winds up *confined* to the maxima (peaks and troughs) of one of these ripples. The waves from around the entire

orbit add up to a new pattern, also a highly regular wave field, which
dictates the path that the droplet can take.

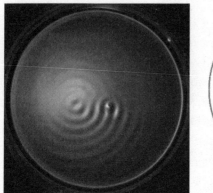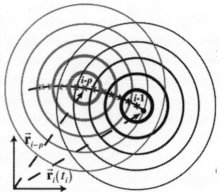

Left: Image of walking droplet in a rotating bath, navigating a corrugated wave
field. *Right*: A diagram showing the simple geometrical considerations explaining the corrugated wave field. Reprinted with permission from Fort et al. (2010)
"Path-Memory Induced Quantization of Classical Orbits," *PNAS* 107:41.

These are the sort of surprises we get from wave memory. We start
with a wave that the droplet causes, and add enough of these waves up,
in a given droplet trajectory, and we get a new wave field that supports
the droplet taking that trajectory. The wavelength of the emerging
wave field depends, itself, on the original wavelength of the wave that
the droplet creates. Because of this, the allowed radius for the droplet is
quantized to half-multiples of the Faraday wavelength.

Because memory can be hard to think about, I want to belabor the
point. Imagine that the particle is going around in a circle, each time
creating an identical standing wave that persists, centered at the droplet impact point. As we add up more and more of these waves, they
cancel out in some places, reinforce in others, and a pattern starts to
emerge. This pattern is the emergent wave field, completely dependent
on the trajectory of the particle. For a particle going around in a circle, the emergent wave field becomes clearer and more pronounced, the
more waves we lay down:

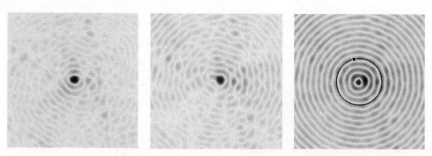

Wave field of circular orbiting droplet building up over time, after (from left to right) 5,000, 10,000, and 20,000 time steps. The circular trajectory of the particle is shown in the rightmost image, but it is present in all of them.

How close is this analogy to real quantum phenomena? As it turns out, very close indeed. We are not exactly dealing with Bohr's atomic model here, because when we rotate a fluid system we get something called the Coriolis force, which is analogous to forces acting on electrons in a magnetic field (called the Lorentz force).

If you have a uniform, smooth magnetic field and send an electron zipping into it, this electron will be affected by the field in a simple way: it will start to curve to the left or right. Eventually, this curve will loop around in a circle. The stronger the magnetic field, or the faster the electron, the tighter the circle. But the circle cannot have just any radius. In fact, the allowed radii form equidistant concentric rings, exactly like the droplet experiment we just described.

This is called Landau quantization, after the Soviet physicist Lev Landau, who discovered many fundamental quantum phenomena. The different radii that charged particles are allowed to adopt in a magnetic field are called Landau levels. Now, take a look at these two mathematical formulae:

$$\frac{1}{\pi}\left(n+\frac{1}{2}\right)\lambda_{dB} \qquad \frac{5}{3}\frac{1}{\pi}\left(n+\frac{1}{2}\right)\lambda_{F}$$

The math here is not hard. The formula on the left is the one that Landau found for the allowed radii for an electron in a magnetic field. The symbol π is just pi, that number they tell you in school: 3.1415.... Dividing 1 by it gives you 0.318 or so. The symbol n is just 1 or 2 or 3 or..., and λ_{dB} is the de Broglie wavelength, the mysterious wavelike property that de Broglie said is associated with all matter.

Fort and coworkers fit the observational data of the droplet orbits in a rotating bath, and came up with the formula on the right. The only difference is the prefactor of 5/3, which arises as a fitting parameter. Multiplied by $1/\pi$ it equals .526 or so. The symbol λ_F is the Faraday wavelength, which defines the waves that the droplet sets down, and is a characteristic of the system as a whole.

These mathematical formulae are almost identical, and so in the case of Landau levels, the analogy is strong. It gives us something very close to the modern theory of particles in a magnetic field. In the droplet case, at least, the meaning of the Faraday wavelength and the origin of the quantization are entirely clear. This is why those who publish papers on HQA often have cause to use some version of Fort's phrase:

> Although this effect is unintuitive in quantum mechanics, it receives a simple interpretation in the hydrodynamic analog.[3]

Finally, let us consider another droplet system with even more quantization. Researchers in Couder's group thought up a clever way to subject their droplets to a space-filling external force, what in physics is called a potential. They "loaded" the droplet with a small amount of ferrofluid, which is a fluid that is full of magnetic nanoparticles. Then, they applied a magnetic field above the oil bath. The field was zero in the precise center of the bath, and it got progressively stronger toward the circumference. This is a linear spring force, so-called because the further the droplet strayed from the center, the stronger the forces pulling it back became.

Such a setup is called a central harmonic potential. It is harmonic because systems subjected to this potential oscillate back and forth, just like a pendulum; they are called simple harmonic oscillators. Couder and Fort's group attacked this problem early on because one of the canonical results of all quantum mechanics is to solve the Schrödinger equation in exactly this kind of potential.

In the quantum case, if we set up the system in two spatial dimensions, we imagine a particle moving in a potential that is shaped like a satellite dish. The solutions are well-known, and worked out in elementary textbooks. They involve a double quantization, with two quantum numbers $n_x = 1, 2, 3, \ldots$ and $n_y = 1, 2, 3 \ldots$, that emerge from the two spatial dimensions. We can therefore index every solution to the Schrödinger equation in this system by its value of n_x and n_y. These states we would write $(1, 2)$ or $(3, 5)$ or $(4, 1)$, and so on.

The results of the droplet experiment were quite remarkable. Published in *Nature Communications* in 2014, the researchers reported that in a specific memory regime, the droplets switched chaotically between a variety of stable orbits. There were simple one-loop orbits like circles and barbells, as well as lemniscates (figure eights), trefoils, and even more complicated periodic loops.

The wave fields accompanying each orbit were unique and served as a kind of fingerprint of the motion.

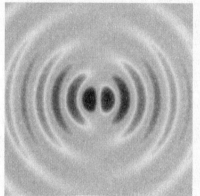 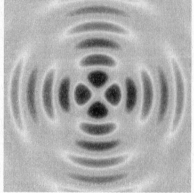

Left: The wave field of a droplet traveling in a circular path. *Right*: The wave field of a droplet traveling in a figure eight. Credit: Matthieu Labousse.

To understand this system, we may look at two main variables: $\overline{L}_z$ is the average angular momentum of the droplet, whereas $\overline{R}$ is the average distance from the center. It turns out that each orbit exists for discrete values of both $\overline{L}_z$ and $\overline{R}$. So this system also has a double quantization, and we can index it with those values, scaled in the appropriate way, to get states labeled as $(1, 1)$, $(2, 3)$, and so on.

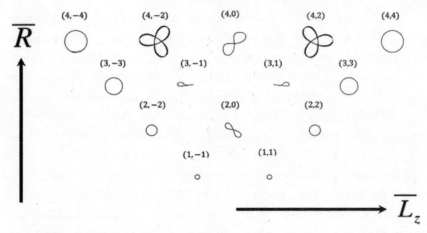

Chart showing double quantization of orbital states in a central spring force. As mean radius increases, the first number goes up. As mean angular momentum increases (or decreases), the second number goes up (or down). Adapted with permission from Matthieu Labousse.

Hydrodynamic quantization is by no means limited to these two experiments. Because it depends on such general features of the droplet system, quantization is observed repeatedly in many different experiments. We will see several more examples as we proceed.

Part V

What We Talk About When We Never Know What We Are Talking About

Quantum mechanics is that subject where we never know what we are talking about.

—Peter Holland[1]

Chapter 18

Schrödinger's Equation

Now we start getting the payoffs I promised. Remember that Fourier's heat equation looks like

$$h_t = h_{xx}$$

and this equation serves to model how heat flows in time. The variable h can be named anything else, and of course this has no influence on the mathematics. What's in a name? So for dramatic effect we are going to rename the variable and call it ψ instead. Naturally we still have the heat equation: $\psi_t = \psi_{xx}$.

Now, let us define i to be a fixed, unchanging number, with the following property: when we multiply i times itself, the answer is -1. If you remember the rules of multiplication, *any* number multiplied by itself is supposed to be positive, so our definition of i is a little strange. As it turns out, extending our idea of numbers in this way is extremely useful and so mathematicians invented a new kind of numbers with this property. These are the *imaginary* numbers.

There was originally a debate about imaginary numbers: Are they "real"? How should we interpret them? But nothing much depends on our acceptance of imaginary numbers, as I will show in a second. The really breathtaking thing is what happens when you take i, and multiply it into one side of the heat equation, like this:

$$\psi_t = i\psi_{xx}$$

Ladies and gentlemen, this is the Schrödinger equation. It looks just like the heat equation, but with imaginary curvature.[1]

So now that we see the extremely close similarity between the equations that Fourier and Schrödinger found, let us get rid of the imaginary numbers. This is not hard to do; a couple algebra tricks allow us to remove i, and what we have instead are two differential equations, with two variables f and g, both of which depend on each other. In the trade this is called a *coupled differential equation*.

So this, too, is Schrödinger's equation:

$$g_t = +f_{xx}$$

$$f_t = -g_{xx}$$

Both of these equations are *almost* diffusion equations, but not quite. The diffusion equation needs the same variable on both sides of the equals sign. It reads $u_t = u_{xx}$, where u is modeling something, like prices, dye in water, or heat, that diffuses. This means the rate of change in time of u at any point depends on its own curvature, which is defined by its neighbors. Remember about keeping up (and down) with the Joneses.

The Schrödinger equation is different. Here we have the rate of change in time of g depending not on its own curvature, but upon the curvature of f. The variable g changes in time based upon what f is doing in space. So at a given point, g will get smaller if f is larger than its neighbors at that point, and g will get larger if f is smaller than its neighbors there. Returning to our Joneses metaphor, it's almost as if you saw how the person across the street was living, and felt so embarrassed that he was so much richer than his neighbors that you spent your own money.

But that person is also watching you, because the variable f returns the favor, but in the opposite way. It changes in time based on the *negative* curvature of g. This is a whole different thing. If you put a negative sign in front of the curvature, you flip the direction of change. In this case, instead of approaching the average over time, the point diverges from the average. If the value at some point is above the average of the surrounding values, then as time proceeds, the value at that point will be *more* above the average. So whatever is being modeled by this does not melt and even out differences, like heat and anything that diffuses. Instead, it does the opposite: it concentrates and intensifies differences. If heat worked like this (the equation would be $u_t = -u_{xx}$), then a single piece of lava would melt more and more of the world, all the while getting hotter and hotter in its interior, and a single piece of ice would freeze more and more of the world, all the while getting colder and colder. Where the two extremes met would be a wall of touching lava and ice, and that configuration could be perfectly stable.

Concentration processes are known in physics, but they are comparatively little studied. They are also called negative or backward diffusion, and seem like curiosities because they not only violate Fick's law of diffusion, they also seem to spontaneously create order. The second law of thermodynamics says that entropy (which we can associate loosely with disorder) always increases. Maxwell and Boltzmann showed why this happens based on irreproachable arguments. How is it, then, that order can emerge spontaneously? How do concentration processes work? Without getting too much into details, here's one example.

There is something called Ostwald ripening that occurs in many physical systems.[2] It happens, for example, when you leave ice cream for a long time in the freezer. Eventually it gets grainy and full of little ice crystals that weren't there before. This is a concentration process ruining your dessert. If you looked at the distribution of ice crystals over time, they would appear to be concentrating and increasing the amount of order. The explanation for this lies at a deeper level than the one that seems mysterious. A molecule in a crystal is more stable when it is on the interior than the exterior, because it is surrounded and cannot escape. Therefore larger ice crystals, which have more molecules bound within their interiors, are more stable than smaller ones. All of the ice crystals are constantly losing and gaining molecules, but the smaller crystals will tend to lose more of their volume than they gain, and the larger crystals gain more than they lose. The smaller ice crystals are spontaneously evaporating, and their water molecules are dispersing through the ice cream, and binding on the larger ice crystals. Thus, through an underlying diffusion process, we get a concentration process.

The Schrödinger equation seems to be modeling a two-way process of diffusion and concentration, locked in a perpetual dance. We can learn a little bit about these processes by looking at the solutions to the equation. Because the Schrödinger equation is actually *two* equations, with two variables f and g, we will have two solutions. They look like waves, which is why ψ is called the wave function.

This is one set of solutions, set up for one dimension of space, for one instant in time. As time passes, these solutions travel together, in the same direction (right or left). Their traveling speed depends strongly on something called the wave number, which counts how many oscillations the wave makes in space. The more oscillations, the higher the wave number and the faster the waves travel.

There are whole families of these solutions. Here is another pair of solutions with a higher wave number, but a smaller amplitude. If the wave above moves with a speed of 3, the wave below moves at a speed of 9. The speed depends on the wave number squared.

Now, like the heat equation, the Schrödinger equation is *linear*. This means that if you can find two solutions, you can always add them together to find a third solution. Imagine we added together our two solutions above, to form a third solution. This is what each component would look like:

This idea of adding waves together is, of course, familiar. It is just Fourier synthesis. So the trick Fourier used to solve the heat equation in 1807 is also the same trick that physicists use to solve the Schrödinger equation today. We can add together as many simple Schrödinger waves as we want to produce a highly complicated shape, and we know how the shape will behave because it is formed of many simple behaviors we understand.

We can almost imagine a physical process underlying this equation, but not quite. We are standing at the boundary of the unknown, grappling with ignorance. Nobody really knows what the Schrödinger equation means. Instead what we have are many interpretations.

So is this it? Have we reached the end of the road? Not quite. The formalism we are considering now is supposed to underpin "one-third of the world's economy."[3] It's supposed to be "the most successful theory in the history of science."[4] So there must be a close link between the wave function and the real world. What is it?

The answer is: we transform ψ using a mathematical recipe called Born's rule. It is also a magical recipe, because we don't know why it works. It forms another layer of the quantum formalism, and results in mathematical structures that actually have physical meaning. Let's see if we can find something to hold on to there.

Chapter 19

The Dice of God

Born's rule is named after Max Born, who worked at Göttingen (and ultimately Edinburgh). He was a lifelong family friend of Einstein, and his wife sent Einstein plays that she wrote. Born is credited with correctly interpreting Schrödinger's equation by stating that ψ should be turned into a probability distribution. As such, the Schrödinger equation should tell us about *probabilities*. This means that, according to Born, the Schrödinger equation is not modeling a physical but rather a probabilistic process. How we interpret *this* idea is a matter of central significance which we come back to in Chapter 20.

Born's rule is where probability enters microphysics. Importantly, the rule has no obvious physical derivation. It is a procedure, a recipe, an algorithm, a ritual, that transforms ψ in a specific way, and that results in a new set of numbers that we can treat as probabilities. Physicists don't like to call their procedures rituals, however, so they call Born's rule a *postulate*. When you see the word *principle* or *postulate* in physics, it generally means there is no known deeper explanation. So the justification for Born's rule is this: if we assume it is true then lots of good things happen.

As we know, probabilities must always be between zero and one. They cannot be negative—that doesn't make sense. However, because both the diffusion and the concentration parts of ψ can be and often are negative, the first step in the Born rule is to modify ψ so that we always have a positive number. We don't care about the technical details here; suffice it to say the procedure is roughly equivalent to multiplying ψ by itself.[1] We will write this as $|\psi|^2$.

Now, the next thing we must do is scale $|\psi|^2$ (make it bigger or smaller), so that when we add up every single possible value for $|\psi|^2$, the sum is equal to one. Remember, this is a general feature of all probability distributions. When you add up all the probabilities for all the different things that could happen, you always get the number one. This is equivalent to rolling a die and then asking, "What is the probability I will see one of the faces?" Of course you will see one of the faces if you just rolled the die. That's certain. In the case of the particle, this is equivalent to assuming the particle exists, and then asking, "What is the probability we will measure it somewhere?" Of course you will measure it somewhere, since it exists. That is also certain, so it has probability equal to one.

So we scale the wave function to give us a probability distribution. This process of scaling is called *normalization*, and a wave that has been normalized is called *normed*. This procedure is common to statistical theories in general; even the Maxwell-Boltzmann distribution of molecule velocities needs to be normed. We should keep in mind, however, that quantum theory normalizes the wave function through a different chain of reasoning than the one used in thermodynamics. In quantum theory, we are doing this because it works, not because we have a physical motivation. The weirdness of this whole process is what led de Broglie to speak of "the fictitious ψ wave of statistical significance, which was arbitrarily normed."[2]

Even though Born's rule is arbitrary, we can now talk about the real world. Generate $|\psi|^2$, norm it, and look at the result. Where the result is small, you have a small chance of finding the particle there; where it is high, you have a high chance. Not only that, but the probability distribution has some familiar properties. Imagine, for example, that we have a bell curve distribution in space. The particle is likely to be found near the center of the curve, and less likely off to the sides.

As time passes, this distribution will spread out in *exactly* the same way as the examples we already considered. The probabilities themselves diffuse. Just like heat in a bar, dye in water, or different gases as they mix, the probability of a particle to be somewhere melts and spreads out. We shouldn't be too surprised to find that this happens. After all, the Schrödinger equation is extremely close, algebraically, to the diffusion equation.

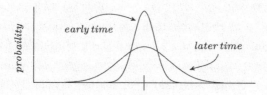

Quantum theory as we have explained it thus far has a double structure. At the base are the solutions to the Schrödinger equation, the diffusion-concentration process, whatever that is. Then, above that, is the probability distribution. The link between them is Born's rule. There is one more feature of the theory we need to talk about, and that is what happens after a measurement.

The theory was invented to model tiny bits of matter that are unimaginably small and fast. Atoms and their components, like electrons and protons, are called *particles* because they are always observed to be sharply localized in space. But the Schrödinger equation is a wave equation, and so we have both of these things together, the particle-wave duality.

Adding the process of measurement to this theory thus creates a tripartite structure. Schrödinger's equation is linked with probabilities through Born's rule, and the probabilities are linked with actual localized flashes via measurement. We can draw it heuristically using the diagrams we already have used, plus one for the localized particle.

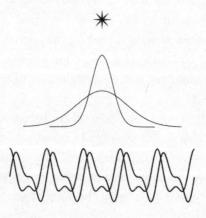

John Nash, the mathematician whose life was dramatized in the film *A Beautiful Mind*, based on the book of the same name, compared this setup to the "herbal medicine used by witch doctors." He talked about "quantum theory," in scare quotes, to underscore its dubious nature, and he said:

> We don't REALLY understand what is happening, what the ultimate truth really is, but we have a "cook book" of procedures and rituals that can be used to obtain useful and practical calculations (independent of fundamental truth).[3]

In the summer of 1957, Nash made an attempt to crack the magic recipe. He wanted to "find a different and more satisfying under-picture of a non-observable reality."[4] However, early in 1958, his mental state rapidly deteriorated into what he described as "paranoid schizophrenic."[5]

Decades later, Nash blamed this attempt to understand quantum theory for his passage into madness. He called it "possibly over-reaching and psychologically destabilizing."[6] Clearly, the quantum mystery is more than a philosophical donnybrook, and a blemish on the face of physics. It is also a menace to public health.

In the chapter discussing Young's explanation for diffraction, we mentioned the double-slit experiment. It was voted by readers of the *Science Times* as the fifth most beautiful experiment in the history of physics. We will now explain number one, which is just the quantum version of Young's original idea.

Recall Young's original setup: There is a ripple tank of water, with a paddle that oscillates back and forth. The paddle sends out regular plane waves, which propagate toward a wall with two slits in it. The waves are blocked by the wall everywhere, except at the slits, where they pass through. Because the slits are narrow, the waves radiate outward from each one in a circle. As they fan out into the space beyond the slits, they interact with each other.

The interaction is very simple. You just take the height of one wave and add it to the other one. Because these waves are oscillating in both space and time, however, the result you get by doing the sum is not very simple. You observe interference: If waves of the same height coincide, they add up to an even higher wave. But if waves of opposite height coincide, they cancel each other out and the water is still at that point, as though there were no wave present at all. We have another wall set up beyond, and we measure the height of the wave that hits it at every point. In this way we observe a pattern of interference that is very easily explained by the wave theory.

To prepare for the quantum case, let's imagine what would happen if we shot bullets at a wall with two slits in it, and then had a big piece of black paper on the other side. The bullets are stopped by the first wall unless they pass through the slits. Maybe they bounce and scatter a little there, but if they pass beyond, they go right through the paper, leaving a hole. If we open one slit (say the left one), as we shoot eighty, then eight hundred, then eight thousand bullets, the pattern we see build up on the paper will look like this:

Here, the peak of the blur is behind the open slit, in this case the left slit. Of course, the situation is symmetrical. If we opened the right slit, the pattern that built up would be off to the right. The shape of the blur is the same, whether we have one or the other slit open. In other words, there is nothing special about the slits, they are identical. Although the point is clear, just for fun we can draw the result here:

If we opened both slits, the one on the left and the one on the right, and fired the same number of bullets, the pattern we would see build up is just one big blur: the peak on the left and the peak on the right add together:

Now, the quantum double-slit experiment is more like the bullet experiment than the water-wave experiment of Young, because particles are always observed as highly localized objects, like bullets. So rather than sending waves toward the slits, we can fire electrons off one by one. If we record where each particle lands (for example, it marks a photoelectric plate with a dot of light), then we can watch the electrons build up gradually. If we do this for the left-hand slit, we get exactly the expected thing for bullets: a general blur that builds up and is peaked behind the left-hand slit.

And, if we open only the right-hand slit, we see again what we expect: the blur is peaked behind the right-hand slit. But then, when we open both slits at the same time, we get something astonishing. The particles still land one by one, but the picture they eventually draw will be described by the wave theory.

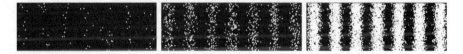

This is called *single-particle diffraction*. It is much more mysterious, at least today, than traditional diffraction with waves. We can understand how water-wave diffraction works; there is nothing strange about that at all. If light is a wave, then diffraction of light also makes perfect sense. But electrons, we know, are particles, so if they diffract according to the wave theory, then they must also be waves. This was de Broglie's matter-wave hypothesis, "without doubt one of the boldest, deepest, and most far-reaching ideas in the whole development" of quantum theory.[7]

We see this same thing when we do the experiment with very, very weak sources of light. This was one of G. I. Taylor's first experiments. As a grad student in 1909, his advisor, none other than Bohr and Rutherford's advisor J. J. Thomson, suggested that Taylor look into the diffraction of single photons. He took a series of exposures of the shadow of a needle, the first of which was with a direct exposure of light from a gas lamp that instantly established the expected diffraction pattern. The next set of exposures had an increasing number of smoked glass screens in the way. The last exposure took a long time, with so many screens between the light source and the needle that the intensity on the receiving end was roughly as bright as a single candle shining a mile away. Taylor then went sailing, since it took about three months for enough light to get through. The ultimate result was the same as the original exposure.[8]

If light is also particulate, then only one photon at a time got through the smoked glass. This essentially proves Einstein's speculation in 1905 that the wave theory of light describes time averages of processes, and that something more complex was going on, much faster than we can see. In classical diffraction experiments done with light, a huge number of photons are recorded in a very short time, and the average pattern is built up almost instantly, obscuring the fact that something more complex is going on. By sending particles through one by one, we slow the process down to something on the human timescale, and the complexity is revealed.

The probability distribution becomes obvious over many trials. When we are doing an actual experiment, the measurements are the primary thing. The scintillations appear here or there, randomly. But we can predict the degree of randomness, where the particles will likely wind up, with the probability distribution. The statistical regularities become evident with more and more observations. As the law of large numbers takes over, the result becomes completely determined by the probability distribution.

But where the heck does the distribution itself come from, and why do the particles obey a distribution that we get from wavelike interference? Feynman had this to say about single-particle diffraction:

> In reality, it contains the *only* mystery. We cannot explain the mystery in the sense of "explaining" how it works. We will *tell* you how it works. In telling you how it works we will have told you about the basic peculiarities of all quantum mechanics.[9]

Chapter 20

The Measurement Problem

It's true that we have just encountered the basic peculiarity of all quantum mechanics, but it is somewhat latent because we did not emphasize it. So here we will do that, using a different and simpler example.

To that effect, we will return to our example with the iron bar, but this time we are going to make it an iron pipe, closed on both ends. Inside the pipe we will say there is a single electron, and we will look at the Schrödinger equation to understand the behavior of this electron. This is called the *infinite square well*, and it is one of the foundational toy examples used in early courses on quantum mechanics.

There are an infinite number of solutions to the Schrödinger equation, given the setup we have described. Each solution has a different energy. Here we draw the two lowest energies: the *ground state* on the left, and the *first excited state* on the right. At the bottom is the pipe, above it is the wave function, and above that is the probability distribution obtained from Born's rule.

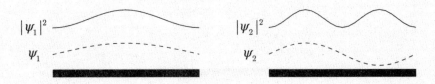

Take a look at the probability distribution of the ground state, on the left. It tells us that the electron has a higher probability of being found in the middle of the pipe than off to the sides. It will essentially never be found at the closed ends of the pipe; the probability there is zero. The distribution of the first excited state on the right has an electron that will never be found at the closed ends, *or* at the middle of the pipe.

The wave function ψ_1 is oscillating, but not in any way we can describe physically, because the oscillation involves complex numbers. Still, we can get an intuitive grasp without pinning down the technical details. Just imagine that the shape is rotating like a jump rope, into and out of the page. The crucial point is that ψ_1 changes with time.

The same thing is happening for ψ_2; it oscillates like a jump rope, but at a different speed. We need to imagine ψ_2 moving like a jump rope four times faster than ψ_1. The rule is, look at the number of places a person can jump, two in this case, and square that number, and that is how much faster than the ground state the jump rope is moving.

However, the probability distributions in *both* cases do not change with time. The jump ropes can be in any position, and if we apply Born's rule we get the same distribution. The term for this is stationary, or time-independent. This is important, so I repeat: the probabilities for electron position do not change with time in either case.

The "only mystery" we are seeking to understand is called *quantum superposition*, and it is made evident through the phenomenon of wave interference. We already understand wave interference; it's the basic driver of Young's explanation of diffraction. The same thing happens here, only with rather different consequences.

We can always add together two solutions to the Schrödinger equation to get a new one, so that is what we will do here. We want to add together the two examples we just considered, to get a new example. We will call this new wave function ψ_3. We simply add the two previous wave functions together. The shape we get, before either of the jump ropes have started moving, is this:

Because ψ_3 is the combination of two jump ropes shaped differently, and moving at different speeds, it will not have a simple motion. Anyone who wants to jump inside ψ_3 will have to pay attention and move around as the shape of the jump rope changes. This means that, after we apply Born's rule, the shape of the probability distribution will also change in time. We cannot get the probability distribution by adding together other probability distributions! We can only get it from Born's rule, applied to adding together wave functions. Now, as time passes, the distribution sloshes back and forth, and so the probability of finding the electron described by ψ_3 looks something like this:

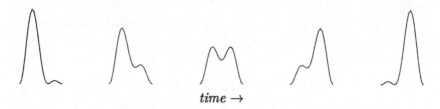

time →

So we had two probability distributions, $|\psi_1|^2$ and $|\psi_2|^2$, and neither one of them change with time. But when we add together ψ_1 and ψ_2, we get a probability distribution $|\psi_1 + \psi_2|^2$ that *does* change in time. Remember that each one of these wave functions describes a *single* energy state. So when we add two wave functions together, we are thinking about the wave function that has *both* energy states present *at the same time*. We know that both energy states are present because the probability distribution that changes in time correctly predicts our observations. The math works.

We can now connect this with the double-slit experiment. In that case, we get the correct statistics by adding together two different wave functions, one for going through the left slit and one for the right slit, and then applying Born's rule. Just as in the example with the pipe, *both* eventualities need to somehow be present, in order to correctly model what we observe. So people say that the electron must, somehow, go through *both* slits!

This is the riddle of superposition. We know it is there because Born's rule gives rise to the correct statistics, through interference. Two different wave functions interface with each other. We need *both* energies to model ψ_3 in the infinite square well. We need *both* slits to correctly model the wavelike diffraction pattern that the electrons eventually make on the screen. So we need two different and contradic-tory energies to be present at the same time in our electron, or we need two separate locations for the electron simultaneously.

Now we can state one of the biggest question marks of modern sci-ence. It's called the *measurement problem*. Imagine that we prepare an electron in a pipe, so that the wave function ψ_3 generates the statistics. Then, we measure the energy of the electron. We will see *either* the ground energy or the first excited energy. We will never, ever see a mixture of both of them. Every measurement only gives a single one of these possibilities. Say we get the ground state. If we do the measure-ment again, any number of times, we will always observe the ground state. So where did the other possibility go? We needed it to model the statistics originally, but after we check the energy, we don't need it anymore.

The measurement of the original system did something, and this something is *not* described by the Schrödinger equation. We had one system described by ψ_3, which was a solution to the equation. We took a measurement, and now suddenly we have another solution, ψ_1. It seems that all the possibilities, which we needed to model the system to begin with, collapsed into a single actuality. This is what people mean with this phrase you might have heard before: "collapse of the wave function."

The most famous example of this is Schrödinger's cat. It's a thought experiment that assumes we have a cat in a diabolical box (what he called a "hell machine") with an instrument that can crack a vial full of poison gas and kill the cat. The instrument only goes off if a single atom under-goes a quantum transition, from high energy to low. If the atom is in the high energy state, the cat is still alive. If it's in the low energy state, then the vial has been smashed and the poor cat is dead. Of course, as you might suspect, the best we can do is describe the *probabilities* that either one or the other energy state will be observed. So, in general, we have a superposition of two states, one with high energy and one with low. But this means that we have a superposition of cats, one being alive and the other not so much. Opening the box, we know we will see only one thing.

This idea of Schrödinger's was inspired by a dialogue with Einstein, who proposed an even more striking idea. Instead of smashing a vial,

Einstein's decaying atom triggered dynamite. This is less catchy than a cat because everything is less catchy than cats, so Einstein's Bomb didn't become a thing. But nevertheless it poses the very same conundrum: How on earth are we to interpret a bomb that is both exploded *and* unexploded?

So now we have this problem. Superpositions are predicted, but never observed. The particle has *one* energy when we measure it; it has *one* position. But we cannot predict which one. Quantum mechanics is fundamentally and irreducibly statistical. It gives us probabilities only. We need these other properties in a superposition to get the probabilities correct. So evidently we need the properties that we don't observe. What are we to say about these unobserved superpositions of properties? If we need them, why do we only observe one single property?

We have a differential equation that works, but why it works is a mystery. Out of this mystery there has effloresced—variously colorful and drab, outrageous and sensible—an entire field of interpretations. Let's go back to our original list from the beginning of this book, naming them, and adding a few items.

Interpretation	Oversimplified formula
Copenhagen	The moon is not there when nobody looks at it.
Many Worlds	An entirely new universe is created each time something happens.
Von Neumann–Wigner	The mind creates the physical world.
Consistent Histories	The universe has more than one past.
Relational	Objects are not real, but their relationships are.
Transactional	The future affects the past.
Ensemble	The wave function generates a statistical distribution in the usual sense.
Pilot Wave	The particle is driven by the wave function.

This list is not exhaustive. We will explore some of these interpretations in detail later on. The point for now is simple: there are multiple interpretations of the Schrödinger equation. The math works, but nobody knows exactly why. In the place of knowledge and understanding, there are many hypotheses. As Max Jammer put it:

> Never in the whole history of science has there been a theory so successful like quantum mechanics in describing and predicting phenomena at the microscopic level and never has there been a theory so problematic like quantum mechanics with respect to its ontological or epistemological interpretation.[1]

Chapter 21

Positivism, Again

It is an interesting historical fluke that quantum mechanics was not discovered once, but twice. The first time was in June 1925, when Heisenberg had an epiphany on the tiny island of Heligoland. The second was Schrödinger's "late erotic outburst" of creative genius, six months later in the Alps. Heisenberg's ideas came to be called matrix mechanics, and Schrödinger's, wave mechanics. These were two different roads to the same city, but the sights seen along the way, and perhaps the temperaments of those taking one road or another, led to very different opinions of the city.

Shortly before this happened, physics was in disarray. Wolfgang Pauli, a close collaborator of Heisenberg and Bohr, said: "Physics is very muddled again at the moment; it is much too hard for me anyway, and I wish I were a movie comedian or something like that and had never heard anything about physics."[1] And earlier Einstein had written to Ehrenfest, "I suppose it's a good thing that I have so much to distract me, else the quantum problem would have got me into a lunatic asylum."[2]

Heisenberg was an outdoorsman who skied and took long hikes, and he spoke of this period in pastoral language. He compared the situation in quantum physics during the winter of 1924–1925 to a nature walk he took with some friends from Krueth to Lake Achen. Mist and fog descended on them; they lost the track and wound up in rocks and brambles, got separated, could see nothing, and were calling out to each other. Then, a patch of fog lifted and they all saw a steep rock face bathed in sunlight. An instant later the fog closed again. Yet that glimpse was all they needed to get their bearings and read the map. Ten minutes later, they were reunited, looking down on a sea of clouds.

It was Heisenberg's work during this time that led him to make this dramatic analogy. He was all alone, on a small island in the North Sea, at well past one in the morning, when he glimpsed the sunlight through the fog. This story has been told a hundred times. Here it is again.

He spent the winter in Copenhagen, working at the institute that Bohr headed there, before returning in the summer of 1925 to Göttingen and his work with Born. He got terrible hay fever and asked for a leave of absence, traveling to Heligoland, a one-kilometer-square island about forty-six kilometers off the German coastline. His face was so swollen from allergies when he arrived that his landlady thought he had been in a fistfight. He sat on the balcony, looking onto a view of village, sand dunes, and sea. He took walks and long swims. He worked. In a few days he "jettisoned all the mathematical ballast" that was confusing the matter, and clarified his problem with the same kind of assumption that served Fourier so well: make a theory out of only what you can observe. His idea led to calculations, which, as he did them, worked out. He became so excited he started going quickly and making errors. By three in the morning he knew it was right. He was "deeply alarmed" and could not sleep, so he climbed a rock jutting out into the sea and waited for the sun to rise.

The paper Heisenberg wrote after his epiphany has this for its very first sentence:

> The objective of this work is to lay the foundations for a theory of quantum mechanics based exclusively on relations between quantities that are in principle observable.[3]

This move was inspired by the fact that Bohr's atomic model had fallen on hard times. It worked well for hydrogen, "but it was a complete fiasco in its depiction of helium."[4] Since helium is the second element in the periodic table after hydrogen, the best model that physics had of the atom could stand on two feet, but it could not walk. Bohr's model had been introduced over a decade ago, and Heisenberg thought it "sensible to discard all hope of observing hitherto unobservable quantities, such as the position and period of the electron."[5]

Heisenberg made much out of the condition that quantum mechanics must be founded on "quantities that are in principle observable." And this intuition is understandable in light of the overemphasis in the Bohr model on electron orbits. Such orbits were primary assumptions—the physics community had worked with them since 1913—but were they really needed? Pauli was in agreement with Heisenberg, speaking with his characteristic acerbity of "weak men, who need the crutch of defined orbits and mechanical models."[6] Given the decade-long run of Bohr's atomic model and its many problems, it could even be said that Heisenberg's move was required.

One year after his happy visit to Heligoland, Heisenberg was invited to give a talk in Berlin. At that time, it was the "stronghold of physics in Germany." Planck, von Laue, Nernst, and Einstein were all there. The colloquium where Heisenberg was to speak was an institution since the time of Helmholtz, and attended by the entire department. It was Heisenberg's first time meeting many of the great and famous physicists of the previous generation. He made an impression, and after the talk was invited to walk home with Einstein.

The exchange was described by Heisenberg in a book originally published in 1969 and known in English as *Physics and Beyond*. The fact that it purports to record a conversation that happened four decades earlier should immediately give us pause, and Heisenberg himself admitted as much. At the very beginning of the book he quotes Thucydides on oration: "I have found it impossible to remember the exact wording." Furthermore, many of the chapters are in dialogue form; they read like an echo of Plato, and that raises its own tangled textual questions ("the dialogue, as Galileo well knew, is itself a most insidious literary device"[7]). The fact that Heisenberg is making things up is quite clear by occasional phrases, when he writes, "I probably said," or "at this point I must have replied." Nevertheless, the position taken here is recognizably Einstein's, and the debate seems plausible.

The very first thing Einstein objected to was Heisenberg's positivism.

> "But you don't seriously believe," Einstein protested, "that none but observable magnitudes must go into a physical theory?"

> "Isn't that precisely what you have done with relativity?" I asked in some surprise....

> "Possibly I did use this kind of reasoning," Einstein admitted, "but it is nonsense all the same.... It is quite wrong to try founding a theory on observable magnitudes alone. In reality the very opposite happens. It is the theory which decides what we can observe."

> I was completely taken aback by Einstein's attitude.... "The idea that a good theory is no more than a condensation of observations in accordance with the principle of thought economy surely goes back to Mach, and it has, in fact, been said that your relativity theory makes decisive use of Machian concepts."[8]

Then follows a discussion about Mach's principle of thought economy, the role of theory, reality, perceptions, and observations in science, which concludes by Einstein saying to Heisenberg: "I have a strong suspicion that, precisely because of the problems we have just been discussing, your theory will one day get you into hot water."[9]

Heisenberg did not listen. The advice of the elder scientist, who had made exactly the same moves when he was the same age as Heisenberg, and to such great success, likely seemed hypocritical. A couple years later, in Heisenberg's famous paper discussing the uncertainty relations, his concluding thoughts leave no doubt whatsoever. He writes:

> In view of the intimate connection between the statistical character of the quantum theory and the imprecision of all perceptions, it may be suggested that behind the statistical universe of perception there lies hidden a "real" world ruled by causality. Such speculations seem to us—and this we stress with emphasis—useless and meaningless. For physics has to confine itself to the formal descriptions of the relations among perceptions.[10]

Comte could have written this sentence. In fact, he did. Check it out:

> In the positive state [of knowledge], the human spirit recognizes the impossibility of obtaining absolute notions, renounces the search...to know the hidden causes of phenomena, in order to apply itself to discover...their effective laws, that is to say their invariable relations of succession and resemblance. The explanation of facts...is thenceforth nothing more than the link established between different phenomena.[11]

Mach would have been proud because he said the same thing:

> When all experiences are regarded as "effects" of an external world extending into consciousness...[this] gives us a tangle of metaphysical difficulties which it seems impossible to unravel. But the spectre immediately disappears if one conceives...that only the discovery of *functional relations* have value to us, and that what we want to know is merely the interdependence of experiences. We then see that reference to unknown, not given, primary variables (things-in-themselves) is entirely fictitious and idle.[12]

Heisenberg called thinking about the real world behind the complex of perceptions to be *"unfruchbar und sinnlos,"* useless and pointless. Mach called it *"fiktive und müßige,"* fictitious and idle. This is the essence of positivism, and it led to the claims, of Comte and Mach, respectively, that we could never know the true nature of heat and that atoms were not real.

These early positivistic spasms could have been a passing thing, but they were not. From this beginning, denial of hidden natures attended the quantum theory with an uncanny intimacy. Consider the following quote by Bohr, who wrote in 1929 that

> In our description of nature the purpose is not to disclose the real essence of the phenomena, but only to track down, as far as it is possible, the relations between the manifold aspects of our experience.[13]

Positivism was so pronounced in the early days because Bohr and his followers believed that quantum theory forces us to make a renunciation. Remember when I flagged this word when Comte used it! Comte thought that true knowledge began with the renunciation of the search for hidden essences, of primary causes. In Copenhagen, what was renounced was the idea that the real world, out there, was composed of anything stable that existed by itself.

The Copenhagen interpretation was all about "renunciation," a word that became practically a term of art. As Pauli said: "This solution [complementarity] is bought by renunciation of the unambiguous objectifiability of natural processes."[14] And Heisenberg wrote that the atom

> has no material characteristics.... As Bohr has emphasized...
> knowledge of color...only becomes possible by renouncing
> knowledge of the atomic and electronic motions...while knowl-
> edge of the atomic and electronic motions in turn forces re-
> nunciation of knowledge of color.[15]

Pais said that the uncertainty principle forces "renunciations of first principle" of the "information obtainable in a given experimental arrangement," and among this information is "a fully causal description involving the localization of the electron."[16] In other words, quantum physics forces us to renounce speaking and even thinking about things we cannot, in principle, observe. And Bohr himself spoke of "the necessity of a final renunciation of the classical ideal of causality and a radical revision of our attitude towards the problem of physical reality."[17]

This view, remember, is called antirealism. It overlaps with positivism so much that the two approaches to knowledge are practically coextensive. We should therefore not be surprised to find many links between the Copenhagen interpretation and positivism. Some links are indirect. For example, as Heilbron pointed out, the two positions had the same sets of opponents. Planck, von Laue, Einstein, and others fought against both. Other links, however, are direct. One of Bohr's greatest defenders was Pascual Jordan, who said he went into physics to "help resolve the discrepancy he saw between Mach's teachings and the old quantum theory."[18] He saw quantum theory in an essentially positivist light, claiming that "The physics of atomic processes really is not a description of objective, isolated, factual situations, but of regularities of observation processes."[19]

Leading historians in the field see positivism very strongly represented in Copenhagen, and say things like "where it really matters Bohr invariably lapses into positivist slogans and dogmas,"[20] and that "the Copenhagen interpretation...rests on two central pillars—positivism" being the first.[21] Philosophers speak of the "Bohrian-positivist alliance," in particular with the logical positivists, which took their inspiration from Mach.[22] Fine recounts:

> Among Bohr's peers it seemed not only natural to associate complementarity with logical positivism...but also honorable to do so. It has become less honorable. Still, it is no less obvious that Bohr's doctrine incorporates a positivist treatment of physical concepts and their significance, dressed in the Copenhagen style.[23]

The positivism of Copenhagen was not considered honorable among all of Bohr's peers. Einstein is the obvious example. There is also Schrödinger, who wrote a letter in 1940 to Arthur Eddington, explaining that he was unimpressed by "alleged grand philosophical revelations" created by quantum physics, calling Heisenberg and his cohort the "enthusiastic champions of that new positivistic outlook."[24] He was unimpressed, he explained, because he was over a decade older than these champions and had read all of Mach before knowing a single word of Bohr. Schrödinger also said this about the Copenhagen interpretation's treatment of the measurement problem:

> The reigning doctrine rescues itself by having recourse to epistemology. We are told that no distinction is to be made between the state of a natural object and what I know about it....Actually—so they say—there is intrinsically only awareness, observation, measurement.[25]

Other physicists, working slightly later, also observed the same dynamic. John Bell said that

The founding fathers of quantum mechanics rather prided themselves on giving up the idea of explanation. They were very proud that they dealt only with phenomena: they refused to look behind the phenomena, regarding that as the price one had to pay for coming to terms with nature.[26]

This retreat from explanation into positivistic exclamations about the limitations of knowledge is what caused Einstein to say about the Copenhagen spirit that "this epistemology-soaked orgy should come to an end."[27]

In 1989, the physicist David Mermin put the whole matter this way: "If I were forced to sum up in one sentence what the Copenhagen interpretation says to me, it would be *Shut up and calculate!*"[28] This phrase became at once a pithy summary of the orthodox view, and a rallying cry for dissidents who refused to shut up. If it sounds familiar here, that is because I already used it to describe Comte's attitude toward the hidden nature of heat. It really does summarize the positivist credo: talk of a hidden reality is fictive and useless; only functional relationships have any value for us.

Chapter 22

The Choice

We have now reached the very heart of the matter. Every sentence of this book has been leading to this point, and every sentence to come will flow from it.

In the early nineteenth century, the mathematical theory of heat transfer had a few defining features. It used a differential equation connecting change in time to spatial curvature: $h_t = h_{xx}$. The equation worked for an unknown reason, and two different interpretations purported to explain the hidden nature of heat. Positivists came along and asserted, through seemingly sophisticated epistemological pronouncements, that this hidden nature was unknowable and that anyone who inquired in that direction was leading science backward. It turned out they were wrong and that the heat equation was a diffusion equation, and as such fundamentally statistical. It encoded the statistical dynamics of an inconceivably rich hidden process.

Now, as we have seen, in the early twentieth century, the theory of quantum mechanics had the same defining features. It used a differential equation linking change in time to complex spatial curvature: $\psi_t = i\psi_{xx}$. That equation worked for unknown reasons, which today we still do not know. It had several interpretations even in the 1920s, and today has more than seems reasonable. Positivists attended the theory closely from the beginning, making the same claims about atoms that Comte made about heat. Finally, just like thermodynamics, quantum mechanics is fundamentally statistical.

The critical, obvious question is this: *Are quantum statistics the same as thermodynamic ones?* This is the only thing worth pondering, for anyone who is not a professional physicist or philosopher, because it casts every other dilemma in its crucial light.

The large-scale features of thermodynamics and quantum mechanics are identical. We have the benefit of history in one case, but not the other. Comte said the hidden nature of heat would never be known, and he was proved wrong. The Copenhagenists said the hidden nature of atoms would never be known, and they might be right. In other words: there is a place where the clear similarity in stories breaks down. The mathematical connection, the multitude of interpretations, the positivism, the fundamentally statistical character—it is all there. What we do not understand is *how* the statistics of the two theories are related, or even *if* they are. We know how the story with heat turned out, but we are living through the quantum story today and have not reached its denouement.

Literally everything depends on this one crucial choice. Some do not see that they are faced with this choice, and I have not seen a single popular book on quantum theory give it the weight I am sure it deserves. In physics departments students are usually not even told that there is a choice to begin with. Or if they are, they are told that the matter was already settled, and so there actually is no choice. The matter is not settled. This is the most significant point in the entire subject of quantum interpretations, which makes it by extension extremely important for future science and the entire human project.

The way this is usually expressed in the literature, and the terms used by Einstein and his colleagues, was as a choice between quantum mechanics being *complete* or *incomplete*. If the theory is complete, then the quantum statistics are *not* generated by any more fundamental, deeper hidden process. Then there is no explanation for the probability distributions we get from the wave function. The statistics are irreducible, pure, primary; they will never go away. On the other hand, if the theory is incomplete, then quantum theory is just like thermodynamics used to be. It is statistical for the same reasons and we will one day find the deeper explanation, just as we did with the heat equation.

The terms "complete" and "incomplete" are a little infelicitous, since they are easily confounded with their overtones of finality. The situation is made a bit worse by the fact that sometimes Bohr and Heisenberg appeared to talk this way. Heisenberg called quantum theory *endgültig*: final, definitive.[1] Karl Popper translated this as *the end of the road*. But if we think that Bohr and Heisenberg and Born and Pauli and the other leading lights of the Copenhagen interpretation actually thought physics itself had reached an end, we will get rather confused. Indeed, there are some who have interpreted quantum completeness in this light, and there are some who, on the back of this situation, think that the terms "complete" and "incomplete" were never clearly defined and have no real meaning today. Both of these views are wrong. Bohr and Heisenberg explicitly mention the future of physics in their writings, speculating about new discoveries and knowledge. They also express, very clearly, many times, precisely what the completeness debate was about. As Niels Bohr put it:

> The question at issue has been whether the renunciation of a causal mode of description of atomic processes involved in the endeavours to cope with the situation should be regarded as a temporary departure from ideals to be ultimately revived or whether we are faced with an irrevocable step towards obtaining the proper harmony between analysis and synthesis of physical phenomena.[2]

To someone familiar with the issues, this run-on sentence is just a core expression of the problem. As is usually the case with Bohr's writings, what he is asking here can be put much more simply: Do we need to renounce classical physics because atoms do not partake of cause and effect, or will we one day discover a hidden deterministic process that explains the statistics? Bohr believed quantum physics was complete. What the Copenhagen interpretation stated, and argued for with unrelenting force, was that physics had passed into a new phase in which statistics and probability were now the absolute best we could do. This idea carries with it many deaths: the death of causality, of realism, of explanations themselves.

Einstein resisted this view his entire life. He is the one who originated the incompleteness critique, and as he put it:

> Assuming the success of efforts to accomplish a complete physical description, the statistical quantum theory would, within the framework of future physics, take an approximately analogous position to the statistical mechanics within the framework of classical mechanics. I am rather firmly convinced that the development of theoretical physics will be of this type; but the path will be lengthy and difficult.[3]

Statistical mechanics is the entire body of mathematical techniques that applies probability arguments to enormous numbers of microscopic entities. It was founded by exactly those people who worked out the kinetic theory of gases, including Clausius, Maxwell, Boltzmann, and Gibbs. So this is the essence of Einstein's position. He believed that the quantum mechanics would one day be explained by a deeper, deterministic layer of reality we have yet to discover, exactly as happened with thermodynamics.

It is usually said that such a theory would be constructed of "hidden variables." The term hidden variable is also somewhat infelicitous, since it has been misconstrued to mean something forever unobservable. For example, Rovelli says that these "variables are hidden in principle: we can never determine them. This is how the theory gets the name."[4] Wiseman claims that "hidden variables are, by definition, empirically inaccessible."[5] This is totally wrong and misconstrues the significance of hidden variables as they have been discussed for a century.

To put it in a formula: hidden variables pertain to future physics. They are either not contained within the current formalism, or currently inaccessible to the experimenters, or both. This is very different than being inaccessible by definition. The best explanation of hidden variables I have seen was given by David Bohm. Those who believe quantum mechanics is incomplete also believe

> that results of individual quantum-mechanical measurements are determined by a multitude of new kinds of factors, outside the context of what can enter into the quantum

theory. These factors would be represented mathematically by a further set of variables, describing the states of new kinds of entities existing in a deeper, sub-quantum-mechanical level and obeying qualitatively new types of individual laws. Such entities and their laws would then constitute a new side of nature, a side that is, for the present "hidden."[6]

In Euclid and Ptolemy's theories of vision, the visual fire is a hidden variable. The ancient scientists supported different versions, either discrete lines or a continuous cone.[7] The properties of the visual fire allowed for the calculation and explanation of visual facts: distances, foreshortening, and recognition itself. The theory of vision *was* the interior formal structure of the hidden variables, along with dynamical rules for interactions with things like mirrors and lenses.

An even better example is that of atoms, because the dispute about thermodynamics is "the classical precedent for the controversy about hidden variables in quantum mechanics."[8] In Boltzmann's time the atomic hypothesis was extremely fertile and full of explanatory power. True, atoms could not be directly perceived, but the concept was strong enough to serve as the foundation of multiple branches of science. In the theory of heat, the molecular motions, indeed the very existence of molecules to begin with, are the hidden variables. These concepts are completely foreign to the heat equation. Not only are they not necessary to derive the equation, they are an actual impediment to it. As we saw, Fourier originally conceived of heat as a many-body problem, and then he gave up and considered heat flow in a continuous body, without any assumptions concerning the nature of that body.

Moreover, the history surrounding heat illustrates Bohm's definition of hidden variables perfectly. The heat equation absolutely is an approximation, not describing a thing, but an emergent statistical phenomenon. The real things, particles in motion, reveal themselves at the "sub-heat level." Moreover, only recently did that deeper level become accessible to experimenters. Today, nobody would say that the hidden variables explaining the heat equation are hidden "by definition." Back in Boltzmann's day they were hidden in practice, and the debate precisely was whether they were hidden in principle, by definition.

As we shall see, Bohr and Einstein argued this issue back and forth every time they saw each other, into old age. Their debate continues to this day. It is, without a doubt, the most important scientific riddle of our time. The choice between completeness and incompleteness is historically and philosophically the most important choice—quite literally the crux of the whole matter. When you start thinking about the issue in this light, many obscure things become clear.

Part VI

Interlude

Chapter 23

Parable of Comets

What are comets? What did our ancestors on the plains of Africa think of these awesome celestial visitors? Clutching their hand axes and looking up at the heavens so suddenly and drastically changed, what stories did they tell their children?

Comets are the original harbingers, the very definition of a portent. The technical term for a comet's visitation is an *apparition*, the same word we use for seeing ghosts. All peoples of all times witnessed apparitions. The Chaldeans, who lived in Babylon, began practicing astronomy in the third millennium BC, and they recorded many apparitions on clay tablets stippled with cuneiform.

In the West, it was Aristotle who first explained comets in a coherent framework. As was well-known in the fourth century BC, the world was made of four elements, arranged in nested spheres like an onion. Earth was the heaviest and sunk to the bottom. Then it went water, air, fire. Above fire, the celestial region began, with the moon, the crystal spheres of planets, and beyond them the sphere of fixed stars, a thin shell with all the stars on it, rotating in eternal harmony.

Aristotle expounded the subject in a book called *On the Heavens*, but he does not mention comets there. That is because he thought the celestial sphere is perfect and unchanging. All the things like comets, shooting stars, the northern lights, earthquakes, rain and lightning, rivers, and so on, those are subject to change; they are temporal, imperfect, and so clearly below the moon. This is why Aristotle spoke of comets in a different set of notes, *Meteorology*, which mainly treated geology and the weather.

That book explained comets, shooting stars, and the Milky Way as atmospheric phenomena with a common cause. You see, sometimes the Earth exhales vapors, which rise into the higher spheres. Moist, cool exhalations rise into the sphere of air, where they form into water vapor and clouds. Dry, hot ones keep going, up and up, into the sphere of fire, where they ignite. Shooting stars are what happens when the exhalations are suddenly consumed by fire in a line. Comets, however, have more and longer-lasting fuel. The element of fire smolders in them, and because the celestial spheres are rotating, the spheres of fire and air are in turbulent motion that catches the comets and glides them along. If a comet was a sort of shooting star, the Milky Way itself was explained as a kind of comet.

Geocentrism is probably the only other scientific system that wedded error and success so thoroughly. "Aristotle's theory of comets was one of the most widely accepted and long-lasting theories in the history of natural philosophy."[1] For a very long time, it had few critics. Seneca the Younger is one notable example. Around 65 AD, he noted that some comets were observed for up to six months. Since when did the weather last this long? Indeed, Seneca criticized all previous cometary theories for weak reasoning, and concluded that comets were a mystery. He said:

> The day will come when, by study pursued through several ages, the things now concealed will appear with evidence; and posterity will be astonished that truths so clear had escaped us.[2]

But the world preferred answers to riddles. Aristotle's theory did a good job of explaining comets if you did not actually look at them, and its seeming adequacy prevented astronomers from doing just that. This, and the difficulty of studying such a rare phenomenon, contributed to keeping a false theory from 340 BC perfectly intact for a very, very long time.

Fast-forward about two thousand years, to 1531. Copernicus has almost completed his book *De revolutionibus*, in which he rejects Aristotle on so many points, but not comets. Europe is undergoing an intellectual rebirth, and three bright comets appear three years in a row. Like everyone in the Northern Hemisphere, Peter Apian and Girolamo Fracastoro, one thousand kilometers away from each other, look up at the apparition. But unlike everyone else, they notice something absolutely crucial, a keystone fact: as the comet moved, the tail always pointed away from the sun. Aristotle had never mentioned this. It took almost two thousand years for anybody to see it. How astonishing that a truth so clear escaped the ancients.

By 1577, when a great comet appeared in Sagittarius, Brahe's instruments were twenty times more accurate than what Copernicus had used, and his triangulation measurements proved that comets traverse the orbits of Venus and Mercury. This shattered Aristotle's cozy theory of crystalline spheres, and refuted the idea that comets were but a bird's flight above the clouds. (Regiomontanus had claimed this in 1472, after overestimating his triangulation angle by a factor of 7,200.) It had taken two millennia, but Aristotle's grip on cometary theory was broken. A new era had dawned.

The Church did not care about comets, and this left astronomers free to speculate about their vexing properties. "Within less than a decade after the appearance of the 1577 comet, about one hundred treatises and pamphlets were published in Europe discussing comets."[3] Everyone seemed to have an opinion.

According to Brahe, the comets were traveling planets, sent by God to inspire horror. Hevelius believed they were formed of exhalations from all the planets, which coagulated into a disk and started moving. Kepler said the comet moved in straight lines and cleansed the ether by attracting waste exhaled from the planets. Even the great Galileo weighed in, saying that comets didn't even exist and were optical illusions like halos and sunbeams. The followers of Descartes said that comets were dead stars, the most solid objects in the universe. Hooke, on the other hand, believed they were soft magnetic bodies that could dissolve into the ether. In short, "no other subject in the whole of astronomy was as controversial as the nature of comets."[4] All was confusion.

While the telescopes were better every year, upon each apparition the images were always the same. A bright nimbus, a glowing haze. Still, astronomy sharpened its tools. The micrometer, invented in 1638,

allowed measurements eight times more accurate than before. Around
the same time, logarithms were invented. They allowed complicated
multiplications and divisions to be reduced to addition and subtraction,
which "by shortening the labors, doubled the life of the astronomer."[5]

From the time of hearing about them, Newton had been interested in
comets. He took notes on varying theories, and he speculated about the
tail. He tried to fit existing astronomical data of apparitions to straight
lines, following the majority view that comets were temporary visitors
to the solar system. But a comet appeared in 1680 that changed everything, and opened the modern era of cometology.

John Flamsteed, the first Astronomer Royal, produced some of the
most accurate data of his time. He tracked the comet through November until it disappeared going toward the sun. Then, in December, another comet appeared departing from the sun. Flamsteed proposed that
they were actually the same. With Flamsteed's data making it possible, Newton took up the calculation. He concluded that Flamsteed was
right. The cometary path was parabolic, and Newton made a drawing
that has become one of the most famous in the history of the subject,
the first true turning point of a comet but also of the entire study of
these mysterious bodies.

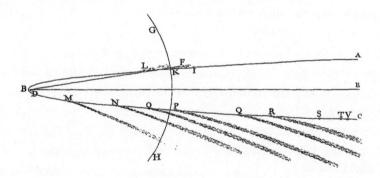

Now we get to Pierre-Simon, marquis de Laplace. We already met
this great French natural philosopher in our study of caloric. One of the
most exceptional comets ever discovered passed extremely close to the
Earth on July 1, 1770, a mere two weeks after it was seen for the first
time. It was only 2.25 million kilometers away, almost six times the
distance to the moon. In 1805, Laplace realized that this near miss

provided a unique opportunity to compute the mass of comets. He computed the effect of the comet on the Earth's rotational period, and showed that comets are not the size of planets, as astronomers had been imagining since the time of Brahe. Comets are tiny.

This had significant psychological effects, because since the beginning of time, humanity had feared apparitions. Pliny the Elder believed they were connected with political unrest and death, and that a comet had caused the war between Caesar and Pompey. Of Caesar's assassination, Virgil wrote in the *Georgics*: "Never did fearsome comets so often blaze."[6] The comet of 1402 was said to have caused the war of independence in Wales. The same great apparition in 1577 that Brahe triangulated was also studied in the Observatory of Istanbul by the sultan's astronomer, Taqi al-Din. He reported good tidings for the war with the Safavid Empire, but that same year several dignitaries died and there was a plague. The sultan then ordered the observatory itself to be destroyed. After the appearance of the great comet of 1618, the German schoolmaster Gotthard Arthusius said the end-times were near, and wrote for ten pages about the disasters associated with comets, including earthquakes, floods, famine, plague, war, treason, and, worst of all, high prices. The Lutheran bishop Andreas Celichius said comets were nothing but "the thick smoke of human sins...kindled by the hot and fiery anger of the Supreme Heavenly Judge."[7] To this, Andreas Dudith made the memorable retort: "If comets were caused by the sins of mortals, they would never be absent from the sky."[8]

With the realization that comets are small, as Laplace himself said, "The light of science has dissipated these vain terrors which comets excited in the ages of ignorance."[9] But what *are* comets? What causes their magnificent tails? The biggest questions were still seemingly untouched because the thoughts needed to answer them were yet unthought.

It was not until 1866, almost two hundred years after Newton's calculation, that William Huggins began analyzing the light reflected from Tempel's comet. He realized that comets contained a strong spectrum of carbon, and also hydrogen, iron, manganese, lead, magnesium, and sodium. In the very same year, a long study of the origins of meteor showers, called the radiant points, revealed that they coincided with the paths of comets. It became clear that meteor showers were actually the debris of comets falling to Earth. The Swift-Tuttle comet was the parent of the Perseid meteors, Tempel-Tuttle for the Leonids, and so on. In 1875 Arthur Wright, a professor at Yale, examined the chemical composition of a newly recovered meteorite. He collected gas from the rock, passed an electric arc through the gas, and observed the very cometary spectrum that Huggins had seen. The association of shooting stars with comets that Aristotle had imagined so long ago was therefore proved, in a way he could never have imagined.

At the same time came the theory of light. In *A Treatise on Electricity and Magnetism*, published in 1873, Maxwell showed that electromagnetic waves should produce a force, and he calculated the pressure exerted by light. Ten years later, in 1882, the Irish physicist George Francis FitzGerald studied the pressure of the sun's rays on idealized particles and found that the sun's pressure could create the comet's tail. Twenty years later, the pressure of light was definitively proven in extremely delicate experiments, and in 1903 two physicists recreated a miniature comet tail using fungus spores, separated from falling sand and streaming away from the source of light.

This is how astronomy discovered the solar wind, and how the mystery of tail formation was finally dispelled. It had taken a new theory of light, the atomic hypothesis, and an understanding of how light and atoms interact.

The fourth question had been answered, but the fifth still remained. What was the actual physical constitution and origin of the comet? This remained a matter of study and debate for decades. Finally, in the 1950s, at the very dawn of the space age, the riddle of comets was solved. Fred Whipple, an astronomer at Harvard College Observatory, proposed what has been called the "dirty snowball" model. Comets, we now know, are loose aggregates of rocks and dust and volatile ices called a rubble pile.

Immediately after Whipple's work, Gerard Kuiper and Jan Oort gave us the origin of comets, in what is now called the Kuiper belt and the Oort cloud. These are the farthest, largest structures in the solar system, giving rise to short- and long-period comets. The Kuiper belt is a kind of enormous asteroid belt, starting at the orbit of Neptune and extending very far out. The Oort cloud has never been observed, but if it is real, it contains an unthinkable number of possible comets, surrounding our sun out to three light-years.

From time to time, gravitational interactions with the outer planets bump one of these icy wanderers from its orbit. It begins a voyage through the endless pure void, during which one star in the ocean of stars slowly grows larger and brighter. After decades, even millennia, the little rubble pile is falling into a hard solar wind that vaporizes both dust and ice. The vapor is blown backward into a tail sweeping for hundreds of millions of miles. Then, the bipedal mammals on a tiny planet close to the sun can look up at this visitor from a silent wasteland, marvel at God's creation, and think the whole thing is about them.

Chapter 24

Diffracting Droplets

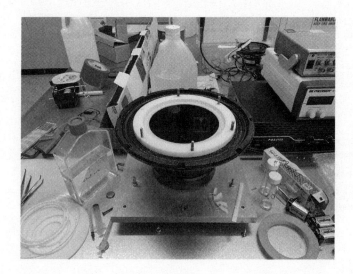

Imagine you come into a lab and find a setup like this. It has a bottle of silicone oil, and the reservoir to contain the oil is resting on a speaker. A machine controls how loud the speaker is, and what pitch (frequency of vibration) it plays.

Turning the thing on, you can hear the hum of the vibration, though at low frequencies, you cannot see any evidence that energy is being pumped into the oil. Taking a toothpick, you can lift some oil out of the bath and drop it down, but the drop of oil does not disappear. It stays there, bouncing with a vague trembling motion just beyond your eye's ability to resolve. The vibrating waves surrounding the droplet are the same.

Then, if you slowly begin to increase the vibrational forcing, the energy in the bath increases, and suddenly everything changes. The droplet picks a random direction and starts scudding across the bath. Before, the droplet was bouncing vertically; now it bounces vertically *and* horizontally. The droplet has become what is called a "walker." Droplets are generally between 0.25 and 0.5 of a millimeter in radius, and walkers move between 10 to 20 body lengths per second. These little guys are cruising. The wave field of such a walker has a very different shape than that of a bouncer.

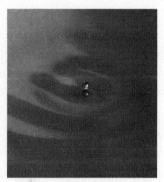

Different wave fields for bouncing (left) and walking (right) droplets. Credit: John W. M. Bush, MIT.

Why does a bouncer spontaneously start moving in a straight line? The explanation lies in something called *gradient-driven motion*. A more familiar term for gradient is *slope*. Think about dropping a little ball onto a big ball. The little ball will bounce straight up only if you manage to drop it on the absolute top of the big ball. If you miss, even by a little bit, the little ball will bounce off the big ball at an angle. The gradient at the absolute top of the big ball is zero. Elsewhere, it is nonzero.

Bouncers turn into walkers because, in a Goldilocks zone of just the right droplet size and vibrational forcing, the bouncing state destabilizes. Small variations are no longer erased, but amplified. If the droplet does not fall exactly in the center of its standing wave, the fluid surface below it will not be exactly flat, and so the gradient will not be zero. Because of this, the force of the air pushing on the droplet to make it bounce again will have a preferred direction. This is how a walker is born.

The walking state is itself stable, meaning that walkers can move about on their parent bath indefinitely. For most parameters in the walking regime, if the walker gets perturbed a little bit (by an ambient air current or the vibrations of a passing subway train, for example), it will briefly wobble before settling back into its straight-line walking state. And so, once you have droplets moving in straight lines, there is no end of experiments you can do.

John Bush has argued that the next step taken by Yves Couder and Emmanuel Fort was a leap of genius. They saw that the droplet system unified both particles and waves, and so might have some bearing on microphysics. Making the connection between this fluid system and quantum mechanics looks obvious in retrospect, but it was not an obvious step at all. Many fluid dynamicists would have simply started studying this new dynamical system to understand how it worked, without connecting it to the larger historical and scientific context.

Couder and Fort made this connection, and decided to do the double-slit experiment. They constructed a barrier for the droplets by placing strips on the bath floor to change the fluid depth. The parameters of the system depend crucially on this depth, which means that with a non-uniform depth, you get different system states depending on where the droplet is bouncing. Waves in the deep part of the oil will be long lived, whereas waves in the shallow oil will disappear very quickly. But remember, the walker regime is a Goldilocks zone. It depends on both the droplet size and the vibrational forcing. The walker needs high memory to maintain its delicate balance. The system parameters in the shallow region of the bath do not allow for walkers to even exist.

As a result, droplets cannot walk over walls of this kind. If the wall is thick enough, droplets cannot even walk *onto* such walls. In effect, these shallow regions become barriers. Walkers coming close to such a shallow region will actually be reflected away from it because the wave field changes, and the droplets bounce off that. Or, as Couder and Fort put it: "The waves guide the droplet and the trajectory is thus defined iteratively by a type of dynamical echo-location."[1]

Using this same idea, gaps or apertures can be created for walkers to pass through. You can then start looking at what happens when the gap is close to the size of the Faraday wavelength. The situation starts resembling single-particle diffraction. This behavior in itself is interesting,

and the subject of several research papers. But what is more interesting right now is what happens when you make two equally sized gaps in the wall, comparable to the Faraday wavelength, and the slits have separation distance 4.2 times the Faraday wavelength. In this case, the setup is very similar to the double-slit experiment.

The statistics that Couder and Fort counted were a decent fit to the diffraction predicted by a wave in Young's famous double-slit experiment. And they went further, explaining the change in the droplet's path by an argument that touched on Heisenberg's uncertainty principle. In this case, of course, causality was fully maintained. The experimenters showed how the "momentum of a walker becomes ill-defined when the transverse extent of its wave is spatially limited." Because that is precisely what is happening when the walker enters a slit, they concluded: "The uncertainty principle inherent to the Fourier-transform of a wave is here responsible for a corresponding uncertainty affecting the motion of the material particle."[2]

Walkers going through slits obey a kind of uncertainty principle, not because causality breaks down, but precisely because causality is maintained.

These results stirred up a lot of interest. Among those who began looking into the pilot-wave system was none other than Niels Bohr's grandson, Tomas Bohr. He enlisted the help of the experimentalist Anders Andersen, and with several others at the Technical University of Denmark and the Niels Bohr Institute, they recreated the experiment and tried to reproduce Couder and Fort's results. They could not do it. Not only that, they showed how the French team had taken only seventy-five droplet trajectories for their data, and accused them of sloppy statistics. They said this "cast strong doubt on the feasibility of the interference claimed by Couder and Fort."[3]

It is important to note that these experiments are extremely sensitive. If you do not control every aspect of the setup very precisely—droplet size, the uniformity of the vibration, forcing frequency, ambient air currents, oil depth, and so on—then there is little hope for reproducing somebody else's results. Similarly, if another lab does not report the relevant values in their experiments, reproducing their results will be like throwing darts blindfolded.

The double-slit was done again in 2018, by Giuseppe Pucci and Daniel Harris, working in John Bush's group at MIT. This work isolated the bath from air currents, constructed a droplet generator capable of repeating droplet radius within 1 percent, controlled oil depth to within two hundredths of a millimeter, and other things besides. They did not find the same interference pattern as Couder and Fort, but they did see diffraction effects. Moreover, these effects were accurately predicted by the accompanying theoretical model, developed by Louis Faria.

Two years later, Clive Ellegaard and Mogens Levinsen published the most comprehensive and thorough study to date. They showed, as Bush's team had done, that "the randomness in the original experiment is an artifact of lack of control" and that the droplets do experience diffraction effects.[4] However, they explained these effects as arising from waves scattering off the back of the slit not used by the droplet. They concluded that the system was "extremely sensitively dependent on so many parameters" and that it had

> clear effects of interference resulting in different, but always sharp and reproducible, angular distributions of tracks between single- and double-slit experiments.... These narrow slits result in much more complicated and fascinating structures making them worth a study by themselves though it is not quantum physics.[5]

What should we make of the droplet trajectories, which behave nothing like bullets, and yet nothing like light either? One option is to just give up entirely. This is what *Quanta Magazine* did, in a 2018 article called "Famous Experiment Dooms Alternative to Quantum Weirdness." After the work by Andersen et al. came out, the writer spoke of the "old dream" of explaining quantum mechanics as a causal theory of particles interacting with waves, and explained that the new finding "has crushed this dream."[6]

If bouncing droplets cannot reproduce the double-slit phenomenon, they must have nothing to say about quantum mechanics! This is a very strange position to take, completely disconnected from every other relevant fact. Forget that for a hundred years we have been told that a classical system could never be both a particle and a wave. Forget the fact the droplets are quantized. Forget that the analogy to Landau levels is almost mathematically perfect. Forget all the other analogies that we will discuss later. One analogy is imperfect; therefore bouncing droplets are doomed.

Why would we leap to such an extreme position? If a baby failed to walk at the normal time, would anyone just throw up their hands and say the baby is doomed? And speaking of babies, I am reminded of the German phrase, first printed in a book called *Appeal to Fools*: Don't throw the baby out with the bathwater.

If we don't think in terms of clickbait headlines, it is clear that these droplet trajectories are actually highly encouraging. They *prove* that single-particle diffraction is possible if you link the particle to an ambient wave field. The droplets undergo trajectory modification owing to wave interference, and this is a fully intelligible effect of classical fluid mechanics. Admittedly, the analogy does not generate the same diffraction that we see in electrons and photons, but why would we even demand this? The point is that we do get some form of single-particle diffraction, which has been a big mystery for a hundred years.

As we have seen, many illustrious minds have decided that single-particle diffraction is impossible to understand classically. But once you have seen these experiments, how likely are you to buy into the claim that this experiment is *absolutely* impossible to understand and that "a few hours of brain racking"[7] should convince you of this?

There is another very important fact, as compelling as any we have seen so far. I hinted at it at the very beginning of this book, and now we will talk about it in depth. The bouncing droplets were discovered in 2005, but they are not new. They are old. The idea of oscillating particles interacting with waves is, indeed, an "old dream," because an almost identical system was proposed in the 1920s, by Louis de Broglie himself.

Part VII

Wrong Turning

In the past decades, physicists have not succeeded in solving any of the open problems in the foundations of their field; indeed it's not even clear we are getting closer to solving them. In this situation it makes sense to go back and look for the path we did not take, the wrong turn we made early on that led into this seeming dead end. The turn we did not take, is resolving the shortcomings of quantum mechanics.

—Sabine Hossenfelder[1]

Chapter 25

The Double Solution

As we saw in Chapter 6, de Broglie's doctoral research "lifted a corner of the great veil." This came after many years of reflection, during which he had generated a good deal of research momentum. Because momentum does not just disappear, we may profitably ask: What was de Broglie thinking about *immediately* following his marvelous prediction of matter waves? He was, in fact, attempting to resolve the wave-particle dualism that his thesis revealed to be such a serious problem. De Broglie published several notes between 1924 and 1927 on this subject, and his thinking oriented itself by degrees to what he eventually named the *theory of the double solution*. He spelled out his thoughts in detail in a May 1927 paper, titled "Wave Mechanics and the Atomic Structure of Matter and Radiation." As others have done, we will refer to this as the "Structure" paper. The ideas expressed in this paper are truly remarkable, in multiple ways, and not least for their relevance to the bouncing droplet system of HQA.

As an introduction to "Structure," let us consider Wolfgang Pauli's opinion of it. Pauli worked closely with Bohr and Heisenberg, and was well-known by 1927, in part for writing, six years earlier, an encyclopedia entry on relativity theory that his teacher Sommerfeld was too busy for. Instead of writing a mere article, Pauli wrote what remains to this day one of the best books on the subject. This impressed everyone, naturally, but what really shocked people was that Pauli was only twenty-one at the time. Over the years, he revealed himself to be a ferocious critic of physical ideas, and this earned him the nickname *die Geissel Gottes*, the Wrath of God. This is what Pauli wrote in a letter to Bohr in August 1927:

> In the last number of the *Journal de Physique*, a paper by de Broglie has appeared.... De Broglie attempts here to reconcile the full determinism of physical processes with the dualism between waves and [particles].... It is very rich in ideas and very sharp, and on a much higher level than the childish papers by Schrödinger.[1]

To begin the paper, de Broglie asked whether, in order to represent the motion of a particle like an electron or photon, we must use "a continuous solution...analogous to those used in the optics of Fresnel." The problem with such an approach was that it "obviously in no way accounts for the atomic structure of matter."[2] This is, of course, the very old problem we examined in previous chapters. To solve it, de Broglie "boldly postulated"[3] that there were two connected systems. One was a particle: Euclidean, Newtonian, Einsteinian discreteness; the other was a wave: Ptolemaic, Hookean, Fresnelesque continuity. The critical assumption was that these two systems were in direct correspondence with each other. There were strict mathematical relationships between them that had to be satisfied at all times.

The particle was considered to be a point of enormous, possibly infinite, intensity, what is known in the trade as a singularity. Also, the particle was considered to be oscillating, as he wrote, with a "periodic value which seems intimately tied to the very existence of matter."[4] This oscillation of the particle was a critical feature of de Broglie's picture. (Sound familiar?)

Furthermore, the particle was at the center of the wave which extended through all of space, and which was also oscillating. The particle touched the wave at all times, and the two of them interacted in a specific way, which de Broglie had first discussed in his thesis. He called this interaction *the harmony of phases*. The harmony of phases meant that the oscillation of the particle was in sync with the oscillation of a wave. There is another word for two oscillations being in sync; it's called resonance. (Sound familiar?)

In his analysis de Broglie derived a guidance equation for particles. That is, he derived a law that described how particles are supposed to move through space. The central idea of this guidance equation is *gradient-driven motion*. The particle was conceived to be at the center of the wave, but the wave itself determined the motion of the particle because the velocity of the particle would be proportional to the gradient of the phase of the wave. (Sound familiar?)

The double solution theory is exceedingly similar to the bouncing droplet system. The similarities are much more than skin deep, too. In 2015, John Bush drew the analogy between HQA and de Broglie's model.[5] He compiled what has been called a Rosetta stone, allowing researchers to translate back and forth between the two pictures. The table, slightly adapted, looks like this:

	Walkers	**de Broglie**
Pilot wave	Faraday capillary	Matter waves
Driving	Bath vibration	Internal clock
Spectrum	Monochromatic	Monochromatic
Trigger	Bouncing	Zitterbewegung
Frequency	Faraday	Compton
Resonance	Droplet-wave	Harmony of phases
Wavelength	Faraday	de Broglie
Motion	Gradient of wave	Gradient of phase

The similarities, as well as the differences, are most fascinating here. De Broglie did not know where the pilot wave came from, though this is most obvious in the droplet system. The trigger for the waves is the bouncing action of the droplet in HQA, while in de Broglie's theory it is a *zitterbewegung*, or "trembling motion." We will discuss this more in Chapter 43.

Since the final row, the gradient-driven motion, is so crucial to both theories, let's explore it a little more. Remember, a gradient is a fancy way of saying slope. Imagine that you are on a skiing slope. There are six directions that you could be pushed at any moment by the snow beneath you: left or right, forward or backward, up or down. (This last one takes some imagination.) The ski slope can be depicted in a topographical contour map, where each contour line traces out a constant elevation of the mountain. If you walked along one of those lines, you would not go up or down in elevation. When the elevation does not change uniformly but in a complicated way, a skier will not go in a straight line.

Below, I have made two imaginary ski slopes, each with two skiers. On the left is a perfectly uniform bunny slope. It is for babies. If the skiers do not turn, but simply follow the path of least resistance, they will always maintain the same distance from each other. Now look at the heliskiers on the right. They are both starting at the top of the same mountain, a little apart. Yet if they ski down following the path of least resistance, they will very soon lose sight of each other. Both will end up in the riverbed, but on completely different sides of the mountain.

This is the essence of de Broglie's idea: there is an extended wave that acts like a ski slope, but it is so subtle that only extremely small things can feel it. One important difference to keep in mind, however, is that the skiers are traversing features of the landscape, while de Broglie's particles, just like the bouncing droplets, always lie *at the center* of the extended waves. The particle and the wave are intimately connected— where one goes the other goes. When the wave is constant, like the bunny slope above, the particle just moves in a straight line. When the wave is more complex, it can change as the particle moves, just as the snow changes beneath a skier who is moving, and so the particle's direction and speed will vary from time to time.

The idea is a rich one. Two skiers that start close together can end up far apart, but notice how the converse is also true: two skiers that start far apart can end up close together. They will be channeled into the low places of the mountainside. De Broglie proposed that the shape of the wave field would channel particles into different places, and that the wave field was, in some sense, a physical object because it could interact with the material world around it.

It is little known, among not only physicists but also historians of physics, that de Broglie's famous matter-wave hypothesis was actually a side effect of his real program, which was to develop a new, non-Newtonian form of dynamics that was capable of encompassing both particles and waves. In his own words:

> Since the motion of the [particle] was connected with the evolution of the wave phenomenon at whose center it was, it would depend on all the circumstances that the wave phenomenon would encounter in its propagation through space. For this reason, the motion of the [particle] would not follow the laws of classical mechanics... [but] would undergo the influence of all the obstacles which would influence the propagation of the wave phenomenon with which it is linked, and thus the existence of interference and diffraction would be explained.[6]

Single-particle diffraction is supposed to be "the only mystery" of quantum mechanics. Indeed, Feynman's declaration in his famous lectures about never being able to explain this experiment is always quoted. But the explanation is actually not hard in this framework, and de Broglie actually explained "the only mystery" in 1927. This has led to some puzzlement, and snarky comments, among physicists sympathetic to de Broglie:

> Feynman's claim about the impossibility of understanding this experiment... is flatly false. It had been known

to be false since 1927, when de Broglie first presented the pilot wave theory....How did Feynman manage to get so confused?[7]

Let us return to the analogy of the skiers. Imagine there are hundreds of skiers starting down from a mountaintop. They are not allowed to turn by themselves, but only to follow the path of least resistance. Beyond this you are not told anything about them. You are not told where each one of them starts, and you are not allowed to watch them ski. But you are asked where they will wind up. The best you can do is look at the topology of the mountain, and see how the slopes are likely to channel the skiers. A broad channel will catch a lot of skiers and so you can say the probability is high that a skier will wind up in the channel. A narrow channel has a lower chance of catching and channeling skiers.

Your predictions might not be great at first. But if you keep flying a bunch of skiers up the mountain and having them ski down over and over, with time your predictions will get better and better. Eventually, they will become perfect because with enough repetition, the topology of the mountain wins. It is the only thing that matters, so long as the skiers start at the top in a uniformly random manner, and are not allowed to turn by themselves.

In the double-slit experiment, the waves would be blocked by the wall, but could pass through the two slits just as water waves do. The result would be the typical interference pattern. The particle goes through one of the slits, but it still feels the influence of the other slit through the mediation of the wave. Thus where the particles wind up on the screen will be the result of a continual channeling by the wave field, just as skiers are continually channeled by the topography of a mountainside.

In this picture, clearly, trajectories exist, because the particle exists at a specific location in space from each time to the next as it moves. The particle goes through one slit or another; it does not go through both slits. The wave goes through both slits and interferes on the other side, changing the gradient the particle navigates, and thus the particle's trajectory. The probabilities emerge organically and in the same way that probabilities emerge in every other branch of science: we do not know the initial location of the particle, and very small differences matter a great deal. Therefore, the best we can do with particles is to say where each one is likely to land. There is nothing mysterious about this, it is just like rolling dice, and the probabilities could, in principle, be eliminated.

Perhaps the most remarkable thing about this version of the double-slit experiment is that it leaves our commonsense picture of the universe almost untouched. Matter is still real and exists in a specific place. The objective world is real and does not depend on measurements. Observations do not change the fundamental nature of what is observed. Classical determinism is also preserved: physical laws operate at all times, and God does not play dice.

Chapter 26

This Subtle Doctrine

In 1926, when de Broglie was developing his ideas about the double solution in Paris, the seed of the Copenhagen interpretation was already planted in Bohr's mind. His approach could not have been more different. That year Schrödinger visited Bohr's house for several days of what turned into an exhausting debate. Returning from this meeting, Schrödinger touched on Bohr's basic intuition in a letter he wrote to Wilhelm Wien:

> Bohr's...approach to atomic problems...is really remarkable. He is completely convinced that any understanding in the usual sense of the word is impossible. Therefore the conversation is almost immediately driven into philosophical questions, and soon you no longer know whether you really take the position he is attacking, or whether you really must attack the position he is defending.[1]

Do not pass too quickly over the crucial part of this description. Bohr was *completely convinced that any understanding in the usual sense of the word is impossible.* Bohr's work of 1913 had inaugurated the quantum theory of the atom by imagining mechanical orbits, both like and unlike the solar system. These orbits were considered real; they could change their shape and radius. But the idea simply did not work

190

for any atom other than hydrogen. The orbits themselves were suspect. Bohr was coming to believe that the difficulties of quantum theory, with which he had struggled incessantly for over a decade, were so serious because the true lesson had not been learned. *No* classical model could ever be constructed of the atom and its pieces.

Bohr abandoned classical space and time, an extreme move that several physicists resisted strongly. Schrödinger was among them, saying that

> Bohr's standpoint, that a space-time description is impossible, I reject at the outset. Physics does not consist only of atomic research, science does not consist only of physics, and life does not consist only of science. The aim of atomic research is to fit our empirical knowledge concerning it into our other thinking. All of this thinking, so far as it concerns the outer world, is active in space and time. If it cannot be fitted into space and time, then it fails in its whole aim, and one does not know what purpose it really serves.[2]

But for Bohr, this inability to situate atoms in space and time grew and grew in importance. It holds the tiny core that gave rise to the world tree of the Copenhagen orthodoxy. Once we accept it, the question becomes: *Why* is any understanding in the usual sense impossible? Early in 1927 he spent over a month "skiing in the Norwegian mountains around Gudbrandsdalen."[3] It was around this time that the solution to the quantum mystery came to him. He called it *complementarity*, and it was the main idea of his life. Bohr developed and "struggled to clarify" this idea until the very end.[4]

The idea of complementarity is simple and general, and vague. Einstein said that "despite much effort which I have expended on it, I have been unable to achieve the sharp formulation of Bohr's principle of complementarity."[5] Unsurprisingly, scholars on the subject also find it confusing, and "what exactly Bohr's complementarity means continues to be an enigma."[6] The situation has continued for so long that the philosopher Don Howard said that "now there are signs of growing despair...about our ever being able to make good sense out of [Bohr's] philosophical view."[7] And while we are on the topic of Bohr and despair, Scheibe said that "anyone who makes a serious study of Bohr's interpretation of quantum mechanics can easily be brought to the brink of despair."[8] In light of this we will have difficulty in expressing things exactly or finding any essence. The best we can do is talk around the issue.

Attempts to precisely define complementarity start out seeming clear, but quickly become problematic the more you think about them. For example, Murdoch summarized Bohr's vision as follows. *Any two ideas are complementary if (and only if) they are different in meaning, together they completely describe a thing, and they are mutually exclusive.*[9] So perhaps the visual and auditory components of a movie are complementary? Clearly, Bohr's idea can be applied to far more than just the quantum realm. If it had emerged from film theory, we can only speculate about how things might have been different. As it turned out, complementarity emerged from thinking about atoms, and so it assumed an air of profundity.

Complementarity emerged from the theory of atoms because atoms had come to be modeled by waves. The details are, in fact, related to Fourier analysis. And so our work on the heat equation has also prepared us to understand this as well.

Recall that Fourier's idea was to decompose any mathematical function into a spectrum of simple waves. The original function exists in the time or space domain, and the waves, which form the fingerprint of the original function, exist in the frequency domain. To refresh your memory, the last diagram we gave on this subject looked like this:

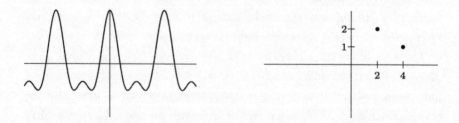

There are only two waves added together in this diagram, but there really is no limit to the number of waves we can add together. It could be infinite. So let's do that and see what happens. We'll consider an infinite number of waves of the same amplitude, with wavelengths on either side of zero. After a wavelength cutoff we have no waves at all, so the frequency domain will be zero beyond the cutoff. Then we obtain the following pair:

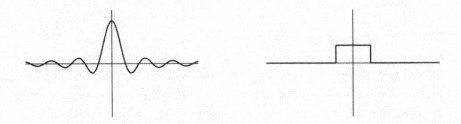

Let us consider the famous bell curve that we met in the chapter on probability. This is also called a Gaussian function. Interestingly, the Fourier transform of one Gaussian is another Gaussian.

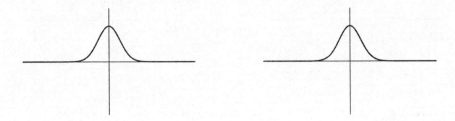

Now that we have refreshed our memory about Fourier analysis, we can see how it is applied to quantum mechanics. With the heat equation, we spoke of the space and frequency domains. In quantum mechanics, we speak of the space and momentum domains. The theory is structured such that a particle's momentum waves form its spectrum, the "fingerprint" of its position wave.

The wave on the left is going to be the function describing the position of the particle, where it exists in space. As we know, it is a probability distribution. The shape on the right will be a function describing the momentum of the particle. That, also, is a probability distribution. But because these are Fourier transform pairs, if we want to use the formalism to describe a particle that is very highly localized, then we need to add up the correct momentum waves. We will get a narrower shape in the space domain, but the momentum domain will be wider.

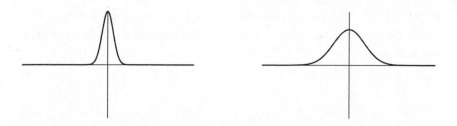

And the converse is true: if we want to describe the momentum of a particle more precisely, we can narrow the function in the momentum domain. But this is a spectrum, a fingerprint of the probability of the particle's position. When we add together all the momentum waves, we get a space domain that is wider.

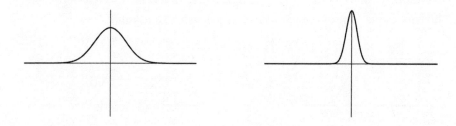

As you can see, there is an inescapable trade-off, a tug-of-war back and forth. The narrower one gets, the wider the other gets. It's almost as though there is some conserved quantity, some number that needs to always stay the same. Like how you can squeeze a balloon tight in one region, but another part of the balloon needs to expand to keep the volume of air constant.

Indeed, this is the case. If we expressed the width of the space domain as Δx and the width of the momentum domain as Δp, multiplying them together can never be bigger than a certain number. The relationship that Heisenberg discovered is that when you multiply the width of the space and momentum domains together, the answer must always be bigger than a certain number that, surprise, involves Planck's constant:

$$\Delta x \Delta p \geq \frac{h}{4\pi}$$

This is the famous uncertainty principle. A narrow probability distribution means we have high certainty, and a wide distribution means we have low certainty. So Heisenberg's principle says that the more you know about where a particle is, the less you know about its velocity. If your knowledge of a particle's position is very precise at one instant, then you will have almost zero knowledge of velocity, and so the particle could be practically anywhere in the next instant.

According to Bohr and Heisenberg, the application of Fourier analysis to atomic theory, which results immediately in the uncertainty trade-off, has very profound consequences for our entire picture of physical reality. Complementarity is the formulation of the special logic that Fourier analysis imposes upon us.

Bohr thought it more than a mere coincidence that "limitations of our concepts coincide so closely with limitations in our possibilities of observation."[10] Position and momentum are the two most common complementary pairs mentioned, and he said that "a sentence like *we cannot know both the momentum and the position of an atomic object* raises at once questions as to the physical reality of two such attributes of the object."[11] To answer this question, he says, we must look at the conditions underpinning our use of the very concepts of "position" and "momentum." Bohr's entire program was to examine how we use these concepts and to show that any attempt to define them precisely with experiments will always run into the uncertainty principle. To give a flavor of his style, here is a diagram he used, outlining the double-slit experiment.

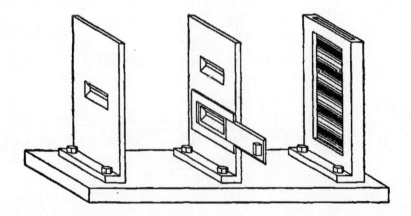

Here, the screens are all bolted down, so by clever machining we can make the slits as narrow as we like and thus get an arbitrarily precise position measurement for the particle going through the first slit. Let's say the particles are electrons. We know exactly where the electrons that pass through the first slit are located. However, because the screens are bolted down, we cannot get any momentum measurement at all. How fast and in what direction is the electron moving immediately after it interacts with the first slit? The experimental arrangement is unsuited to yield this information.

The second screen has two slits. Because we are completely ignorant of the electron momentum as it heads toward the second screen, we have no way of knowing whether it goes through the top or bottom slit. When we do not know this information, on the third screen we get an interference pattern, which is a wave behavior. The only way we have of assuring that the electron goes through the top slit is to close the bottom slit. But if we do that, the interference pattern—the wave behavior—disappears and reverts to particle behavior on the third screen. And so we see that position and momentum, particles and waves, are complementary. They cannot appear together with arbitrary sharpness.

If we seek to gain more knowledge about the momentum of a particle, we might imagine a different experiment, somewhat like in the following diagram:

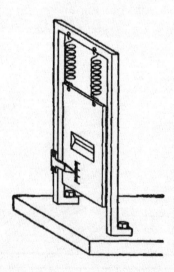

Here, the whole screen and slit are suspended by weak springs, so that if the electron goes through the slit, it will create an effect on the screen, from which we can form "an estimate of the momentum transferred."[12]

However, now we have an opposite problem, because the screen will move, making a position measurement less precise. Bohr also said that at times measurements cause disturbance, so when we shine a light on it to read the pointer, it will cause "an uncontrollable change in the momentum" of the screen such that

> there will always be, in conformity with the indeterminacy
> principle, a reciprocal relationship between our knowledge
> of the position of the slit and the accuracy of momentum
> control.[13]

This basis of this argument is the experimental apparatus, which is large and real and constructed using things like lenses, lasers, filters, screens, and all that. Macroscopic objects that humans can pick up and put down. Bohr asserted that all analysis must be founded on classical, human concepts, and experiments at the human scale, because otherwise we cannot even talk to each other or describe "what we have done and what we have learned."[14] This has been called the "doctrine of the indispensability of classical concepts"[15] and it underpins the "foundational status of wave-particle duality"[16] in the Copenhagen interpretation. Rosenfeld attributed the necessity of classical concepts to the size of our sense organs, to the fact that all phenomena we know directly are macroscopic. Assuming that human thinking is forever tied to macroscopic objects and that such objects are fundamentally different from quantum objects leads, according to Bohr, naturally, inescapably, and directly to complementarity. Bohr was making this same argument in an interview the day before he died.

After he developed complementarity, Bohr's method of analysis was always the same: he tried to solve problems "by applying epistemological analysis to fundamental concepts and demonstrating the limits of their applicability."[17] He developed these ideas to show that there was never any way, even in principle, of escaping this trade-off, this back and forth, this complementarity, between position and momentum, between coordinates and causality, between particles and waves. Yet, according to Bohr, the combination of concepts of space-time coordination and those of causality "into a single picture of a causal chain of events is the essence of classical mechanics."[18] This combination is precisely what quantum mechanics does not allow, and therefore the classical modes must be transcended.

This is why wave-particle duality became such a foundational issue in Copenhagen. "Bohr felt that the wave-particle duality should be made the focal point of all treatments"[19] precisely because it drew out the necessity of complementarity so clearly. In Heisenberg's view, Bohr thought of the

> dualism between waves and corpuscles as the central point of the problem, and [wanted] to say, "That is the center of the whole story, and we have to start from that side of the story in order to understand it."[20]

Bohr himself extended the idea by applying complementarity beyond physics. He spoke of the complementarity between mercy and justice in jurisprudence, between sentiment and reason and psychology, between vitalism and mechanism in biology. He said there was a complementarity between truth and clarity, between free will and determinism, quality and quantity, intuition and logic.

The import of this idea for him was truly enormous. Bohr believed that we live suspended in contradictions, for nature is a system of glorious contradictions. This became Bohr's idée fixe. Rosenfeld recalled that, on one of those "unforgettable strolls" with the great man,

> Bohr declared, with intense animation, that he saw the day when complementarity would be taught in the schools and become part of general education; and better than any religion, he added, a sense of complementarity would afford people the guidance they needed.[21]

In 1947, Bohr was awarded the Order of the Elephant, a very high award in the Danish system normally only given to royalty and heads of state. As his grandson Vilhelm described it: "You get to hang an elephant around your neck. And, when you go to the Queen and important parties you have to wear that thing, and when you die you give it back to the Queen."[22]

Members of the order have their coats of arms displayed on a wall in Frederiksborg Castle. Bohr designed his own; it had a yin-yang symbol and the Latin motto: OPPOSITES ARE COMPLEMENTARY.

Coat of arms of Niels Bohr.[23]

Chapter 27

Solvay

From the moment they were developed, de Broglie's double solution and Bohr's complementarity were on a collision course. Their encounter decided the future of physics.

Throughout most of 1927, de Broglie had every reason to be optimistic. He submitted his "Structure" paper to the *Journal de Physique* on New Year's Day. Four months into the year, a paper by Davisson and Germer was published in *Nature* that showed electrons scattering from a crystal of nickel and diffracting, just like waves.[1] The matter-wave hypothesis conceived by de Broglie and supported by Einstein was now experimentally proven. One month later, "Structure" was published, and Einstein presented a paper to the Prussian Academy of Sciences in which he proposed that photons are guided by a "ghost wave," thus maintaining the classical idea of a trajectory for particles. This was, in spirit, the exact same idea that de Broglie had been developing for years. By June, de Broglie was in correspondence with Hendrik Lorentz about giving a talk at the fifth Solvay Conference on Physics at the end of the year. This has a poetry to it, as de Broglie had begun his scientific career at nineteen years old by tagging along with his brother Maurice to the very first Solvay Conference.

The young man who, only a few years before, had obtained his PhD with a wild hypothesis that nobody understood was now accepted as an expert by the leading physicists of his time. The other speakers at this conference were some major names—Compton, Heisenberg, Born, Schrödinger—and the attendees would be most of the others. Planck, Pauli, Bohr, Einstein, Lorentz, Curie—they would all be there. The invitation must have been very meaningful for de Broglie, both professionally and personally. It was certainly the highest profile event of his career so far. Yet, in preparing his presentation, de Broglie was confronted with a problem. His "Structure" paper was, as Pauli said, full of ideas and very sharp, but it was also full of difficulties and very hard. De Broglie himself acknowledged this, saying of his double-solution program:

> It did not take long to realize that its justification would run into very serious mathematical difficulties....I did not feel myself capable of resolving these difficult mathematical problems.[2]

The choice he made for the talk was to present a much simplified version of his ideas. Rather than go into the double-solution theory, which intimately connected particles and waves and was full of suggestive physical ideas, he presented something called pilot-wave theory. The last section of de Broglie's "Structure" paper was called "L'onde pilote," the pilot wave. Here, de Broglie writes that

> If one does not want to invoke the principle of the double solution, it is admissible to adopt the following point of view: we will admit the existence, as distinct realities, of the material point and the continuous wave...and we will take as a postulate that the movement of the point is determined as a function of the phase of the wave....We then conceive of the continuous wave as guiding the movement of the particle. It is a pilot wave.

In other words, rather than deducing the guidance equation and gradient-driven motion from physical ideas that link the particle and the wave mathematically and suggest how they may be interconnected phenomena, let us assume that particles and waves exist as separate entities, somehow. If the wave pushes the particle around, then we get

the same experimental predictions. De Broglie thought of this as nothing more than a provisional attitude. He thought it would eventually be
necessary "without doubt, to *reincorporate* the corpuscle in the wave
phenomenon and we will probably be led to ideas analogous to those
developed above."[3] In short, he did not mention any of the driving ideas
or the physical intuition behind the results he obtained. He ignored the
harmony of phases, the philosophical justifications, and especially the
mathematical difficulties arising when you assume each particle has its
own wave field that affects all the other particles. Instead, he started
with the results, and went from there. It did not turn out well, and
as de Broglie himself admitted: "I did not perceive at the time that,
by adopting this kind of line of reasoning, I had greatly weakened my
position."[4]

In the preface to *Phenomenology of Spirit*, Hegel wrote,

> The bud disappears in the bursting-forth of the blossom, and
> one might say that the former is refuted by the latter; simi-
> larly, when the fruit appears, the blossom is shown up in its
> turn as a false manifestation of the plant, and the fruit now
> emerges as the truth of it instead. These forms are not just
> distinguished from one another, they also supplant one an-
> other as mutually incompatible. Yet at the same time their
> fluid nature makes them moments of an organic unity in
> which they not only do not conflict, but in which each is as
> necessary as the other; and this mutual necessity alone con-
> stitutes the life of the whole.[5]

For quantum theory, the "thirty years that shook physics," beginning
in 1900, were the bud.[6] These years were seething with early activity
and breakthroughs that began the long work of slow, careful development. The end of the beginning, when the bud had fully burst forth into
the blossom, came in 1927 at the fifth Solvay conference in Belgium.

This is probably the most famous scientific meeting of all time, and it is considered by some to be "when the new quantum mechanics was fleshed out once and for all."[7]

The Solvay conferences were founded by Ernest Solvay, a chemist and industrial magnate who patented a special process for producing sodium carbonate. This chemical has many uses but is most recognizable as a builder in detergent. It softens water and allows soap to work much more effectively. Solvay used his great wealth to further science in many ways, the best known being the creation of the invite-only conferences that bear his name, and continue to this day.

The fifth Solvay conference was on electrons and photons, and it is well-known in the scientific community as a turning point in the history of quantum theory. It has been called a "crossroads" for quantum mechanics. It was also a turning point in de Broglie's career. The list of attendees includes almost every significant physicist of the age.

Back, from left: Piccard, Henriot, Ehrenfest, Herzen, de Donder, Schrödinger, Verschaffelt, Pauli, Heisenberg, Fowler, Brillouin. *Middle*: Debye, Knudsen, Bragg, Kramers, Dirac, Compton, de Broglie, Born, Bohr. *Front*: Langmuir, Planck, Curie, Lorentz, Einstein, Langevin, Guye, Wilson, Richardson.

This is probably the most famous conference photograph in the world. There was, in fact, a video camera at the gathering, brought there by the American chemist Irving Langmuir. One can find the silent movie on YouTube, with a helpful voiceover on who's who.

Most of the video shows the scientists talking with each other, smiling, smoking, and joking around. They point at things, wave their hats, and ham for the camera. The end of the film, however, shows everyone coming out of the building and down the stairs, and they are almost unanimously pensive, serious, and subdued. The narrator says: "Note the changed demeanor as the participants exit from the meeting."

When the general discussion began, Lorentz, the chairman of the conference, opened with a question: "Could one not maintain determinism by making it an article of faith? Must one necessarily elevate indeterminism to a principle?"[8] He was the grand old man of physics, who had only one year of life left to him. It was not so easy for people like Lorentz and Einstein to give up what appeared to them a central pillar of science. Along with others, such as Schrödinger and de Broglie, this group comprised "the unrepentant realists."[9]

They were faced off against those, including Bohr, Born, Pauli, Heisenberg, and Dirac, who had accepted that quantum mechanics takes away "the whole foundation for causal spacetime description"[10] (Bohr), and that the "indeterministic foundations will be permanent"[11] (Born), and that the entire idea of a visualizable picture underpinning atomic processes was "crap"[12] (Heisenberg).

De Broglie gave the first talk of the second day, in the afternoon. The usual story that is told goes like this:

> According to legend, as [de Broglie] outlined his ideas he was heckled loudly by the brash Wolfgang Pauli....De Broglie took Pauli's interruptions with dignity...but after the lecture was over people were more interested in Pauli's questions than de Broglie's answers.[13]

This is completely wrong in almost every way. We know what actually happened because the conference was carefully recorded, and scholars have gone over the proceedings in detail. There was an exchange between Pauli and de Broglie; it may even have been a difficult one. But it happened in the discussion period, after de Broglie's talk. And if you just think about it, the idea of heckling is ludicrous. These people were all extremely sophisticated, and nobody would have disrespected the chairman of the conference, Hendrik Lorentz, in this way. It is also untrue that people were more interested in Pauli's questions. The exchange was quite brief, and nobody else chimed in.

Retelling this drama amounts to little more than repeating false but delicious gossip, with its inevitable inventions and authorial embellishments ("de Broglie took Pauli's interruptions with dignity"). But the main error of this account is not one of commission. It is of omission. It leaves the true story untold. Another exchange took place during the discussion section of de Broglie's talk that was probably the most crucial moment in the entire conference.

During most of the talks, Einstein was silent. Listening, observing. Some flavor of his attitude is conveyed in a note that survived. Ehrenfest wrote on a piece of paper and passed it to his friend: "Don't laugh! There is a special section in purgatory for professors of quantum theory, where they will be obliged to listen to lectures on classical physics ten hours every day." Einstein passed the note back: "I laugh only at their naiveté. Who knows who would have the laugh in a few years?"[14]

During the discussion dedicated to de Broglie's presentation, Einstein finally rose to speak. It was here that he made his main contribution to the conference. It was, to adopt Schrödinger's appraisal in another context, "brief but infinitely far seeing." He began by apologizing for not having entered deeply enough into quantum mechanics, and then put his finger on precisely the heart of the issue.

He proposed a thought experiment, which we need not repeat because in its essence we have already described it with the example of an electron trapped in a pipe. The pipe can be as long as you like, an astronomical unit even, but when we look for the electron we will measure it in a single place. Einstein says that we have a simple choice:

1. The theory only gives statistical information emerging out of collections of individual processes.
2. The theory claims to provide a complete description of individual processes.

We have, of course, met this choice before. Under the first conception, we would think of Schrödinger's equation as giving us a probability distribution, while remaining silent on precisely the things we want to know most: the hidden variables that generate the distribution. In Nash's words, "We don't REALLY understand what is happening."[15]

On the second conception, the Schrödinger equation *is* the particle. It says everything there is to say, which means the electron *really is* spread out over ninety-three million miles (or any arbitrary distance). Or, perhaps more accurate would be to say, it has no specific location. The idea of location is itself suspect, until we have made a measurement. Then, the entire electron is localized to a small space and the wave that was smeared over huge distances is no longer there. As Einstein pointed out, for the wave to disappear immediately everywhere "assumes an entirely peculiar mechanism of action at a distance."

In other words, if we decide that quantum mechanics is complete and the wave function says everything there is to say about the particle, then the wave function has to be real in some sense, because the particle is real. Then how does this wave function "know" to instantaneously collapse everywhere in ninety-three million miles (or even far greater spatial extent) all at the same time? The theory of relativity says that it takes about eight minutes for the fastest possible thing to travel ninety-three million miles. So if we say quantum mechanics is complete, we create a conflict with another pillar of physics.

Einstein concluded:

> In my opinion, one can remove this objection only in the following way, that one does not describe the process solely by the Schrödinger wave, but that at the same time one

localises the particle during the propagation. I think that Mr de Broglie is right to search in this direction.[16]

You might think that, with Einstein's support, de Broglie's ideas gained some influence. But this was not what happened. As he remembered his talk at Solvay:

It received hardly any attention.... Only Einstein encouraged me somewhat on the path I wished to tread. But I was faced with redoubtable adversaries: Niels Bohr and Max Born, scientists of world renown. And there was also the group of young researchers who formed the Copenhagen School, amongst them, in particular, Pauli, Heisenberg and Dirac, who were already authors of remarkable works. They interpreted the duality of corpuscles and waves by the theory of complementarity recently proposed by Bohr and...concluded by abandoning any clear picture of a wave or a particle. I was very distressed. I found Bohr's complementarity quite obscure and I did not like abandoning physical images which had guided me for many years.[17]

Einstein had said to everyone gathered that he thought de Broglie was searching in the right direction. This support was meaningful to de Broglie, who decades later remembered it. But it was not enough. He felt that Einstein had encouraged him to continue in the way he had started, "without giving a clear approval of my endeavor."[18]

The year 1927 had started with victories and great expectations for de Broglie, but it was the year everything unraveled. Einstein had withdrawn his own double solution paper only weeks after he presented it. De Broglie had fatally weakened his position at the conference by not having faith in the double solution, despite the mathematical difficulties, and deciding to present the pilot-wave theory only. Objections were strong and support was weak. De Broglie came home to Paris with Einstein, and as they parted, Einstein told the young man: "Carry on. You are on the right road."[19] But de Broglie did not carry on. As he said:

I came to the conclusion that...the theory of the pilot-wave
was untenable. Not daring to return to the double solution because of its mathematical difficulties, I became dis-
heartened and aligned myself with the purely probabilistic
interpretation of Bohr and Heisenberg.[20]

Note the adjectives in the quotes above: *very distressed...disheart-
ened.* This is what happened to de Broglie's momentum. He did not do
anything as important as his early work ever again. The matter-wave
hypothesis and the double solution were vital, imaginative, insight-
ful. But he stopped pursuing his own ideas. He adopted the viewpoint
developed by Bohr and Heisenberg, that the electron could not be localized, that measurement forced the electron to assume a specific position,
that the universe was fundamentally random. De Broglie participated
in the burying of his own theory. At the age of thirty-five, his career as
a creative scientist was essentially over. This is why de Broglie has been
called "one of the most tragic figures in the history of science."[21]

Chapter 28

The Bohr-Einstein Debate

The fifth Solvay conference was also where the most famous debate of twentieth-century science began. Much of it occurred in the Hotel Metropole, where all conference attendees were staying. Einstein would emerge in the morning with a new thought experiment, an objection to the uncertainty principle. It would be discussed over coffee and crois-sants and then on the walk to the conference building, and in between the sessions. Heisenberg and Pauli conferred among themselves and then conferred with Bohr, and by the end of the day at dinner at the hotel, Bohr explained why Einstein's latest critique could be answered.

This happened, again and again, and it convinced Bohr, Pauli, and Heisenberg that they were on firm ground, "and Einstein understood that the new interpretation of quantum mechanics cannot be refuted so simply."[1] Ehrenfest wrote one of the most vivid descriptions of the debate shortly after it occurred, recounting his experience to Goudsmit and his other students:

> Brussels-Solvay was fine!...Bohr towering completely over everybody. At first not understood at all...then step by step defeating everybody. Naturally, once again the awful Bohr incantation terminology. Impossible for anybody else to sum-marize. (Poor Lorentz as interpreter...summarizing Bohr.

209

And Bohr responding with polite despair.) (Every night at
one a.m. Bohr came into my room just to say one single
word to me, until three a.m.) It was delightful for me to be
present during the conversations between Bohr and Einstein.
Like a game of chess. Einstein all the time with new examples...to break the uncertainty relation. Bohr from out of
philosophical smoke clouds constantly searching for the tools
to crush one example after the other. Einstein like a jack-
in-the-box: jumping out fresh every morning. Oh, that was
priceless. But I am almost without reservation pro Bohr and
contra Einstein...!!!!!!! Bravo Bohr !!!!!![2]

At the next Solvay conference, held three years later in 1930, Einstein posed a problem for Bohr's view that greatly disturbed him. This
is how Rosenfeld remembered it:

I shall never forget the vision of the two antagonists leaving
the Club: Einstein, a tall, majestic figure, walking quietly,
with a somewhat ironical smile, and Bohr trotting near him,

very excited, ineffectually pleading that if Einstein's device would work, it would mean the end of physics. The next morning came Bohr's triumph and the salvation of physics; Bohr had found the answer.[3]

The Einstein-Bohr debate seems to have been characterized by what we might call mirror asymmetry.[4] The first asymmetry is presented strikingly by the above image: On the one hand, Einstein, "majestic," secure in his position, quiet, wearing an "ironical smile." On the other hand, Bohr, "trotting near him" and "ineffectually pleading." The other asymmetry, however, reverses the roles. Bohr "triumphs" over Einstein in the conceptual arena, which is, after all, the one that matters. Or at least, he answers the challenge posed. Because of this apparent triumph, Bohr is more influential than Einstein, and actually directs the development of quantum mechanics.

Both asymmetries persist together because Bohr could not convince Einstein of his point of view no matter how much he tried. In Heisenberg's words: "We knew in his heart he was not convinced."[5] Once, Bohr told his antagonist that other concept pairs, such as subject/object, and thought/sentiment, are complementary. This was meant to illustrate the general validity of Bohr's epistemological discovery, but one can hardly imagine an argument less likely to impress Einstein. For as long as he lived, his position never wavered a single millimeter. He was as immovable as an obelisk. And this rock-solid intransigence of Einstein's continued to distress Bohr.

Bohr's triumph, to Einstein, was an illusion because Bohr had not answered Einstein's actual concerns. He had merely shown that basing his analysis on the uncertainty relations was logically consistent. Einstein was unable to disprove Bohr, but that did not make Bohr right. Complementarity resulted in a view of the world that Einstein described as "the system of delusions of an exceedingly intelligent paranoiac, concocted of incoherent elements of thoughts."[6] Because such a theory is logically possible, should it be considered true?

No matter how many people sided with Bohr, Einstein remained completely secure because of his strong and unwavering intuitions about the nature of science and the natural world. If others were drawn in by the illusion, let them sleep. As he wrote to Schrödinger in 1935:

The Heisenberg-Bohr tranquilizing philosophy—or religion?—
is so delicately contrived that, for the time being, it provides
a gentle pillow for the true believer from which he cannot
very easily be aroused. So let him lie there.[7]

Einstein continued to fight against the "tranquilizing philosophy"
for the rest of his life. To him, the Copenhagen physicists did "not
understand what a risky game they are playing with reality."[8] Einstein's unmoving resistance inspired others who thought something was
wrong with the story they had been told, and his powerful voice of
dissent lives on. He always wrote with exceptional directness and elo-
quence about these issues. Everything he said about God not playing
dice, about bombs being exploded and unexploded, his quips about
disappearing moons and ectoplasmic beds, it was all designed to resist
Bohr's influential claim that no further story could be told. Einstein
made this point over and over, and here is yet another version:

It seems certain to me that the fundamentally statistical
character of the theory is simply a consequence of the incompleteness of the description....It is rather rough to see that
we are still in the stage of our swaddling clothes, and it is
not surprising that the fellows struggle against admitting it
(even to themselves).[9]

Whenever they were in the same place, the two old friends immediately dove straight into the old argument. Sometimes the engage-
ments left emotional marks, as when Bohr left a debate "in a state of
angry despair," lamenting, "I am sick of myself."[10] Other times, they
were brief and humorous, as when they speculated about which one
of them Spinoza would have sided with. (Probably Einstein.) Abraham Pais witnessed some of the debate firsthand, and he described it
perfectly:

I listened as the two of them argued. I recall no details but
remember distinctly my first impressions: they liked and
respected each other. With a fair amount of passion, they
were talking past each other....It had not taken long before

> I grasped the essence of the Einstein-Bohr dialogue: comple-
> mentarity versus objective reality.[11]

By "objective reality" Pais means realism, summarized nicely by Einstein's claim that "what we call science has the sole purpose of determining what *is*."[12] Of course Einstein would resist complementarity, which had grown from the instinct that we cannot rely on descriptions based in space and time.

Pais wrote biographies of Einstein and Bohr; he knew both well. He said that the struggle over quantum physics was "Bohr's inexhaustible source of identity. Einstein appeared forever as his leading spiritual sparring partner."[13] And he said that Einstein spoke about relativity "with detachment, about the quantum theory with passion. The quantum was his demon."[14]

Back and forth they sparred, neither ever convincing the other. Nevertheless, Bohr's advantage steadily grew larger, until only a handful of physicists would even engage Einstein on these issues. Ultimately, Bohr convinced most everyone else, in part because of a book published in 1932 that had given him a decided advantage.

Chapter 29

The Proof of a Martian Anthropologist

Born in 1903 in Budapest, János Lajos Neumann von Margitta (called Johnny by his friends) was one of a generation of remarkable Hungarians. The men of this group were all born within a few years of each other, all of Jewish descent, emigrated from Europe to the United States, and all achieved great things. The sudden flowering of so much brilliance from a single place has been called the Hungarian Effect, and subject to some study. The group is sometimes jokingly called "The Martians" because they seemed to come from another planet.

Not only was von Neumann the greatest of the Martians, but during his life, nobody could match him for raw intellectual firepower. It could be that he had the most efficient brain to ever exist. When he was thirty years old, von Neumann was invited to be on the founding faculty of Princeton's Institute of Advanced Study, along with Albert Einstein and one other (Oswald Veblen). Those who knew both von Neumann and Einstein attested that von Neumann was much intellectually sharper ("quick and acute" is the phrase Wigner used).[1] This is easy to believe, considering the many stories of his prodigious gifts that are often repeated. Here is a sample of them.

By the time he was six, he could multiply (some say divide) two eight-figure numbers in his head. He mastered calculus when he was eight years old. He "quickly outgrew children's games...and took to observing those around him with the disquieting, detached manner of an anthropologist."[2] Gábor Szegő, one of the foremost mathematicians of his generation, was moved to tears when he first met the teenage von Neumann and witnessed his talents. George Pólya, one of the Martians and another leading light in his time, recounted this anecdote:

> There was a seminar for advanced students in Zürich that I was teaching and von Neumann was in the class. I came to a certain theorem, and I said it is not proved and it may be difficult. Von Neumann didn't say anything but after five minutes he raised his hand. When I called on him he went to the blackboard and proceeded to write down the proof. After that I was afraid of von Neumann.[3]

If von Neumann's mental speed and technical virtuosity were astonishing, his memory was equally so. Around the age of ten he tore through all forty-five volumes of a world history edited by Wilhelm Oncken (it was the centerpiece of a private library his father had purchased), and "was able to recite whole chapters verbatim decades later."[4] As he lay dying of cancer, possibly caused by his work on the Manhattan Project, von Neumann's brother read to him from a book they had enjoyed as children (some say Goethe's *Faust*, others Dickens's *A Tale of Two Cities*). When his brother paused to turn the page, von Neumann would recite the lines that followed. This amazing recall allowed him to become a world expert on a wide variety of topics. A Princeton professor once agreed to attend a party on the condition that von Neumann would not discuss Byzantine history, since, as he said, "Everybody thinks I am the world's greatest expert in it, and I want them to keep on thinking that."[5]

I end with quotes from two of the Martians, both friends of von Neumann. The mathematical physicist Eugene Wigner said: "Whenever I talked with von Neumann, I always had the impression that only he was fully awake, that I was halfway in a dream."[6] And finally, Edward Teller put it this way:

> Von Neumann would carry on a conversation with my three-year-old son, and the two of them would talk as equals, and

I sometimes wondered if he used the same principle when he talked to the rest of us.[7]

In short, John von Neumann was a real-life Will Hunting, the genius played by Matt Damon in the film *Good Will Hunting*. Like Leonardo da Vinci, who is "widely considered one of the most diversely talented individuals ever to have lived,"[8] the example of von Neumann reveals to us the unbelievable heights to which the human brain is capable of ascending.

In 1927, the year that the fifth Solvay conference was held in Brussels, von Neumann was twenty-three. Around this time he became interested in atomic problems, something that was called "very fortunate indeed"[9] for quantum mechanics itself. Two years later von Neumann had completed his analysis, which was published three years after that, in 1932, as *Mathematical Foundations of Quantum Mechanics*. The book is a landmark for many reasons, not least its startling clarity of exposition, and has been extremely influential. It was here that von Neumann addressed the problem of hidden variables with full rigor for the first time.

Like Euclid, von Neumann used the axiomatic method, stating clearly all of his assumptions and then deriving everything from those axioms alone. Introducing the problem, von Neumann notes that if quantum mechanics could be explained by hidden variables, "such an interpretation would be a natural concomitant of the general principle that every probability statement arises from the incompleteness of our knowledge."[10]

The way von Neumann thought about hidden variables is *exactly* as we have laid out in the thermodynamic chapters. He notes that the state of a gas is specified by six variables for each atom (three each for position and velocity), giving an unthinkably huge number. In practice, however, we use only two variables: pressure and temperature. The kinetic theory of gases shows how these are statistical ideas that arise as complicated functions of the overwhelming number of kinetic motions that actually define the state of the gas.

Von Neumann asked whether a similar situation occurred in quantum mechanics, and he concluded that any theory that tried to supplement quantum mechanics with hidden variables could not have the same theoretical structure. It would have to be a fundamentally different theory. He wrote that "an introduction of hidden parameters is certainly not possible without changing the present theory in essential respects."[11]

This sane conclusion would not be very discouraging to Einstein or anyone who believed that quantum mechanics was incomplete. Why should we believe that the theory deeper than quantum mechanics should have the same mathematical architecture? The opposite happened in the case of heat: Fourier's differential equation was explained by Maxwell and Boltzmann using an entirely different set of mathematical concepts that relied on mechanics and statistics.

In the 1930s, knowledgeable physicists understood this state of affairs. As Max Born wrote, the significance of von Neumann's work was that

> no concealed parameters can be introduced with the help of which the indeterministic description could be transformed into a deterministic one. Hence if a future theory should be deterministic, it cannot be a modification of the present one but must be essentially different. How this should be possible without sacrificing a whole treasure of well-established results I leave to the determinists to worry about.[12]

This is hardly the knockout blow that the Copenhagen camp would like it to be. In no way does it show that "to hope for hidden variables is as ridiculous as hoping that $2 \times 2 = 5$."[13] It does not show that quantum mechanics is complete. All it tries to prove is that any future theory that might explain quantum mechanics will not be quantum mechanical.

However, this was not the way von Neumann's proof was read or used. The Copenhagen physicists made much of the passages in the book that confirmed their beliefs. True, von Neumann said that "it would be an exaggeration to maintain that causality has thereby been done away with."[14] But then he went on to say that "mindful of such precautions, we may still say that there is at present no occasion and no reason to speak of causality in nature."[15] The second part is exactly what Bohr and Heisenberg were proclaiming with their relentless rhetoric of finality and inevitability. Indeed, their proclamations are what led von Neumann to address the question in the first place.

By not paying too much attention, Bohr's followers could use von Neumann's work as a "proof" that Bohr was right. And this is what they did. "Gottingen-Copenhagen physicists often referred to John von Neumann's 'impossibility' proof as a conclusive strike against hidden variables."[16] Very quickly, von Neumann's proof became another feather in Bohr's cap.

The Austrian philosopher Paul Feyerabend spent some time in Copenhagen, and he witnessed the following very revealing event:

> Bohr came for a public lecture.... At the end of the lecture he left, and the discussion proceeded without him. Some speakers attacked his qualitative arguments—there seemed to be lots of loopholes. The Bohrians did not clarify the arguments; they mentioned the alleged proof by von Neumann, and that settled the matter. Now I very much doubt that those who mentioned the proof, with the possible exception of one or two of them, could have explained it. I am also sure that their opponents had no idea of its details. Yet, like magic, the mere name "von Neumann" and the mere word "proof" silenced the objectors.[17]

For the community of practicing physicists, the debate was over. The impossibility of getting underneath quantum mechanics was *proven* by von Neumann, among the greatest mathematicians of the age. As the years passed, the generation of Heisenberg and Pauli considered Einstein to be a deplorable reactionary, unable to change with the times, overly tethered to outdated assumptions. A calcified fossil who "had difficulty" understanding the Copenhagen interpretation.[18] Maybe even his brain had calcified and he was now senile.

This is not a joke. The charge of senility was actually made, and not just against Einstein but anyone who agreed with Einstein and yet could not be dismissed as a quack. Carver Mead, an emeritus physicist at Caltech who also got his undergraduate degree there, remarked, "That's what they always say. By the time quantum mechanics came around he was getting a little senile." And Clauser, another Caltech alum who studied physics a few years later, recounts:

> On many occasions, I was personally told as a student that these men had become senile.... This gossip was repeated to

me by a large number of well-known physicists from many different prestigious institutions.... Under the stigma's unspoken "rules," the worst sin that one might commit was to follow Einstein's teaching and to search for an explanation of quantum mechanics in terms of hidden variables.[19]

Even Einstein knew this was the way he was viewed: "Even the great initial success of the quantum theory does not make me believe in the fundamental dice-game, although I am well aware that our younger colleagues interpret this as a consequence of senility."[20]

And so the Göttingen-Copenhagen physicists got their way. The implications of quantum theory were established. There is a "pure" probability that does not arise from patterning of deeper structure, and our universe is driven by it. Our capacity for knowledge is bounded by complementarity. There is no objective world that exists apart from observations. There are no trajectories. There are no sub-quantum facts.

Dissent lived on, underground, audible only to those already attuned to hear it. For the rest of the world, the battle was over. Completeness had won.

Part VIII

Interlude

Chapter 30

Parable of Miasma

Bad air! It creeps in through open windows, under doors, and even down your chimney. It rises from bogs and marshes and all foul places. From sewers and cesspools it creeps in, the pollution, and it weakens, it sickens you. For the unlucky, it can even kill!

This was the most accepted theory for disease transmission going back to Hippocrates, in ancient Greece, and all the way into the 1800s. The technical term for this was *miasma*. It was believed that the invisible, creeping poison spread through the air from general foulness. It was thought to grow worse at night, when the Earth exhaled pestilence, and so people closed their windows against night air. (Malaria, which means "bad air," is spread at night, by mosquitoes.) You could not control the miasma; you could not even see it. But you knew it was there because it stank.

Between 1845 and 1860, there was a global outbreak of cholera. It is called the third cholera pandemic, and was the most deadly of all pandemics in the 1800s. It originated in India, and quickly spread to Asia, North America, Europe, and Africa. Before it was done, over a million Russians were dead. In the year 1853–1854 alone, more than twenty-three thousand people died in Great Britain, about half of them Londoners.

It was assumed that cholera spread—somehow—through the air. The miasmatists, as those who supported this theory of disease transmission were called, were the dominant force during the London pandemic in the 1840s. The theory had folklore and tradition on its side, since most people assumed that illness spread from the foul smells of the city, and it had many prestigious supporters, among them the sanitation commissioner, the city demographer, as well as most doctors, scientists, and members of Parliament.

The miasmatist point of view was forcefully expressed by Florence Nightingale, the first nurse and a great advocate and transformer of the role of women in Victorian England. In her *Notes on Nursing*, published during the nightmarish outbreak, she wrote,

> The very first canon of nursing...the first and last thing... the first essential...without which all the rest...is as nothing...is this: TO KEEP THE AIR HE BREATHES AS PURE AS THE EXTERNAL AIR.[1]

"Nightingale believed that cholera, smallpox, measles, and scarlet fever were all miasmatic in nature."[2] So if you were stupid enough to breathe through a rotten banana peel for a time, you could get cholera and god knows what else. Of course, the miasma theory is as garbage as the garbage it fears, and so attempts to address the pandemic were predictably worsened by the miasmatists. The most influential of them was the head of the Board of Health, one Edwin Chadwick. He devised a plan to reduce miasma in homes by outlawing "any foul or offensive Ditch, Gutter, Drain, Privy, Cesspool or Ashpit." In other words, no longer were Londoners allowed to keep "great heaps of turds" in their cellars.[3] Where was the waste to go? Straight into the Thames. The river quickly became one of the most polluted in the world.

Because cholera is a bacterium spread through drinking wastewater, and people drank the water from the Thames, this was exactly the worst possible thing to do. The outbreak worsened considerably, and "a modern bioterrorist couldn't have come up with a more ingenious and far-reaching scheme."[4] And yet, the authorities were pleased with the effectiveness of their new law; they even celebrated the tons of waste being dumped into the water despite being fully aware that people drank it. Why? They thought they were improving public health. As Johnson wrote, it seemed like madness to destroy the Thames in this way, and indeed, it was a kind of madness, the madness that comes from being under the spell of a Theory. If all smell was disease... then any effort to rid the houses and streets of miasmatic vapors was worth the cost, even if it meant turning the Thames into a river of sewage.[5]

Like all false theories, miasma had serious problems. Just to take one example, consider that class of London poor called toshers, or sewer-hunters. They did what has been called "quite likely the worst job ever": exploring the sewers of London, wading through human excrement while scavenging for valuables. This could be quite lucrative and the goods they brought up were called "tosh." The toshers explored by light of kerosene lanterns, and every couple weeks one of them would come across a particularly dense patch of methane gas and "the hapless soul would be incinerated twenty feet below ground, in a river of raw sewage."[6]

If the miasma theory was correct, the toshers should have had the shortest life span of anyone in the city. After all, they spent their days breathing "the most noxious—sometimes even explosive—air imaginable."[7] And yet the leaders of the gangs, men in their sixties, seventies, and eighties, had been doing the same job their whole lives and were in excellent health. As one contemporary author puzzled:

> Strange to say, the sewer-hunters are strong, robust, and healthy men, generally florid in their complexion, while many of them know illness only by name.[8]

The way the miasmatists made sense of this sort of thing was with an appeal to another invisible, accommodating variable: "inner constitution." Of course foul air had different effects on people, because people themselves are different. The sewer-hunters did not die upon minutes of entering London's plumbing, and were evidently hale and hearty to the man, because they had very strong inner constitutions. In general, however, the poor had the same variety of inner constitutions as the wealthy. Many of them were not so strong, and miasma could sicken them. And in this sense the situation with the poor supported the miasma theory: they lived amid filth and squalor, with its accompanying bad smells, and they died of illness in far greater numbers than the middle and upper classes.

During this time, there was a physician named John Snow. As a child he had excelled at mathematics, and as a country lad of twenty-three, he had walked two hundred miles alone to pursue a medical career in the city of London. He wished to be more than a country apothecary, and placed in the first division of students at the University of London. He was calm and emotionally flat with his patients, and he was gifted with a prodigious memory for past cases, an analytical inclination, and no small part of ambition. In a short number of years, Doctor John Snow rose to the height of Victorian society. His researches into ether and chloroform had made him the most sought-after anesthesiologist in London. He attended hundreds of operations each year, and in 1853 he was called upon to administer chloroform to Queen Victoria upon the birth of her eighth child.

Snow started doubting the miasma theory and investigating the cause of cholera. His interest was triggered by reading about how John Harnold, a crewman on a German steamer that docked in London in September 1848, came down with the only case of Asiatic cholera in the country. It was known that this variant had recently spread to the continent, and was active in Hamburg itself. Harnold died within hours, and the new lodger named Blenkinsopp, given the dead man's room, came down with Asiatic cholera within a week. From there, the disease spread through the entire country, killing 50,000 people.

It strained the miasma theory to imagine that bad air has suddenly arisen in a single room to breathe Asiatic cholera into a man who had arrived, the day before, from a country stricken with that disease. Far more plausible was that cholera had been *communicated*, because it was caused by some currently invisible physical mechanism. Harnold brought the cholera from Germany to England, and then somehow gave it to Blenkinsopp.

This was the theory of contagion, which had struggled against the miasma theory for some time as a minority view. It had a few supporters; for example, there was an editorial in the *Lancet* in the early 1830s that proposed cholera to be a kind of poison independent of "all condition of the air...one that makes mankind the chief agent for its dissemination."[9] It was this theory that Snow began to suspect was the right one, because he was analytical, and demanded that cause-effect relationships make sense. He wrote:

> Who can doubt that the case of John Harnold...was the true cause of the malady in Blenkinsopp?...And if cholera be communicated in some instances, is there not the strongest probability that it is so in the others—in short, that similar effects depend on similar causes?[10]

Part of the reason Snow doubted miasma was that he was an expert in ether and chloroform. He had thoroughly researched these gases. First he had experimented on frogs and mice in his own lodgings, and then on himself, before being satisfied that he fully understood their operation. In his hundreds of uses of these anesthetics, Snow had seen the "inner constitution" theory disproven utterly. In Snow's firsthand experience, the human response to chloroform was mechanistic and predictable; otherwise it would have been entirely unsuitable for medical use. It was his mastery of a real-life miasmatic gas that prepared Snow to disbelieve what was the accepted wisdom of his entire generation.

In 1849, Snow published an essay titled "On the Mode of Communication of Cholera" where he suggested that the illness was spread through contaminated water. Five years later, on August 31, 1854, cholera came to Snow's own neighborhood of Soho and struck it savagely. Snow was there, he saw everything, and he described it as "the most terrible outbreak of cholera which ever occurred in this kingdom."[11] This was not hyperbole or emotion—it really was an extreme and devastating outbreak. Nearly seven hundred people in a 250-yard radius died in two weeks. If much of the population had not immediately fled, the toll would have been even higher.

On Thursday, September 7, 1854, a week after the outbreak had begun, there was an emergency meeting of the board of governors of the parish. In the middle of it, John Snow came forward and addressed the assembly. He told them that they must immediately close the water pump at Broad Street. Snow had come to his conclusion from a detailed study and many interviews conducted over the previous days. The deaths clustered near the Broad Street water pump, he explained. But what was more, those who lived nearby who had escaped the disease either had their own water source, and so did not drink from the pump, or they did not drink water at all, as was the case with the workers of the Lion Brewery on 50 Broad Street. They had no need for water, since they were happy with their daily allotment of malt. Those who lived farther away, near other water sources, and had yet fallen ill, did so because they somehow drank the contaminated water. They received special deliveries of the favored Broad Street water, for example, or they went to school near Broad Street and often stopped at the pump.

All of this Snow pieced together in an astonishingly short amount of time. As his biographer later wrote about him: "No one but those who knew him intimately can conceive how he laboured, at what cost, and at what risk. Wherever cholera was visitant, there was he in the midst."[12]

Hesitant and unconvinced of Snow's solution, the governors were nevertheless at a loss, and so they followed the esteemed doctor's advice. They had the handle of the pump removed, and this moment has been called a historical turning point.

> For the first time, cholera's growing dominion over the city would be challenged by reason, not superstition.[13]

These sorts of researches and interventions became more standard as the years passed. The trend of thought that Snow started became more accepted, and London adopted fundamental changes to its management of waste and water. Other cities followed, and public health improved greatly across the entire world.

This was all despite the fact that Snow was as ignorant of the true cause of cholera as anyone else. But he did not allow his ignorance to stultify him. He relied on reason and a faith in mechanism, and he struck the target despite being blind. Compare this to the complacency of ignorance exhibited by the *Times*, in an 1849 article devoted to various theories of spread:

> How is the cholera generated?—how spread?...These problems are, and will probably ever remain, among the inscrutable secrets of nature. They belong to a class of questions radically inaccessible to the human intelligence.[14]

Remembering Comte's blunder that primary causes are "questions necessarily insoluble for our intelligence," we can predict how this prediction aged.[15] Indeed, it was made only five years before the cholera bacillus was first glimpsed in Italy, by Filippo Pacini.

Now, of course, we know that John Snow was perfectly correct about contaminated water. We know that cholera is caused by the bacteria *Vibrio cholerae* and the toxin it secretes upon invading the small intestine. Because of many searchers who believed in mechanism despite their ignorance of it, in the space of 170 years we went from considering cholera to be an "inscrutable secret" that is "radically inaccessible to the human intelligence," to sentences like this:

> The bacterium has a flagellum at one pole and several pili throughout its cell surface. It undergoes respiratory and fermentative metabolism. Two serogroups called O1 and O139 are responsible for cholera outbreaks.[16]

Chapter 31

Many Droplets

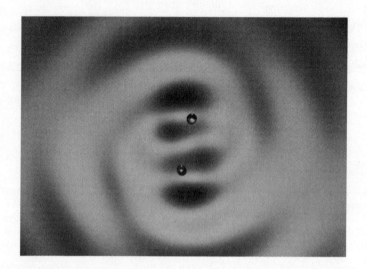

This is a photo of two droplets orbiting in "twin-star-like-motion." Such behavior was observed very early, and reported in one of the first papers on HQA. The droplets' orbits are themselves quantized, which makes sense. One droplet travels in the trough of the wave of the other, and so the orbits can only exist for certain diameters. These diameters change, depending on whether the two drops are bouncing in phase (up, up), or out of phase (up, down), and the quantization itself is explained by the form of the pilot-wave field.

In a previous chapter we described how if you put sodium bicarbonate into a fire and split the light from the fire apart with a prism, you observe two bright bands of yellow, the D lines. Pieter Zeeman was a Dutch physicist who, in 1886, had the idea to put the fire inside a magnetic field. When he added sodium to the flame and looked at the spectrum of light, Zeeman saw something remarkable: each one of the D lines split into multiple, closely spaced spectral lines. The separation between them depended on the magnetic field: the stronger the field, the more the lines moved away from each other.

Because the effect is thoroughly atomic, and therefore quantum, its explanation had to wait for a long time. The quantum theory of Zeeman splitting is relatively complex, because the effect has several different components. However, the simplest form of splitting is actually the best preparation for what comes next. So long as you understand there is much more to the story, I am not leading you astray.

The simplest case is when the spectral line divides in two, like this:

The upper dark rectangle shows a spectral line at a certain position, without a magnetic field. If we add a magnetic field, we get the lower dark rectangle: the spectral line splits in two and separates.

According to Bohr's theory, when an electron goes from a higher to a lower energy, it will emit light at a frequency determined by the change in energy. This frequency of light is what causes the spectral line. When we add a magnetic field, the electrons themselves change their energy levels. This means that when they jump to the same lower energy as before, the energy change will be either smaller or larger than it was originally. In other words, the light emitted will be lower or higher frequency, more red or more violet than before. As we increase the strength of the field, moving farther right in the shaded region, the difference in electron energies grows larger and larger: the two split spectral lines broaden.

We got the analogy to the Landau levels in Chapter 17 by rotating the fluid bath. Recall that this gives rise to the Coriolis force, which has an effect on the droplets that is almost mathematically identical to the Lorentz force produced by a magnetic field on charged particles. (It differs by a scaling factor of two.)

Because of the mathematical similarity between the Lorentz and Co-riolis forces, effects of magnetic fields on charged particles can be recre-ated by rotating reference frames. The two systems will behave in very similar ways. This is why researchers in Couder and Fort's group ex-amined the effect of a rotating fluid bath on orbiting droplets. They ob-served an effect analogous to Zeeman splitting, in a system that gives us physical pictures and clear causality.

Here, we have pictures; we can understand exactly how it works. The droplets can orbit either clockwise or counterclockwise, and this orbit-ing direction can be either with or against the rotation of the bath. If it is *with* the bath rotation, then the droplets experience a boost in their velocity from the bath; this increases the orbit diameter. If the droplets orbit *against* the bath rotation, then their motion is slowed down, and the radius decreases.

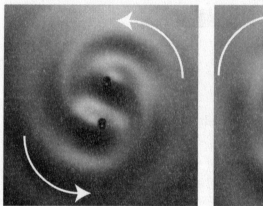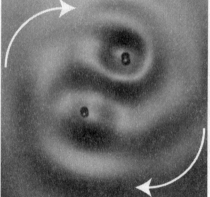

Left: Two droplets rotating against the rotation of the bath; their radius is de-creased. *Right*: Two droplets rotating with the rotation of the bath; their radius is increased. Credit: Yves Couder.

Now, let us make plots of the energy versus magnetic field strength in the Zeeman effect, and of the diameter versus rotation speed in the droplet system. Both show the same change, as the intervention in-creases. Remember, rotation is mathematically analogous in fluid sys-tems to a magnetic field. Both plots show the same type of change in the quantized levels as field/rotation increases. It is a linear increase (or decrease).

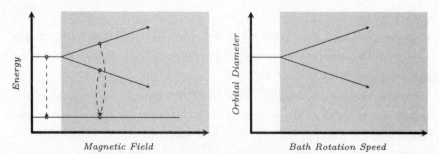

Left: The Zeeman effect. As soon as the magnetic field is turned on, the energy level splits, and the amount of the split depends linearly on the field strength. *Right*: The analog Zeeman effect. As soon as the bath rotation is turned on, the radius of the rotating droplets changes depending on whether the droplets and bath are rotating in the same or opposite directions.

So two droplets can orbit each other, and generate quantum analogies. Let's see what happens when we start adding more than two droplets together. Amazingly, they lock into stable crystalline patterns.

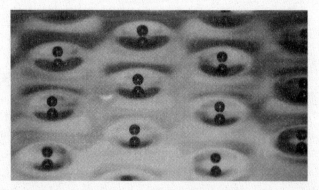

Droplets arranged in a Bravais lattice, all bouncing in sync. Credit: Miles Couchman and John W. M. Bush, MIT.

The droplets are coupled to each other through their respective wave fields, and the global wave field determines where droplets are allowed to be. The study of these lattices started early on, and their mathematical description was recently developed. These structures do not spontaneously dissolve because they "are most stable when each drop bounces in a minimum of the net wavefield produced by its neighbours."[1] This means that the structural geometry of a lattice is a crucial factor in how strong it is.

There are many possible lattice configurations. In 2007, the Paris lab produced eight of the eleven total Archimedean tilings of the plane, using droplets for the vertices. Not all of them, however, were accessible when the droplets were all in phase. Some lattices required a specific spatial alternation between in-phase and out-of-phase bouncing. The image below shows the same lattice at three different times. If you look closely, you can see that some droplets are hitting the bath while others are in the air.

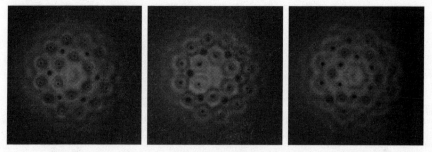

Three different times for the same Archimedean lattice. Credit: Yves Couder.

These lattices are most stable in low memory. As memory is increased, we expect the lattice to destabilize as the droplets become walkers. In the transition to instability, small oscillations of single droplets are passed onto neighbors, and the whole lattice begins to waver in a structured pattern called a *phonon*. Eventually, as memory is increased further, the whole lattice collapses into an irregular pattern.

There are connections with one of the most basic models of condensed matter physics, called the *canonical mass-spring lattice*. This treats molecules as simple masses, linked to their neighbors by springs. Disturbances in the placement of the masses stretch or compress the springs, and then the response can be computed.

The droplets work like this too, but with an important difference. In the mass-spring lattice, a stretched spring always pulls on the attached masses. With droplets, the wave field operates as the spring. However, the field oscillates from positive to negative curvature and back. Thus, whether a droplet will respond to perturbations by being pushed or pulled depends on not only its position relative to other droplets, but also its phase. Droplet lattices are actually *much richer* than mass-spring lattices.

Miles Couchman, currently an assistant professor of mathematics at the University of York, did his PhD on multiple droplet systems with John Bush at MIT. He uncovered many novel behaviors by considering rings of droplets. Such rings can exhibit specific oscillations, both in and out of phase, radial and angular. They can even start rotating, forming a chain of walkers following each other in a circle.

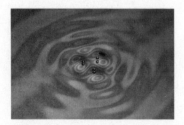 

Forming a larger ring of fifteen to twenty droplets, and slowly increasing memory to approach the instability threshold, leads to a remarkable geometric transition. The rings start in a perfect circle, but with increasing memory begin exhibiting radial, out-of-phase oscillations. At even higher memory, the oscillations grow larger until an instability threshold is passed. Then, the circle quickly transforms into other stable shapes, including squares, pentagons, and irregular lattices.

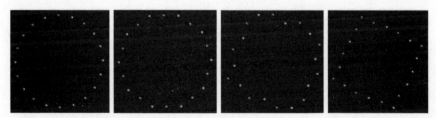

Four different times showing a vibrating circular ring adopting a square formation. Credit: Miles Couchman.

These transitions were rationalized through the mathematics of stability theory, and can be roughly explained with an energy argument.

The average height of the wave field in the circular ring is larger than in the shape that is ultimately chosen, be it square, pentagon, or something more exotic. This means the droplet system is seeking the configurations that have the least amount of energy.

The work we have just considered is only a sample, mostly eye candy, of a research theme in HQA that started in the early days and has continued up to the present. The fact that droplet lattices have some of the properties that we associate with collections of molecules is something to wonder at. What would happen if we could extend this analogy into three dimensions?

Part IX

The Tranquilizing Philosophy

The Heisenberg-Bohr tranquilizing philosophy—or religion?—is so delicately contrived that, for the time being, it provides a gentle pillow for the true believer from which he cannot very easily be aroused. So let him lie there.

—Albert Einstein[1]

All these years it has seemed obvious to me as it did to Einstein and Schrödinger that the Copenhagen Interpretation is a mass of contradictions and irrationality and that, while theoretical physicists can, of course, continue to make progress in the mathematical details and computational techniques, there is no hope of any further progress in our basic understanding of nature until this conceptual mess is cleared up.

—Edwin Jaynes[2]

Chapter 32

The Only Mystery

What happens when we assume quantum theory is complete? What paths do our thoughts travel? It's not very hard to find out since modern physics is replete with examples. After all, completeness won. So let's start with what we know, the double-slit experiment.

If quantum mechanics is complete, then the Schrödinger equation is basically the end of the road. All we can talk about is the wave function, the probability distribution we get from it, and our observed scintillation on the screen. The wave function goes through both slits and its two branches interfere with each other, creating the interference fringes.

In that case, we must agree with Feynman's analysis of the double-slit experiment, when he claimed that it is "*not* true that the electrons go *either* through hole 1 or hole 2," and we should reach this conclusion "undoubtedly."[1] On the same page he doubled down on his argument, saying,

> Many ideas have been concocted to try to explain [the interference pattern] in terms of individual electrons going around in complicated ways through the holes. None of them has succeeded.[2]

Only a few pages later, Feynman tripled down, saying that any attempt to explain the double-slit experiment using electron trajectories was bound to fail. Why? Because any proposed machinery inside the electron that determines its motion "should not be dependent...upon whether we open or close one of the holes."[3] In this argument, Feynman echoed Heisenberg, who said a few years earlier:

> If [a single photon] goes through the first hole and is scat-
> tered there, its probability for being absorbed at a certain
> point of the photographic plate cannot depend upon whether
> the second hole is closed or open.[4]

Therefore, one conclusion drawn from quantum completeness is that we cannot trace a continuous line or curve in space for the electron's path as it goes from point A to point B. This path is also called a trajectory, and if we need to give up trajectories, we cannot keep the related idea of the particle existing at specific points in space, from moment to moment as it moves. That idea is "classical," created by our intuitions of the crude, macroscopic world. A tennis ball has a trajectory, as does an airplane. But particles, no. Their wave nature defeats their particle nature in this respect. As Landau and Lifshitz, the authors of a well-known physics textbook, concluded:

> It is clear that [the results] can in no way be reconciled with
> the idea that electrons move in paths....In quantum me-
> chanics there is no such concept as the path of a particle.[5]

Here we have another renunciation, forced on us by quantum theory, and it is very important to pause at this juncture and to understand what this renunciation means. It's a big deal. First of all, without trajectories, we lose our usual concept of matter. We think of matter, stuff, things, as existing in a specific place, at a specific time. Your body, your bed, raindrops, rocks, trees, moons, stars—they are all things that exist in a place. But, because all of these things are made of particles, and particles do not exist in a place, evidently this idea of existing in a place is an illusion. As Heisenberg said:

> The idea of an objective real world whose smallest parts
> exist objectively in the same sense as stones or trees ex-

ist, independently of whether or not we observe them...is impossible.[6]

The electron is always and forever *observed* as a particle. But if we just concluded that it does not exist from moment to moment as a particle, then what is it about the photographic plate that forces the wave nature of the electron to give way to its particle nature? The orthodox answer is that, because we did not observe the electron at the two slits (by definition we cannot, since the two slits are empty space), the electron was able to keep its wave nature there. It is only at the final screen, where we observe the electron, that the particle nature emerges. Thus, there is something special about the act of observation that converts the wave into the particle.

Pascual Jordan, a coauthor with Born and Heisenberg on one of the founding papers of quantum mechanics, said that when we make an observation at the back screen

> the electron is forced to a decision. We compel it to assume a definite position; previously, it was, in general, neither here nor there; it had not yet made its decision for a definite position.[7]

Because of the mathematics of the theory, it is clear that the wave nature of the electron takes over again, after an observation has been made. So long as we do not make another observation, the electron spreads out and again is "neither here nor there." Absent any measurements, it would seem that the natural state of an electron is to be vaguely smeared out over space. In two (actually very many) places at once. Once we make an observation again and look at the electron, we see it as a particle.

This whole setup is expressed pithily by the physicist Bhatia in what some consider the best summary of quantum physics in five words:

> Don't look: wave. Look: particle.[8]

It is a small step from here to the idea that it is *our* act of looking at the electron that has collapsed its wave nature into its particle nature.

If you find this idea distasteful, you are not alone, for it was one of Einstein's biggest problems with quantum mechanics. Indeed, we can now better understand the objection that he raised about the moon not existing when you don't look at it. The same objection in a more understandable form is what Putnam called "the joke of Einstein's bed":

> Look, I don't believe that when I am not in my bedroom my bed spreads out all over the room, and whenever I open the door and come in it jumps into the corner.[9]

But others did not have the same qualms, and one of them was at Princeton at the same time as Einstein. This was John Wheeler, one of Feynman's teachers, and a central figure in the physics community of that generation. He proposed what he called the "participatory principle." He thought that

> we could not even imagine a universe that did not somewhere and for some stretch of time contain observers because the very building materials of the universe are these acts of observer-participancy.[10]

In other words, when you look at something you actually bring it into being as a concrete thing. Great job! One might ask where observers even come from. But this question is probably disallowed in Wheeler's view because the past is not real. In the participatory universe,

> we have to say that we ourselves have an undeniable part in shaping what we have always called the past. The past is not really the past until it has been registered. Or put another way, the past has no meaning or existence unless it exists as a record in the present.[11]

Clearly, quantum mechanics is very weird. And yet, from this weirdness we can also derive more weird things. Recall that the wave pattern associated with the particle expresses the probability that we will observe the particle there. If we accept that the particle has no trajectory, is not anywhere specific before we look, and that observation compels it to assume a definite position, then the fact that we cannot predict more than a probability for the location of a specific electron becomes much more serious, because we seem to be dealing with a brand-new kind of probability.

Generally, probabilities arise because we don't know all the meaningful information. We say rolling a six-sided die is random because we can't measure all the relevant parameters—the weight of the die, the speeds and angles of the throw, the density and currents of the air, the surface of the table—that would allow us to predict the outcome of any particular throw. Pretty much always, probabilities enter science this way. It is conceivable that, with a powerful supercomputer, we could accurately predict how a die throw would turn up. In fact, you don't even need a computer to do this a little bit, which is why people shave the edges of dice to change the probabilities. The point is that whenever probabilities arise in science, this is because we either don't know all the relevant parameters, or there are just too many of them. In principle, however, we can eliminate the probabilities from our model.

But an electron is not a die, and it is "neither here nor there" before we measure it. So we cannot use its position as one of these parameters. If we cannot use its position, then we cannot use its velocity either, because velocity is a measure of how position changes with time. Then we cannot use its momentum either, because momentum is the product of mass and velocity, and so on. In fact, *all* measurable quantities can be reduced to a measurement of position and velocity, and so it would seem that we cannot actually find *any* parameters for the electron that would allow us to construct a more precise picture of what is going on, and to eliminate the probabilities.

This is astonishing and creates a disjunction between probabilities in quantum physics and in every other branch of science. Max Born, to whom the probabilistic interpretation of Schrödinger's equation is due, was very clear on this question: "No concealed parameters can be introduced with the help of which the indeterministic description could be transformed into a deterministic one."[12]

Quantum mechanics is inherently and essentially a probabilistic theory. Its predictions are irreducibly statistical. That is fine in itself—there are a lot of statistical theories out there. But if quantum mechanics is complete, then quantum probabilities are different than every other type of probability in science. Completeness means we cannot do better than the probabilities; we cannot eliminate them *even in theory*. The concealed parameters, the hidden variables, that would allow this do not even exist. This was the view of Born in Göttingen, it was considered to be definitively proved by von Neumann in 1932, and it spread to become very widely accepted.

What this means for our view of the world is that the world itself is inherently probabilistic. Events, at their deepest level, are random. They seem to "literally have no cause."[13] That is, "quantum mechanics defeats causality." The universe, at its *foundation*, is random. The view today, expressed by one PhD student in a blog post, is that "we absolutely cannot know exactly how something will turn out before it happens."[14] This is not the irresponsible opinion of some trainee, either; it is the assumed truth of our entire scientific culture. As the *Stanford Encyclopedia of Philosophy* puts it: "With the advent of quantum physics...it is now widely assumed that nondeterministic (or probabilistic) causation is possible."[15]

Again, Einstein disliked this, and he had a long debate in letters with Born about it. One of these letters codifies what is probably Einstein's most famous quote, which he said several times in conversation: God does not play dice. In this letter Einstein said:

> Quantum mechanics is certainly imposing. But an inner voice tells me that it is not yet the real thing. The theory says a lot, but does not really bring us any closer to the secret of the Old One. I, at any rate, am convinced that He does not throw dice.[16]

Born did not agree, and his view is that taken by most people historically. It has been established that quantum theory is statistical and that this fact cannot be escaped. Some might try to escape it, but how this could be possible, Born said, "I leave to the determinists to worry about."[17]

And this brings us, finally, back to a quote of Feynman's. In discussing the double-slit experiment, he makes clear that

> Nobody knows any machinery. Nobody can give you a deeper
> explanation of this phenomenon than I have given—that is,
> a description of it.[18]

And so, "nobody understands quantum mechanics." How could they?
It strikes so deeply against our basic intuitions and confounds so fully
the assumptions of classical science that it is among the greatest revolutions in the history of human thought.

In the foregoing I felt it wise to include many quotes, because the
point of view presented is so extravagant that skepticism is the natural
response when first coming into contact with it. This point of view was
and remains so dominant, however, that it has melded with the background of science and much of it has simply become assumed as a given
truth. For example, in a piece about philosophy (not quantum mechanics
at all), Lawrence Krauss writes:

> How one moves from the remarkable and completely non-
> intuitive microscopic world where quantum mechanical in-
> determinacy reigns supreme and particles are doing many
> apparently inconsistent things at the same time, and are not
> localized in space or time, to the ordered classical world of
> our experience where baseballs and cannonballs have well-
> defined trajectories, is extremely subtle and complicated and
> the issues involved have probably not been resolved to the
> satisfaction of all practitioners in the field.[19]

"Probably not been resolved to the satisfaction of all" is quite an
understatement. As we can see, Krauss is here *assuming* that quantum
mechanics is complete. This is the common view, so common it is impossible to avoid. For example, the other day I ran across this little gem
on the Wikipedia page for quantum tunneling: "The wave function of
a particle summarizes everything that can be known about a physical
system."[20] This intense claim is completely unnecessary in the article,
but someone felt compelled to slip it in there as a simple fact. Multiplying examples is trivially easy. Here is a sentence in a book by Max
Jammer: The wave function "was assumed...to be the most complete
possible description of an individual physical system...in the various

versions of the Copenhagen interpretation."[21] And here is another one from a book I was just skimming: "The wave function is a complete description of any physical system and it evolves according to the usual laws."[22] I could multiply examples endlessly at the rate of perhaps one per day just by proceeding with my usual reading and keeping my eyes open. The following paragraphs, cobbled together over a period of time, are proof of this.

I just rediscovered this one yesterday, from a book published in 2020: "I will pound the table and insist that the wave function is the sum total of reality."[23] Here's another one, also very recent: "Quantum mechanics is a theory about the physical description of physical systems relative to other systems, and this is a complete description of the world."[24] I also just found this: "A pure state provides a complete and exhaustive description of an *individual* system."[25] And this: "Bohr and his supporters maintained that quantum mechanics was a complete theory, a claim challenged by Einstein."[26] And this:

> There must, [Einstein] thought, be some sort of undetected "hidden variables" that determine a particle's properties. Science simply had not found them yet. Seven decades later, none has been seen and none is expected.[27]

Heisenberg said in 1958 that "to hope for hidden variables is as ridiculous as hoping that $2 \times 2 = 5$."[28] And Karl Popper wrote in 1982,

> Another source of the present crisis in physics is the persistence of the belief that quantum mechanics is final and complete. And the strongest reason for my own opposition to the Copenhagen interpretation lies in its claim to finality and completeness....It is also a claim that may have been forgotten by now. I think that it has indeed been forgotten: nobody mentions it any longer.[29]

I'm ending with this one: In a discussion hosted by the Institute of Art and Ideas in 2022, the string theorist and popular science author Michio Kaku told a packed house,

> The quantum theory is based on a strange principle: that electrons can be in two places at the same time. Now Einstein hated that principle. He once said that the more successful the quantum theory becomes, the sillier it looks. Well, sorry about that Albert, but it turns out that the world economy is based on the fact that electrons can be in multiple states at the same time until you measure it.[30]

Like I said, it's trivial to multiply examples. Rather than go on, I will just keep a running tally of the quotes I have come across since writing the above. Number of claims that "quantum physics is complete" stumbled upon: fifty-six.[31]

Physicists speaking in public *assume* that quantum theory is complete. I think it likely they don't even realize they are making this assumption, that the views they present are the direct result of this assumption. It is likely that if you ask them point-blank, they will say of course something might underlie quantum theory as we currently know it. Individually, they are likely to take a conservative, plausible position. Yet, without this idea of completeness, and a historical understanding of how it took root, we cannot understand why scientists say—publicly, repeatedly, unapologetically—things like "sorry about that Albert," particles "are not localized in space or time," and this is a "fact."

Chapter 33

Everything a Great and Good Man Could Be

How *did* quantum completeness become so prevalent? How is it that positivism became so dominant in quantum mechanics? When we are trying to understand why the mainstream view, already in the 1930s, was that quantum physics is complete and there are no deeper explanations we can ever find for what we observe, we are in the presence of a mystery. John Bell, one of the clearest thinkers on this subject, remarked "with some heat"

> that among the books he would like to write...would be one tracing the history of the hidden-variable question and especially the psychology behind people's peculiar reactions to it. Why were people so intolerant of de Broglie's gropings?[1]

A physicist of no less insight than David Bohm was at a loss to answer this question. He said:

> I have never been able to discover any well-founded reasons as to why there exists so high a degree of confidence in the general principles of the current form of the quantum theory.[2]

248

We will meet both of these important thinkers in future chapters. They had no idea why everyone assumed quantum mechanics was complete. Perhaps things have since been clarified? Sadly, no. When I have asked philosophers and historians for answers, some have tentative proposals, but most have no idea. And I suspect this is related to Bohm's puzzlement. He could not find any well-founded reasons because there are none. What we have, instead, is a story of animal instincts.

The thesis that quantum mechanics was complete was already the orthodox opinion in the 1930s. The analogy in thermodynamics would have been if Comte dominated all of science for decades and was worshiped as one of the most profound philosophers of his age, to the point that several generations of scientists considered it "obvious" that there was no hidden reality that could ever be found to explain the heat equation.

We know this would have been a disaster for thermodynamics, stultifying generations and freezing progress in a glacier of dogma. It did not happen, thankfully. Has quantum completeness been a disaster for atomic physics? We cannot definitively answer this question. This is the scientific mystery we have inherited.

The denial of hidden causes is quite an unusual position to take for any scientist. Why did the majority of scientists take this position with respect to the atomic world? I believe most of the explanation for this remarkable fact lies in the influence of Niels Bohr. More than anyone else, he was associated with the new atomic theory. He was the grand architect of the Copenhagen interpretation, and the primary philosophical interpreter of the quantum mysteries. As Heisenberg put it: "Bohr's influence on the physics and the physicists of our century was stronger than that of anyone else, even than that of Albert Einstein."[3] This influence flowed from Bohr's amazing personality, his formidable intelligence, his outstanding scientific achievements of the past, his strong physical intuitions, his physical stamina, and his institutional power.

The most straightforward aspect of Bohr's influence was this last one, his power. He began pushing for the creation of a special institute in 1916, the same year that he was appointed as a professor at the University of Copenhagen. In 1921, largely funded by the Carlsberg Foundation, the charitable arm of the famous brewery, the Institute for

Theoretical Physics was founded. This institute became the premier place to study and visit for anyone interested in quantum physics. In the 1920s and '30s, it was practically synonymous with the field itself. Bohr was director of the institute until his death in 1962, so over forty years, and was succeeded by his son Aage Bohr. A few years after his death, on what would have been Bohr's eightieth birthday, the Institute of Theoretical Physics at the University of Copenhagen was renamed the Niels Bohr Institute, the name it bears today.

Human beings are inevitably attracted to capital. They aggregate around centers of capital the way lions, hyenas, and vultures gather around a downed zebra. People need the bounty of physical capital—food and shelter, most obviously—but also of cultural capital: respect, prestige, influence. As the head of the most famous school for quantum physics in the world, with its own buildings, places to meet, places to stay, shining reputation, and ability to mint professors at universities far and wide, Bohr controlled awesome capital. Over the years, hundreds of physicists made the obligatory pilgrimage to Copenhagen, hundreds of papers came out of work done there, and Bohr's influence compounded as his students took on their own students.

As Mara Beller put it:

> Bohr could provide simultaneously intellectual stimulation and help in advancing careers, spiritual fulfillment and down-to-earth fun, material benefits and psychological counsel.... All this was offered at Bohr's institute in Copenhagen.... Bohr could support Heisenberg's and Pauli's visits to Copenhagen much better than the young German physicists could expect to be financed in Germany.[4]

This, by itself, confers authority, and would have put Bohr at an advantage in the quantum debate. Einstein was constitutionally unsuited to head his own school. He was out hunting for big game, by himself, and students would have only slowed him down. And this observation brings us to probably the most crucial advantage Bohr had in the debate: the awesome strength of his personality.

Many people use the word "charisma" when describing Bohr, and it is appropriate. But there are different types of charisma, as different

Copenhagen Conference at the University of Copenhagen Institute for Theoretical Physics (UITF, Universitetets Institut for Teoretisk Fysik). 1930. *First row*: Oskar Klein, Niels Bohr, Werner Heisenberg, Wolfgang Pauli, George Gamow, Lev Landau, Hendrik Kramers. *Second row*: Ivar Waller, Piet Hein, Rudolf Peierls, Walter Heitler, Felix Bloch, Tanja Ehrenfest, Walter Colby, Edward Teller. *Third row*: Oscar Rice, Aurel Wintner, Christian Møller, Mogens Pihl, A. J. Hansen. On the front desk in front of Heisenberg is a toy trumpet to give a fanfare to popular ideas, between Pauli and Gamow there is a cannon to shoot down unpopular ideas, and in front of Kramers a toy drummer boy to provide applause. Credit: Niels Bohr Archive.

from each other as a firefly and lightning bolt. When I say Bohr had charisma, I am not talking about charm. I mean the charisma of the hypnotist, the charisma of the cult leader. A kind of reality-warping personal magnetism that arranges people in orbit like planets circling the all-radiant star.

We will miss the mark if we do not focus here. People believed in completeness *because Bohr told them to.* Murray Gell-Mann said that Bohr "brainwashed" an entire generation of physicists into thinking that the problem of quantum foundations had been solved. This expression is marked by Gell-Mann's wonderful acerbity and independence, but I think it is not quite accurate. Brainwashing implies manipulation contrary to the desires or awareness of the victim. Hypnosis is probably closer to the mark, because there is no victim: it involves an overwhelming of the subject, usually with full awareness and cooperation.

The best example I know is that of Karl Popper, one of the most serious and accomplished philosophers of the last century. He had a deep interest in quantum mechanics and a natural antipathy toward what he saw as the "subjectivism" of the Copenhagen interpretation. He wrote that "the view of the status of quantum mechanics which Bohr and Heisenberg defended was, quite simply, that quantum mechanics was the last, the final, the never-to-be-surpassed revolution in physics," and he condemned such a view, saying, "This epistemological claim I regarded, and still regard, as outrageous."[5] And yet, Popper also said this:

> When I first had the great opportunity, in 1936, of talking to Bohr, he impressed me as the most wonderful person I had ever met or would be likely to meet. He was everything a great and good man could be. And he was irresistible. I felt that I must be wrong about quantum mechanics, even though I certainly could not say that I now understood it, rationally, in Bohr's sense: I did not. But I was overwhelmed.[6]

This irresistible power was experienced by many people; it is recorded over and over. Some charisma enters through the eyes; Bohr's entered through the mind. His conversation was beautiful and dangerous. As Herman Kalckar wrote, "The best of all was when the conversation turned to the so-called 'eternal questions.'...Nowhere was the overpowering intellectual and emotional impact of his personality more irresistible."[7]

Given the philosophical environment of the institute, full of promising young men seeking deep truth and run by a single older man whose personality was overwhelming and irresistible, it is not surprising that Bohr was compared to Socrates. Carl Friedrich von Weizsäcker said, quite seriously: "I think probably Niels Bohr was a reincarnation of Socrates."[8] And Otto Frisch had a very similar feeling. He remembered,

Bohr had invited a number of us to Carlsberg where, sipping our coffee after dinner, we sat close to him—some of us literally at his feet, on the floor—so as not to miss a word. Here, I felt, was Socrates come to life again...and when I cycled home through the streets of Copenhagen, fragrant with lilac or wet with rain, I felt intoxicated by the heady spirit of Platonic dialogue.[9]

Another telling example is this comment made by John Wheeler:

Nothing has done more to convince me that there once existed friends of mankind with the human wisdom of Confucius and Buddha, Jesus and Pericles, Erasmus and Lincoln, than walks and talks under the beech trees of Klampenborg Forest with Niels Bohr.[10]

That's right, John Wheeler, founder of the Unitarian Church of Princeton, compared Bohr to Jesus. But this comment cannot have arisen from a sober judgment of Bohr's human wisdom. Keep in mind that when they were taking these walks, Wheeler was about twenty-three years old and Bohr about fifty. Much more likely, Bohr overwhelmed Wheeler with world-enhancing, category-melting ideas that the younger man had never before imagined. Ideas, of course, have physical effects. A skillful orator can whip a mob into a murderous frenzy. Or, as in Bohr's case, elevate the soul of the listener and induce an unforgettable feeling of intellectual and emotional expansion. For brainy young men, Bohr's conversation was exactly as Frisch described it: an intoxicant.

Even older colleagues, who were Bohr's equals and less likely to fall under his spell, were unanimously impressed by his personality. Schrödinger gives a very interesting character sketch of Bohr. He said that it is a rare man

who achieves such enormous internal and external success, who in his field of work is honoured internationally almost as a demigod, and at the same time remains—I will not say humble and free of presumption—but as shy and timid as a

Theology candidate. Incidentally, I do not necessarily mean this in a positive sense: it is not my ideal of a human being, [but] in comparison to the often very plentiful self-regard that one finds especially in the middling stars of our profession, this attitude is deeply sympathetic.[11]

When Einstein came to meet Bohr for the first time in 1920, the two of them became so enraptured in conversation that they missed their stop on the trolley, got back onto the trolley going the opposite direction, and then missed their stop again.[12] Writing to Bohr after he left him, Einstein said, "Seldom in my life has a person given me such pleasure by his mere presence as you have."[13] He wrote to his friend Paul Ehrenfest, "Bohr was here, and I am just as enamoured of him as you are. He is like a sensitive child and walks about this world in a kind of hypnosis."[14]

Being around Bohr meant to be blinded by the beauty of his soul. Such sensitivity and innocence, such quietness and timidity, combined with such a strong mind with unwavering physical intuitions, left a very great impression on everybody. As Pais put it, "Only those who knew him personally could experience the immense inspiration exuding from his intuitive grasp of physics and his humane personality."[15] The American physicist David Frisch called Bohr the "wisest of living men."[16] The English chemist Frederick Donnan wrote to Bohr that "we all look up to you as the profoundest thinker in science, the Heaven-sent expounder of the real meaning of these modern advances."[17] The physicist Percy Williams Bridgman wrote "that Bohr was now idolized as a scientific god through most of Europe."[18] Feynman wrote about Bohr's visit to Los Alamos, judging that "even to the big shot guys, Bohr was a great god."[19]

Those who studied and lived in Copenhagen thought of Bohr as practically a holy figure. Historians on the subject refer to Bohr's students as "disciples" or even "apostles." Rudolf Peierls said of Bohr that "although often we could not understand him, we admired him almost without reservation and loved him without limits."[20] This sentiment was echoed by Pais, who put it plainly: "I loved Bohr."[21] This sort of thing motivates the judgment that touching and genuine affection, even love, shines in the words of many of Bohr's disciples. Yet these words also reveal an unbounded, intense admiration, often bordering on worship.[22]

Bohr wielded the charisma of the religious leader, the charisma of the guru who, by virtue of his nontransferable and unmediated grasp of mantic truth, promises access to a hidden world. And so, the Copenhagen group was, in a very real sense, a personality cult, centered on the special and wise Bohr. He became a father figure to many young men at critical periods in their development. Most of those who learned from Bohr would never dare, then or later, to challenge his authority. This unwillingness to criticize the father begins in the child's dependence on him, and the family's unspoken norms of acceptable behavior. It ends in cognitive dissonance and episodes of intellectual debilitation.

> Weizsacker's reminiscences about his first encounter with Bohr constitute perhaps the best evidence for the overpowering, almost disabling impact of Bohr's authority. After the meeting with Bohr, Weizsacker asked himself: "What had Bohr meant? What must I understand to be able to tell what he meant and *why was he right?* I tortured myself on endless solitary walks." The question was not: Was Bohr right? or To what an extent was Bohr right? or On what issues was Bohr right? but, quite incredibly, What must one assume and in what way must one argue in order to render Bohr right?[23]

Einstein referred to Bohr as a prophet, and Bohr also jocularly referred to himself this way. The atmosphere, at Copenhagen, we are to understand, was one of "hero worship." For example, during the celebration of the institute's twenty-fifth anniversary, the students all stood on chairs with beer and toasted Bohr, also on a chair, and sang a hymn, changing some of the words to praise "noble Bohr" who "knows the way amidst all false tracks."[24] Those false tracks were, by definition, criticism of Bohr's "Heaven-sent" explanations. When the Copenhagen physicists put on a parody of Faust, as part of the traditional light performance at the end of an annual spring conference at the institute,

"Bohr, fittingly, was represented by God himself."[25] For Bohr's fiftieth birthday, the physicist Hendrik Casimir wrote a humorous poem about a gunfight. The great Bohr is outnumbered three to one and the poem ends:

> Bohr accepted the challenge without ever a frown;
> He drew when we drew...and shot each of us down.
> This tale has a moral but we knew it before:
> It's foolish to question the wisdom of Bohr.[26]

Chapter 34

The Compton Effect

Those foolish enough to question Bohr's wisdom were in for a nasty scrap. This is, I believe, where the true pith of Bohr's influence resides. No matter how beautiful your soul, if you are a fundamentally nice person who lets others get their way, in moments of crisis you will be overturned. I do not know of a single time when Bohr was overturned. As Vilhelm Bohr said about his grandfather, "It's not very well known, but he could be a bully."[1] Arguing against Bohr was not a simple thing to do; it took enormous energy because he subjected opponents to absolutely crushing pressure. Everyone experiencing this pressure, with the possible exception of Einstein, was emotionally marked by it.

For example, Bohr's breakthrough paper in 1913 was written on Rutherford's atom, shortly after Bohr was a postdoc working under Rutherford. Naturally, he sent the paper to Rutherford for an opinion. His advisor wanted to shorten the paper and kept trying to cut passages. Bohr refused. Rutherford insisted. Eventually, Bohr came to Manchester and argued with Rutherford for interminable hours over several long evenings. "An exhausted Rutherford finally gave in," seeing that, in fact, "it was impossible to change anything."[2] He had been swept over completely to Bohr's opinion on the matter.

This strategy of siege was Bohr's main weapon in argumentation. He ground his opponents down into dust, and then he ground the dust. He had the stamina and pain tolerance of a champion bare-knuckle boxer. Beller called this Bohr's "inexhaustible determination."[3]

Heisenberg recounts a time when Schrödinger was invited to Copenhagen and discussed physics intensely "day and night" so that "Schrödinger fell ill, perhaps as a result of his enormous effort."[4] Even then, Bohr would not leave him alone and perched on the side of Schrödinger's sickbed to continue the debate. Similarly, after discussions with Bohr, Rudolf Peierls left the institute "in a state of complete exhaustion."[5] There is a cartoon, drawn by George Gamow, depicting Bohr debating with Lev Landau. In it, "Landau was bound and gagged, with Bohr hovering over him and saying, 'May I get a word in?'"[6] Another famous episode came when Heisenberg and Bohr got into it over the uncertainty paper. Heisenberg said, "I remember that it ended with my breaking out in tears because I just couldn't stand this pressure from Bohr."[7] Most of these stories are well-known, but perhaps the best example of Bohr's grinding power is hardly ever mentioned.

Hendrik Kramers was from the Netherlands, and graduated with a physics diploma in 1916. He wanted to work with Max Born, but going to Germany was out of the question because of the war. Instead, he took a boat through the North Sea, circumventing Germany, and arrived in Copenhagen unannounced, "without introduction, help, or recommendation."[8] Bohr had been a professor for only a few months, and was only nine years older than Kramers. He took Kramers on as his first graduate student. Both Bohr and Kramers began in Copenhagen together, and Kramers wound up spending over a decade there. After obtaining his doctorate in 1919, he was Bohr's assistant for seven years.

In October 1920, Kramers married Anna Peterson, a young Danish woman whose impressive voice and personality had led an older benefactor to pay for her singing lessons in Copenhagen. She was known as Storm, a nickname she got from her running up stairs and constantly "storming around." The following story comes from her, through interviews conducted by Max Dresden with Maartien Kramers, Agnete Kramers, and Jan Kramers. Evidently it was repeated frequently in the Kramers household, "using even the identical language."[9] There are also several other pieces of hard evidence supporting it.

Sometime in 1921, Kramers came home "insanely excited." He told his wife that he had discovered a "striking, unexpected" quantum effect.

This in itself was memorable because Kramers was rarely excited, and he rarely spoke about his work to Storm except in generalities. Kramers had calculated a surprising relationship between incident and scattered light that amounted to a frequency shift.

Kramers did not have a chance to talk to Bohr about it, as he was on a hiking trip for several weeks. But this was for the best, because it gave Kramers a great deal of undisturbed time. As the wife of another physicist there mentioned, "Oskar also, always did his best work when Bohr was out of town."[10] By the time Bohr returned, Kramers had obtained definite results and was eager to publish them. However, there was one problem: if Kramers was right and the frequency shift was observed, it would indicate strongly that Einstein's light particles were actually real.

Almost everybody disagreed with the light quantum hypothesis. Remember, Planck said this would throw back the theory of light "not by decades, but by centuries."[11] Even though Robert Millikan had confirmed a prediction Einstein made on the back of the hypothesis, he did not believe the hypothesis himself, and wrote that "the physical theory upon which the equation is based is totally untenable"[12] and that Einstein's "bold, not to say reckless" hypothesis "flies in the face" of the interference of light.[13]

Bohr's resistance to the photon concept was greater than probably anyone else's. He was famously, trenchantly, permanently against it. He said, in a prize-acceptance speech in 1922, that "the hypothesis of light-quanta, which is quite irreconcilable with so-called interference phenomena, is not able to throw light on the nature of radiation."[14] Because the unexpected quantum result that Kramers obtained implied the existence of light quanta, there was reason to be apprehensive. What happened was worse than Kramers could have imagined.

When Bohr returned from his hiking trip, Kramers presented his work, and the siege began. Kramers and Bohr argued nonstop about light particles. Their arguments lasted all day and into the night. Storm called them "no holds barred."[15] They were "interminable and acrimonious."[16] We can imagine them going at it, walking down streets, drinking coffee, pointing at chalkboards, while the ghosts of Biot and Fresnel, of Brougham and Young, of Newton and Hooke, of Euclid and Ptolemy hover above them.

Given what we know of Bohr's inexhaustible determination, it ended predictably: "Kramers did not withstand Bohr's juggernaut."[17] In the end, Kramers became utterly exhausted and so sick that he checked into the hospital. Around the same time, he came over completely to Bohr's view. We don't know if the hospital stay came before, during, or after Kramers conceded. What we do know is that when he came home, he got rid of his own work on photons. Dresden says he "disposed" of his papers.[18] Think about that for a moment. There was a day when Kramers took his own papers and threw them in the trash, threw them in the ocean, set them on fire. What must he have been thinking and feeling in the midst of this self-mutilation?

The Latin *convincere* means "to convince," and also "to conquer utterly." Both senses are appropriate here. Kramers was convinced. He abandoned his own ideas and adopted Bohr's. He "became violently opposed to the photon notion, and he never let an opportunity pass by to criticize or even ridicule the concept."[19] In 1923 he wrote a popular book with Helge Holst called *The Atom and the Bohr Theory of Its Structure*, which has a handsome photo of Niels Bohr at the beginning. In it, Kramers says that if Einstein's light quantum hypothesis is admitted, "the whole wave theory becomes shrouded in darkness."[20] He mentions several reasons for this, all of which were no doubt used against him by Bohr. The main reason was the standard one: that particles cannot account for interference, something Fresnel's wave theory does with effortless beauty. Kramers's judgment was that "the theory of light quanta may thus be compared with medicine which will cause the disease to vanish but kill the patient."[21] And so, as Dresden said, "his conversion was complete."

In the exact same year that Kramers's popular book came out, the American physicist Arthur Compton published a paper called "A Quantum Theory of the Scattering of X-Rays by Light Elements." Compton discovered the same mathematical formula that Kramers had found back in 1921, and also confirmed it in the lab. These results were recognized immediately as being of the highest importance, and became widely known. The majority of physicists had been fighting against Einstein's photon for almost twenty years. Compton's work was the beginning of the end of their resistance, and Compton was highly celebrated for his accomplishment. The formula describing the scattering of light from matter is today called the Compton effect.

Kramers and everyone else was eventually forced to accept the existence of light particles. This process, for Kramers, was accompanied by a growing bitterness. In 1926, he finally left Copenhagen and became a professor in his own right at Utrecht University, a move that was likely long overdue. Free at last, Kramers "became retroactively angrier and angrier at Bohr."[22] A year after his appointment, in 1927, the Nobel prize in physics was awarded to Compton "for his discovery of the effect named after him."

One day, long after all this had happened, Kramers rediscovered some of his early notes that he had failed to destroy. He gathered them in a folder, and gave it a heartbreaking title: "The Compton Effect."

Chapter 35

Our Quantum "Culture"

The stakes of the Bohr-Einstein debate, "complementarity versus objective reality," could hardly be higher.[1] Multiple authors have called their long duel a struggle for the soul of science. Because science is one of humanity's most important institutions, we should not be surprised to find that a struggle for its soul had many aftereffects on artistic and philosophical culture. Whether Bohr's victory was a Faustian bargain or an exorcism, it has affected our civilization in unexpected ways.

Remember the measurement problem? What makes the wave function collapse? If quantum mechanics is incomplete, this is a fake problem. The obvious explanation goes like this. The particle always exists; it has a specific position and momentum at all times and thus traces out a deterministic trajectory. However, the things we measure, like position, are described by a complicated probability distribution (Schrödinger's equation plus Born's rule) for *physical* reasons currently unknown to us. Sure, we observe the electron in a single place, we observe a single energy, but this is not any more mysterious than when we roll three hundred dice and observe the bell curve of possibilities "collapse" into a single outcome. All of the mysteries surrounding quantum measurement point to some currently unknown *physical* process, which we should investigate.

If quantum mechanics is complete, however, we cannot say the obvious and natural thing. We are forced into what Einstein charitably called "egg-walking" and "unnatural theoretical interpretations"[2] because the electron *is* the wave function. That is, the wave function is synonymous with the particle, and there is "no language deeper than quantum mechanics"[3] that can explain it. Because the wave function is the ground truth of reality, we cannot appeal to hypothetical future knowledge, and now we have a serious problem.

So what causes the electron to be observed as a scintillation? What makes the wave function instantaneously alter its shape? The physicist Eugene Wigner, a close friend of von Neumann's, imagined the following thought experiment to clarify the issue.

Imagine that Alice is in the lab, studying the behavior of an electron. What the electron will do depends on the wave function, and to keep things simple let's say the wave function describes two possibilities in a superposition. You can imagine the infinite square well from Chapter 20, with the two different energies. We need both to describe the statistics, but only one of these will be observed. When Alice does her experiment, we know what will happen. Alice will observe a single energy. Let's say she observes the ground state.

However, this picture does not play well with the existence of multiple observers. Imagine we have another physicist, Bob, who is in the next room. He is considering the wave function not only of the electron, but also of Alice and her experimental apparatus. His observation involves what energy level will be written in Alice's lab journal. There are two possibilities, since there are only two energies. But the wave function that Bob is using will continue to evolve deterministically, even after Alice makes her measurement. Why? Because Bob has not yet made his measurement, and *that* is when the outcome is determined.

Thus, until Bob makes his observation, the Schrödinger equation will show a superposition of two possibilities: in one, Alice wrote in her lab book that the electron was in the ground state; in the other, Alice wrote that it was in the first excited state. But Alice has already written that the electron is in the ground state, and so Bob is predicting the wrong statistics with a theory that is supposed to be complete.

Wigner's solution to this problem was that *human consciousness* is what collapses the wave function. He concluded that "the linear time evolution of quantum states according to the Schrödinger equation cannot apply when the physical entity involved is a conscious being."[4] And so when Alice makes her observation, she has crystallized reality for Bob as well. This has been called the von Neumann–Wigner explanation. Von Neumann implied it in his famous *Foundations* book, and for a little while at least, Wigner supported it explicitly. He said: "It was not possible to formulate the laws of quantum mechanics in a fully consistent way without reference to consciousness."[5]

John Bell memorably mocked this idea when he asked: "Was the world wave function waiting to jump for thousands of millions of years until a single-celled living creature appeared? Or did it have to wait a little longer for some more highly qualified measurer—with a PhD?"[6] And it was this view that Einstein disputed with the joke of his evaporating bed that suddenly jumped back into the corner when he opened the door. These ideas also led de Broglie to say that the Copenhagen interpretation tended toward "idealism," which is a philosophy implying, among other things, that the mental is more fundamental than the physical.

Whatever its actual truth status, the idea that human consciousness creates physical reality was a remarkable gift to those who think "metaphysics" is actually a spiritual term. It originated a tsunami of books on the mind-body connection, as supposedly proven by quantum physics. Back in the 1970s and '80s, this was the main point of contact the public had with the mystery of the atom. Prominent examples include Fritjof Capra's *The Tao of Physics* and Gary Zukav's *The Dancing Wu Li Masters*, both published by physicists in the 1970s and emphasizing the connection between quantum theory and Eastern mysticism. Deepak Chopra published *Quantum Healing* in 1989, a best-seller about the mind's role in healing the body. It uses the word "quantum" incessantly but in the loosest metaphorical sense. The following sentence conveys the general flavor of the treatment: "The discovery of the quantum realm opened a way to follow the influence of the sun, moon, and sea down into ourselves."[7] In 1993 there was *Elemental Mind* by Nick Herbert, which advocated for a "quantum animism"[8] and proclaimed that "quantum mind scientists" will "incorporate quantum thinglessness" when they describe "how mind causes matter to come into being."[9] Then in 2004 there was the documentary *What the Bleep Do We Know!?* about the mind-blowing implications of atomic theory. The film deploys painfully dated computer graphics of spiritual matters, subatomic phenomena, and even a swarm of pink cells with long eyelashes crying in terror as they are decimated by nasty darts of stress. Between such scenes, a group of physicists and philosophers intone: "What makes up things are not more things... but what makes up things are ideas, concepts, information."[10]

Here is another example in this direction, which I have selected completely at random: the book *Quantum Physics and the Power of the Mind* states that our thoughts create our own reality, describing how observing turns pure potential into physical actuality in space and time. Withdrawing your attention, all becomes potential again.[11] Here, the author has applied Wigner's idea to self-help psychology, associating it with the so-called law of attraction popularized in books like *The Secret*. This "law" says that you ultimately create your life through the patterns of your thinking. Many authors have claimed this over the years, and for all I know there may be something to it; but we should be very clear that attempts to make it seem like an established fact because of quantum mechanics rely fundamentally on the idea that consciousness collapses the wave function. And that idea, to repeat, arose from trying to solve a problem generated by quantum completeness.

Nonscientists like to appropriate scientific thought. This has always happened and can only be mitigated by public science education. Physicists, understandably, get touchy about this kind of thing. Bell said it seems to him "irresponsible to suggest that technical features of contemporary theory were anticipated by the saints of ancient religions... by introspection."[12] But the point is, physicists made the process so easy it was hardly appropriation. The professionals said, in all seriousness, that observations create reality.

The von Neumann Wigner interpretation is mostly used today to amaze laypeople, or as a skyhook for spiritualism. Not many serious physicists or philosophers currently support this view, but it is not dead either. One recent example is *Quantum Enigma*, a book by two physicists published in 2006 by Oxford University Press. It tries to show that quantum physics and consciousness are inextricable and has sentences like "Quantum theory... encounters the essence of our humanity, our consciousness"; "quantum mechanics can make Nature appear almost mystical"; and "observation *creates* the reality observed."[13] The book concludes with some grand speculations:

Quantum theory tells us that physics' encounter with consciousness, *demonstrated* for the small, applies to everything.... Copernicus dethroned humanity from the cosmic center. Does quantum theory suggest that, in some mysterious sense, we *are* the cosmic center?[14]

Rick and Morty is a hit TV show about the smartest man in the universe and his sidekick grandson, decidedly less intelligent. Among other things, Rick Sanchez invented portal tech, which allows anyone to hop from one universe to another. The next universe might be identical to this one, only everybody pronounces the word "parmesan" as *par-mee-zee-in.* (Morty: "Gross. I hate it.") It might be a universe where Earth is just a fetid sea sprinkled with butt islands. It might be the blender dimension, a cosmos of tightly packed spinning blades. It might even be a world where pizza is alive and hats wear people.

The grand superstructure of these universes is called the multiverse, and it represents the infinite union of all possibilities. The idea has become popular in fiction precisely for its liberating effect. Philosophers sometimes say that anything conceivable is possible. In the multiverse, anything possible is actual. Therefore, we have a setup where, according to what Doctor Stephen Strange calls "the grand calculus of the multiverse," if you can imagine it, it's real. Or, in the words of Evelyn Wang, the heroine who experiences many universes in the 2022 film *Everything Everywhere All at Once*:

> I can think of whatever nonsense I want, and somewhere out there, it exists. It's real.

Marvel Studios is just one storytelling team that has gone all in on the multiverse. It's a very convenient underpinning for fiction. As Deadpool quips: "The multiverse fills all plot holes." The writers can kill off a popular character for real, and then base an entirely new multi-season TV series on the same character. It needs no explanation beyond a nod to the fact that we are in a different universe than the one where our favorite character died.

But what about the dilution you would expect by imagining that there are an infinite number of such characters? Why should we make the crucial psychological investment in the protagonist? The answer is to use that old trick: you make the characters special. In *Rick and Morty* our protagonist is Rick C-137, and we care about him because he is the Rickest Rick. Another example comes from *Doctor Strange in the Multiverse of Madness*, where we get the following exchange:

RACHEL MCADAMS (aka Christine Palmer):
> You sound a lot like my Stephen right now. He had to be the one holding the knife and then that knife killed a trillion people.

XOCHITL GOMEZ (aka America Chavez, because she can create portals between universes that look like the stars on a certain spangled banner):
> This Stephen is different. He is. It doesn't matter about all the other Stephens. You're not like them.

BENEDICT CUMBERBATCH (aka Stephen Strange):
> Smart kid.

And then they stop talking, thank goodness, and start running from some booming noises.

The ubiquity of the multiverse has not left everyone happy. In 2016, Sam Kriss published an essay in *The Atlantic* called "The Multiverse Idea Is Rotting Culture." Kriss mentioned, among other things, the tens of thousands of people who believe in the Mandela effect, which is the idea that if a group of people are absolutely sure that something happened when it really didn't (like Mandela dying in jail in the 1980s), then perhaps they are all travelers from a different universe that varies from this one in the crucial way. Kriss describes this as "a new name for the phenomenon previously known as 'being wrong about things.'"[15]

What does all this have to do with quantum physics? Well, the idea
of a multiverse arises directly from the doctoral dissertation of one
Hugh Everett III, who in 1956 was at Princeton working with John
Wheeler. Bohr and his coterie did not like Everett's ideas. After a trip
to Copenhagen, one of Bohr's most loyal followers (Léon Rosenfeld)
said that Everett was "undescribably stupid and could not understand
the first thing in quantum mechanics."[16]

The feeling was mutual. Everett called the Copenhagen interpretation "a philosophic monstrosity with a 'reality' concept for the macroscopic world and denial of the same for the microcosm."[17] Everett
considered Bohr's ideas "hopelessly incomplete," but for the opposite
reason that Einstein said the same thing. For Everett, Bohr relied too
much on classical physics. He was correct that quantum mechanics was
complete, but had not really digested what that meant. Everett wanted
to treat quantum theory "in its own right as a fundamental theory
without any dependence on classical physics, and to derive classical
physics from it."[18]

The Wigner's friend paradox should rightly be called Everett's friend,
because he thought it up five years before Wigner. In the first pages
of his dissertation, he took the two postulates of quantum mechanics
(time evolution by Schrödinger's equation, and observation given by
wave function collapse), and combined them with multiple observers in
precisely the way Wigner did to obtain his paradox. But Everett did
not invoke consciousness to solve the problem; his idea was to explore
what happens to quantum mechanics when the collapse postulate is
simply thrown away. In that case, wave functions just evolve all the
time—they never undergo mysterious collapse. This does completely
solve the paradox of Wigner's friend, because we no longer have a magical moment when reality crystallizes out of possibility. The cost of this
solution is the multiverse.

Because the wave function of Alice never collapses, both energy states
she can observe go on existing in the wave function. Because the wave
function is "universally valid" and, according to Everett, "all of physics" can be derived from it, it must be the case that *both* energy states
are real. Why, then, does Alice only see one of them? Because Alice
herself has split into two possibilities. There is an entire universe where
Alice sees the ground state, and another universe, identical to the first

in every respect, except Alice sees the first excited state. When Bob sees Alice has written one or the other state in her lab notebook, that is just because he has tagged along with her into the relevant universe.

This idea is surprisingly popular, and growing more influential every day. It is also very polarizing. The multiverse has been called "gob smacking,"[19] "startling,"[20] "alluring and thought-provoking,"[21] "the ultimate catastrophe,"[22] "the courageous formulation,"[23] and "dangerous to science and society."[24] Reactions are bimodal, clustering around one of two options: either "I hate the multiverse"[25] or "The only astonishing thing is that it's still controversial."[26] Very few people seem to be lukewarm. The closest I have seen is the opinion of one of my best friends: he thinks that the many-worlds interpretation is probably right, because it is so disappointing. (A British take if ever there was one.)

As it turns out, the Everettian multiverse is intimately linked to the idea of quantum completeness. Early in his thesis, Everett listed five different possible solutions to the Wigner's friend paradox. The last two are the most important. Writing "state function" for what we have called the wave function, Everett says:

> **Alternative 4**: To abandon the position that the state function is a complete description of a system.... It is assumed that the correct complete description, which would presumably involve further (hidden) parameters beyond the state function alone, would lead to a deterministic theory, from which the probabilistic aspects arise as a result of our ignorance of these extra parameters in the same manner as in classical statistical mechanics.

> **Alternative 5:** To assume the universal validity of the quantum description, by the complete abandonment of [wave function collapse].

As Everett makes clear, "we shall concern ourselves only with the development of Alternative 5."[27] He then immediately proclaims the completeness of quantum mechanics, saying "the universal validity of the state function description is asserted" and that "all of physics is presumed to follow from this function alone."[28]

Rarely is an interpretation as clear as this on the issue of completeness. After explicitly considering that quantum mechanics might be incomplete in exactly the way that Einstein had been arguing for decades, Everett immediately and explicitly rejects this possibility in favor of the exact opposite. The many-worlds interpretation stands and falls with quantum completeness. The link between the interpretation and completeness was direct and explicit at the very beginning, and it has remained so. For example, Sean Carroll, a physicist and philosopher at Johns Hopkins and a prominent booster of the many-worlds interpretation, wrote that "the entirety of the Everett formulation is simply the insistence that there is nothing else [than the wave function and Schrödinger's equation],"[29] and also that "I will pound the table and insist that the wave function is the sum total of reality."[30]

In 2022, I had the opportunity to visit Carroll in his office and discuss this strong claim. He gave me water in an Einstein mug, and we sat at a table beneath a dozen empty champagne bottles, each one representing a PhD that he had helped mint. To my relief, he did not pound the table. In fact, he showed all the signs of a nuanced epistemology. The way he justified quantum completeness to me was like this:

> Neither Everett nor I nor anybody is saying, "We have confidence that it's the correct theory of everything ultimately." The claim is that thinking the wave function is everything fits the data. That's it. Working as a physicist trying to understand the world, if we have a theory that fits all the data and we don't understand that theory very well—which I would say is the case for Everettian quantum mechanics—the way to bet as a researcher is to try to elucidate that theory, not to change that theory absent any empirical hints about what directions to change it in.[31]

This would not have worked for the heat equation, and we do have empirical hints in the form of the droplets, but never mind. The *Stanford Encyclopedia of Philosophy* makes the exact same point, saying: "It can be argued that the burden of an experimental proof lies with the opponents of the Many Worlds Interpretation, because it is they who claim that there is a new physics beyond the well-tested

Schrödinger equation," and that analysis has shown that "we have no such evidence."[32]

Let's imagine for a moment that this is right. There are two possibilities. Either the universe splits upon measurement, or it is always splitting all the time. Both are discussed. The first possibility does not solve the measurement problem—it just replaces it with a "splitting problem." Why does observation split the universe? The second possibility does solve the measurement problem, but it is also the most metaphysically extravagant proposal in human history.

To see just how extravagant, take a look at yourself. The energy in your cells comes from organelles called mitochondria. They do it using something called the electron transport chain, a variety of protein structures that pass electrons from one site to another. According to the second reading of Everett, whenever an electron is passed, there splits away a universe in which it was not passed. The number of mitochondria in each cell vary from zero for mature red blood cells, to hundreds or even thousands for liver and muscle cells. Your body has about thirty trillion cells, and so it has maybe three hundred trillion mitochondria. There are four different complexes that transport electrons, made of various parts. Complex I, for example, has about fourteen subunits. (Structure determination of these complexes is an active area of research.) There are a huge number of these complexes per mitochondrion, potentially thousands. But let's just say one hundred. Let's also say that each has ten subunits that hold only one particle that can move at any one time. The last question is, how many electrons pass through a specific subunit per second? Let's be extremely conservative and say only one.

Guess what? In the time it took you to read that paragraph, you generated at least nine quintillion—9,000,000,000,000,000,000—copies of this universe, just from your natural metabolism! What are you, some kind of god?

Don't laugh, it's not allowed. This is our quantum culture. One in which "it can be argued" that anyone who disagrees with this must bear the burden of disproving it.

Chapter 36

The Shadow Physics of Our Time

In 2018, a staff member at the Audubon Zoo burned a pastry in a toaster, and some light smoke got into the reptile enclosures. The fire alarm did not go off, and the smoke was not visible. But several Australian sleepy lizards (*Tiliqua rugosa*) reacted very strongly. This led to a paper called "A Wake-Up Call for Sleepy Lizards" that described "increased activity, including swift pacing, repeated escape behaviors, and rapid olfactory-related tongue-flicking" in the lizards that were exposed to the smoke. The lizards not exposed "did not exhibit any such changes in activity or behaviors."[1]

Of course, we know what these sleepy lizards were thinking: *Where there's smoke, there's fire.* This proverb, taken literally, just means that effects have causes.

In philosophy, this idea is called the *principle of sufficient reason*. It says that there is a reason that suffices to explain all things that happen. As such, every fact has an explanation. No matter what you observed in the world, there is a cause for it. The discussion of this principle is as old as philosophy itself, and notable names in the tradition include Parmenides, Aquinas, Spinoza, Leibniz, Hume, and Kant.

The *assumption* of this principle is much, much older than philosophy, even than human beings. The sleepy lizards just demonstrated this. They were born in the zoo and had never encountered a bush fire in their lives. And yet, their first exposure to smoke alarmed them greatly. In other words, the principle of sufficient reason is an evolutionary adaptation that is hardwired into animals because the world is rational.

What Einstein called "the Heisenberg-Bohr tranquilizing philosophy— or religion" attacks the principle of sufficient reason. Remember, quantum events seem to "literally have no cause,"[2] and consequently "quantum mechanics defeats causality"[3] and there is no deeper causal theory explaining where the probability distributions come from. The universe is fundamentally random, and "the laws of physics can be expressed only in probability form."[4] This is, of course, just another way of formulating quantum completeness. But it is often useful to formulate the same problem in multiple ways, so here is a way that I think philosophers would appreciate: *If quantum physics is complete, then the principle of sufficient reason is false.*

The moment you deny that explanations exist, it's the beginning of the end. Because the core principles of thinking and science are actually negotiable now, pretty much anything goes. After all, if the king himself, the principle of sufficient reason, went to the guillotine, the reign of terror has begun and nobody is safe. If I had to pick one dramatic expression of the situation, naturally it would be that scene from *Ghostbusters* when the heroes warn New York City's mayor of what is about to happen:

> We are headed for a disaster of biblical proportions. Real wrath of God–type stuff. Rivers and seas boiling! Forty years of darkness! Earthquakes! Volcanoes! The dead rising from the grave! Human sacrifice! Dogs and cats, living together! Mass hysteria![5]

There has been a growing sense among some physicists that the discipline has taken a wrong turn somewhere. Many highly respected professionals have expressed discontent. For example, Carver Mead wrote in 2000, "It is my firm belief that the last seven decades of the twentieth century will be characterized in history as the dark ages of theoretical physics."[6] His critique of the modern physics culture does not pull its punches:

It's absolutely astounding to my mind. They've built their own coffin and they've built their own trench to put it in, and they don't know it. The way they are going at physics makes the young people, or anybody who had any notion of doing anything real, leave theory. They go over to applied physics or electrical engineering. The people who stay are the math types, and so it reinforces.[7]

This opinion is actually commonplace. Books by physicists and science writers with titles like *The Trouble with Physics*, *Not Even Wrong*, *Farewell to Reality*, *The End of Science*, *Rational Mysticism*, *The Copenhagen Conspiracy*, and *Lost in Math* have been coming out at a steady pace.

Most of these books deal with recent research programs like string theory, and try to diagnose why theoretical physics has completely stalled and made no progress in more than forty years. To give a flavor of the treatment, Jim Baggott calls the multiverse, along with other extravagant theories, "fairy-tale physics."[8] Sabine Hossenfelder roasts the "mindless production of mathematical fiction," a state she calls "sick science."[9] Lee Smolin mocks the names of supposed particles that have never been observed (prefixed by an *s* or suffixed by *ino*): "Sooner or later, tangled in the web of new snames and naminos, you begin to feel like Sbozo the clown. Or Bozo the clownino. Or swhatever."[10] John Horgan predicted that the disputes of physicists "will become more and more like those of that bastion of literary criticism, the Modern Language Association."[11] And he called modern physics "little more than science fiction in mathematical form."[12] David Ferry accuses Bohr of founding a religion, and judges that "the Copenhagen-Göttingen positivist approach has devolved to an empty shell."[13] Peter Woit offers a sober diagnosis: "The lack of useful experimental input has made the traditional organizational system of particle theory seriously dysfunctional."[14]

The books invariably talk about quantum theory at some point because string theory is a variation of the standard model, which is a type of quantum field theory, which is of course based on quantum theory. Evidently, the problems that string theories are trying to solve pale almost into insignificance compared with one of the most fundamental problems inherent in contemporary physical theory—the quantum measurement problem.[15]

If this is right, then it's unsurprising that all is not well at the heights of modern mathematical physics. In the restaurant at the top floor of the skyscraper, we are told that there are rats. Well, nobody who has seen the rats coming through the cracks in the foundation has time to worry about such refined problems. Edwin Jaynes was someone who spent considerable time in the basement. He studied with Oppenheimer and Wigner at Princeton, became a professor at Stanford, and ultimately was the Wayman Crow Distinguished Professor of Physics at Washington University in St. Louis. In his magisterial *Probability Theory*, Jaynes diagnosed the situation as follows:

> In current quantum theory, probabilities express our own ignorance due to our failure to search for the real causes of physical phenomena; and, worse, our failure even to think seriously about the problem. This ignorance may be unavoidable in practice, but in our present state of knowledge we do not know whether it is unavoidable in principle; the "central dogma" simply asserts this, and draws the conclusion that belief in causes, and searching for them, is philosophically naïve. If everybody accepted this and abided by it, no further advances in understanding of physical law would ever be made.... But it seems to us that this attitude places a premium on stupidity; to lack the ingenuity to think of a rational physical explanation is to support the supernatural view.[16]

Jaynes accuses physicists of *failure to even think seriously about the problem of causes*. This condemnation is sadly accurate. Think of complementarity, the multiverse, consciousness collapse, or pretty much any interpretation you like. How well does it explain, actually explain, the particle-wave duality? Or quantization? Interference patterns? Single-particle diffraction? Or literally any other quantum mystery you can think of?

The authors of most quantum interpretations don't even *try* to explain the very things we want to know. There might be some small issues on the fringes that they can comment on, but actually explaining where the physical phenomena come from? Why they are like this rather than that? They wouldn't have the foggiest notion of where to begin. Nobody expects this of them and they do not demand it of themselves because explanation at this level is *assumed* to be impossible. After all, causes do not exist, right?

Clearly, any interpretation that denies the principle of sufficient reason when it comes to individual atomic processes will never be able to offer any explanations that rise to the standards of the 1800s, such as the Maxwell-Boltzmann analysis of the heat equation. These also happen to be the standards of nearly all modern science. The papers on droplets are expected to *explain* things, because the droplets are causal. Real physical analysis rationalizes precisely how particle-wave duality emerges in the droplet system, why droplets undergo single-particle diffraction through slits, how quantization arises in orbiting droplets and rotating baths, and why certain lattice configurations are stable.

Because it is supposed to be complete, however, quantum mechanics is exempt from the rigorous demands of causality. This is why the philosopher Imre Lakatos said, "Bohr and his associates introduced a new and unprecedented lowering of critical standards for scientific theories. This led to the defeat of reason within modern physics."[17] For the exact same reason Karl Popper said the exact same thing: "The general antirationalist atmosphere which has become a major menace of our time... has led to a most serious deterioration of the standards of scientific discussion."[18]

If you defeat the principle of sufficient reason, you defeat reason. Without causes, explanation becomes "explanation," and reality slides away. Then what exists depends on how you look, and there is no hard fact of the matter. This made Bertrand Russell extremely depressed. Toward the end of his life, he resided in Telegraph House, built on the site of a West Sussex semaphore station. There was a study in the tower, with four windows in the cardinal directions. It was midnight and Russell sat at his desk. Fog pressed against the windows. He was thinking about how quantum idealism had become to him a "haunting nightmare, increasingly invading my imagination." Overcome by gloom and darkness, he wrote a short meditation that he titled "Modern Physics."

> Alone in my tower at midnight, I remember the woods and downs, the sea and sky, that daylight showed. Now, as I look through each of the four windows, north, south, east and west, I see only myself dimly reflected, or shadowed in monstrous opacity upon the fog....It seemed that what we had thought of as laws of nature were only linguistic conventions, and that physics was not really concerned with an external world. The revolutions of nebulae, the birth and death of stars, are no more than convenient fictions in the trivial work of linking together my own sensations....No dungeon was ever constructed so dark and narrow as that in which the shadow physics of our time imprisons us, for every prisoner has believed that outside his walls a free world existed; but now the prison has become the whole universe....Why live in such a world? Why even die?[19]

Today we live and die in a scientific climate where something is seriously wrong. Being faced with the starry-eyed speculation that we are the cosmic center because our minds collapse the wave function, or that quintillions of universes popped out of you just now because there is no collapse, anyone with their head screwed on would reply, No, this is absolutely absurd; one of your assumptions somewhere along the line was monkey bananas. But if there are no causes in microphysics, this response is not based on anything and can be no more than an expression of personal disgust. We are thus reduced to toothless exclamations based on taste. Things like Katie Mack's "I hate the multiverse. I know it's a cliché to hate the multiverse, but I hate it."[20] And Steven Weinberg's "I wish I could prove that this continuous splitting of histories is impossible. I can't. But I find it repulsive. Obviously the people who originated it don't find it repulsive. But that's what it is."[21] In this environment, foundational science splinters into battling schools, or cults if you like, and "they all think each other's theory is repulsive."[22]

Just as a refresher, here are some of the actual solutions that have been tried: Perhaps time does not flow in one direction? Perhaps there are endless copies of our universe spawning even now? Perhaps human consciousness creates matter? Perhaps matter is pure potentiality that comes into being upon measurement? Perhaps the universe has more than one past? Perhaps objects are not real, but their relationships are? Perhaps when an atom does something, two different human minds are generated? And so on.

Notice how different these interpretations are from each other ontologically. Each one proposes a different cosmos. To claim that, say, the moon is not there when nobody looks at it, when actually there are endless copies of the moon in endless universes, is to get the nature of things *fundamentally* wrong. This is the sort of error that theologians and philosophers made in the Renaissance. It is like believing that comets are "the thick smoke of human sins," heralding the end-times and bringing about treason and inflation, when, in fact, comets are piles of rock and ice.

This reveals a major difference between the quantum debate and the historical debates we have explored so far. The scientists of those days, working on the problems they inherited, did not mistake the fundamental nature of their subject. Descartes said the comet is a dead star and the tail is a diffraction phenomenon, Newton that it is a planetoid with an atmosphere, and Galileo that it is an optical effect akin to the northern lights, caused by gases rising out of the shadow cone of the Earth. These are all wrong, but they are more right than a wrong quantum interpretation because they are correct about the type of world comets reside in and what an explanation for them should look like.

When Newton and Hooke argued over the theory of light, when Lavoisier and Rumford proposed different theories of heat, when Mach and Boltzmann disputed about atoms, everyone operated with the same bedrock assumptions. They agreed on what kind of world they lived in and what plausible causes could explain the facts. Before quantum physics, science believed in a world that was mysterious, surprising, and devilishly complex, but nevertheless *comprehensible*, because explanations existed for every fact.

If the principle of sufficient reason is false and there are no actual explanations for the facts of atomic behavior, then on what grounds is any quantum interpretation going to be refuted? None can be refuted because there are no causes to adjudicate the matter. This is why the quantum debate seems so interminable and confusing and pointless. Nobody can get a firm hold to pin down an interpretation and disprove it. As Mermin quipped: "New interpretations appear every year. None ever disappear."[23] They never disappear because the only way of killing one off is to find a mathematical contradiction within its implications. But mostly these interpretations are just bloodless ghosts hovering over the Schrödinger equation, nonmathematical stories, so of course no errors can be found.

In formal logic when you try to deduce something from a falsehood, you can get anything at all to be true. This has the coolest possible name: the principle of explosion. Assuming a falsehood to be true means that you will never be able to discover an internal logical contradiction in the subsequent chain of reasoning. You allowed a false proposition through the door, and like some kind of demon worm it gave you infinite wiggle room. In particular, you will no longer be able to make a *reductio ad absurdum* argument, which introduces assumptions and then derives a contradiction to disprove one of the assumptions.

Something similar to logical explosion seems to have happened in quantum foundations. From the identical starting point of quantum completeness, the various interpretations wind up in different places so outlandish that they would seem to be self-refuting. But we cannot actually refute them because no contradiction can be obtained.

Reductio ad absurdum arguments only work if two people share a sacred truth. Anything that contradicts that truth is then "absurd." Such arguments are used all the time in mathematics. If a chain of pristine reasoning ends with the statement that $2 + 2 = 5$, then at least one of the assumptions that started the chain is wrong. Mathematicians take this as definitive, but this is a *cultural* norm as much as a logical one. We could easily imagine a society so intellectually decadent, so married to a false assumption, that a proof ending in $2 + 2 = 5$ was not a *reductio* for them, but instead a profound realization about how mathematics is much richer than previously thought. How mind-blowing! Scholars could spend their careers spelling out the nature of this new mathematics, in which all numbers are actually equivalent in some deep sense, and what was previously considered impossible is actually true. What other things follow, from $2 + 2 = 5$? Do we have the courage to accept them?

This is roughly the stage we are at with quantum foundations. The moment physics denied the principle of sufficient reason it let falsehood through the door. Physicists stopped thinking seriously about the actual problem. Now they produce mindless "mathematical fiction." There are no *reductio ad absurdum* arguments anymore, because absurdity is a matter of taste, and there is no arguing about taste. When trying to understand Schrödinger's equation these days, the only constraint is mathematical consistency. The interpretation just needs to be logically possible.

Unfortunately, when we take every logical possibility seriously we debase our physical intuition. It becomes increasingly difficult to talk to each other as our pictures of the cosmos diverge. I submit the following diagnosis. The root of the illness, the philosophical thorn in the lion's paw, is the denial of the principle of sufficient reason. You simply cannot do science without assuming there are causal explanations. Such a faith is nonnegotiable. As Jaynes put it:

> There is clearly a major, fundamentally important mystery still to be cleared up here; but unless you maintain your faith that there is a rational explanation, you will never find that explanation.[24]

And in a letter to his friend Maurice Solovine, Einstein expressed his faith in the principle this way:

> I have no better expression than "religious" for confidence in the rational nature of reality insofar as it is accessible to human reason. Wherever this feeling is absent, science degenerates into uninspired empiricism.[25]

What if this feeling is not only absent, but actively despised? The answer is evident because we are living through such a period now.

Part X

Interlude

The work of Yves Couder and the related work of John Bush provides the possibility of understanding previously incomprehensible quantum phenomena, involving "wave-particle duality," in purely classical terms. I think the work is brilliant, one of the most exciting developments in fluid mechanics of the current century.

—H. K. Moffatt[1]

Chapter 37

Parable of d'Alembert's Paradox

Lest the longevity of the quantum paradoxes be mistaken for their insurmountability, fluid mechanics has a cautionary tale to tell.

—John Bush[1]

In the winter of 1717, a baby boy was left on the steps of a small church in Paris. The church was called Saint-Jean-le-Rond, and it was a tiny baptistry attached just to the left side of the Notre Dame cathedral. As was customary, the baby was named after the church's patron saint, and so began the life of Jean Le Rond d'Alembert. He was one of France's great natural philosophers, a correspondent of Euler's, coeditor of the famous *Encyclopédie* with Diderot, fellow of the Royal Society, who solved the wave equation in one dimension and, in France at least, gives the fundamental theorem of algebra its name.

In 1749, when he was thirty-two, d'Alembert discovered "a singular paradox which I leave to future mathematicians to elucidate." He was working with the equations of an ideal fluid with no viscosity. In some regimes, these equations describe water surprisingly well. D'Alembert's singular paradox was this: evidently, if a body is moving in an ideal fluid it will experience no drag.

In other words, fish should only need to flip their fins once, and they can glide forever. A bullet shot underwater should be deadly at one hundred, even one thousand miles. These predictions of the equation are bad enough, but it got worse. The math of realistic, viscous fluids, called the Navier-Stokes equation, in certain cases approaches the ideal fluid equations. So d'Alembert's paradox also predicts that a cannonball should follow a perfectly parabolic path, and that the harder the wind blows, the safer your house is.

This problem was well-known throughout the entire 1800s. Nobody had any answer for it. All that was known for sure was that the mathematical theory, which seemed to be based on perfectly sound physical assumptions, could not be trusted. Luminaries like Kirchhoff and Rayleigh tried to solve the problem, but their ideas did not work. Sir Cyril Hinshelwood said of d'Alembert's paradox that it caused a schism in fluid mechanics, between "hydraulic engineers who observed things that could not be explained and mathematicians who explained things that could not be observed."[2]

Fluid mechanics itself suffered because of this. What to do? The equations were wonderful for so many things; indeed they are still taught unchanged to this day, and yet they generate a "patently absurd conclusion."[3]

The future mathematician that d'Alembert imagined was born in 1875, in Germany. Ludwig Prandtl became a professor of fluid mechanics at Hannover, and, in 1904, just shy of his thirtieth birthday, he resolved the problem with a combination of mathematics and experiments. His idea was to look closer at what was happening in the thin layer between the fluid and the body itself. This is called the boundary layer, studied extensively by engineers because it is so important to the dynamics of traveling through fluids.

Prandtl took the idea of a no-slip condition, and derived something surprising. No-slip means that the fluid molecules that are actually touching the body are moving *with* the body. They stick to it. But if this was happening, Prandtl reasoned, then you could expect velocity of the fluid molecules to change rapidly as you moved up through the boundary layer. This rapid velocity change means the fluid molecules in the boundary layer are flowing in unstable sheets that soon break down into countless little whirlpools. This, in turn, causes the boundary layer to separate away from the body downstream, creating a low pressure region in the wake. And this creates a bunch of drag.

Streamlining is so important because without it, the boundary layer will separate away from the body and create a huge wake. A streamlined object the area of a backpack has the same drag as a pencil when traveling through air.

The ideal fluid equations are obtained by throwing away a term of the full fluid equations that only becomes important at very small scales. Reasoning about cannonballs, boats, and other things d'Alembert was interested in led him to believe that he did not need these scales. The mistake he made was in thinking the proper scale was the cannonball, when it was actually the paper-thin boundary layer between the cannonball and the rest of the air.

Prandtl showed how the flows produced in this minuscule layer on the body's surface could generate substantial drag out of seemingly nothing. The solution involved unanticipated dynamics on an unimagined scale. D'Alembert's paradox bedeviled scientists for 150 years, until they looked closer and saw the new variables, hiding within the viscous boundary layer.

Chapter 38

Droplet Statistics

If you take a felt ball and put it in a bowl, it will stay there indefinitely. Only a ballistic child, a strong gust of wind, an earthquake, or some other forceful perturbation would knock the ball out. To escape, the ball has to go over one of the edges of the bowl.

If the ball has some natural motion of its own, it *might* escape. The edges act like a barrier; the taller they are, the more speed the ball needs to escape. If the ball isn't going fast enough, then it has less energy than it needs, and can't ever get out of the bowl. It will just roll back and forth. Eventually, friction will do its work and the ball will become still.

The quantum version of this situation is very different. If a particle is trapped in the microscopic version of a bowl, the Schrödinger-Born description shows that the bowl will act like an insulator. It will hold the wave function inside, thus making the probability function highly peaked inside the bowl. But the wave function extends through the walls and out into the space beyond. That means that, even if the particle does not have enough energy to escape, there is a small chance of measuring the particle outside the bowl.

This prediction arises from a straightforward application of Schrödinger's equation, and has been confirmed in countless ways. When a particle passes through a barrier that, according to classical theory, *should* contain it, we say the particle has tunneled. The phenomenon of tunneling is how we currently explain radioactive decay and nuclear fusion. (The second process powers the sun and thus all of life.) Tunneling underlies photosynthesis, and the construction of biomolecules in all cells. An understanding of tunneling also allows us to construct extremely fine electronic devices. For example, it is used to program the floating gates in your USB flash drive. In fact, tunneling limits how small electronics can get, because if two transistors are too close, they are no longer isolated because electrons can simply tunnel between them. In short, when people talk about quantum mechanics providing new technologies, tunneling is the kind of thing they mean.

Indeed, this is a prime example of quantum phenomena, something where the mystery cannot be "explained." There are no classical explanations of tunneling. Tunneling has been described as "teleportation" or "like a car passing through a wall without damaging itself or the wall." It is frequently claimed that it has "no counterpart in classical physics."[1] The droplets have changed that.

In a paper published in 2009 called "Unpredictable Tunneling of a Classical Wave-Particle Association," Antonin Eddi and coworkers in Couder's group studied the ability of the walking droplets to perform an analogy of quantum tunneling. They used variable bath depth, as we described in the section on droplet diffraction, to confine the droplet in a square geometry. Essentially trapped in a box, the droplet walked inside, bouncing off the walls. Every once in a while, however, the droplet would actually pass over the barrier, and escape.

It turns out that the probability for the droplet to escape the box depends on the same variables in quantum mechanics, and in the hydrodynamic analogy. These variables also come together in the same way, meaning that the probability of tunneling, in both instances, decreases exponentially with the width of the barrier.

This is cool, and I had to mention it, but I did not bring up tunneling to talk about this analogy. Let's just note it and move on to an even more exciting droplet analogy.

One of the things that an understanding of tunneling allows is called a *scanning tunneling microscope*. This is the most powerful microscope in existence today, capable of resolving structure at the atomic level. It does not use lenses, but rather a tip that is atomically sharp. This tip is placed so close to a surface that electrons begin tunneling, and the microscope measures the tunneling voltage. Moreover, through inconceivably fine movements, these microscopes can actually become tools, manipulating the structure of atoms, dragging them across a substrate and placing them in a given geometry one by one.

There is also a specific type of experiment that has been done for decades using this technique. Approximately thirty to eighty atoms are dragged into a circle, called a *corral*. The microscope can then be used to explore the voltage within the corral, generating breathtaking pictures of standing wave patterns. For example:

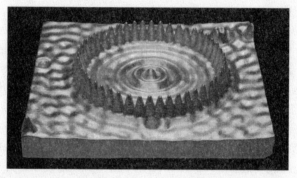

The Well (Quantum Corral) (2009) by Julian Voss-Andreae. Created using the 1993 experimental data from Lutz et al., the gilded sculpture was pictured in a 2009 review of the art exhibition "Quantum Objects" in the journal *Nature*.

What is that wavy stuff in the middle of the quantum corral? This is actually the electric charge at different points in space, and so it stands for the location of electrons. It is a wave, and this is also the exact form of the Schrödinger equation when you confine an electron to a two-dimensional circle.

The idea for doing an analogy of this situation came from John Bush. Dan Harris, now professor of engineering at Brown University, did the experiments as part of his PhD at MIT. They published the full paper in 2013, along with coauthors Moukhtar, Fort, and Couder.[2] The analogy is simple, beautiful, and extremely suggestive.

Recall how the walking state for droplets is stable: small changes, caused by air currents perhaps, tend to be erased and the droplet goes back to its previous state. The main characteristic of the walking state is that the droplet goes in a straight line at a characteristic "walking speed." This speed tends to increase as the memory of the system increases.

As we saw, putting the droplet in a bath with boundaries completely changes the dynamics. A double-slit geometry will result in single-particle diffraction, a square with narrow walls will result in frequent recoil and intermittent, unpredictable "tunneling," and so on. This next analogy creates a circular geometry for the droplet to explore, with barriers that are so wide that tunneling is impossible.

The diameter of the circle is quite small, between 2 and 3 centimeters. Because the droplet radius itself is about 1 millimeter, it is exploring quite a small space. The Faraday wavelength of the droplet's standing wave is about 4.75 millimeters, so a little more than 4 full wavelengths can fit into the whole circle.

For a low level of memory, placing a droplet in this small circular bath results in natural circular orbits. The droplet, affected by its own wave field reflecting from the barriers, begins orbiting the center of the bath. But as we know, things get more interesting as we increase the memory. As the droplet experiences more and more of its own past, it destabilizes from the circular orbit and begins tracing out trajectories that are called "epicycloidal," a fancy term for the kind of patterns you can draw with a Spirograph.

Now, when we increase the memory even more, the trajectory of the droplet becomes completely chaotic. The wave field is deeply complex and continuously evolving. Its form changes continuously in time, and it tends to have a peak amplitude near the point of the droplet's last impact. Any prediction of where the droplet will go from one instant to the next would require a full modeling of the wave field, which in turn requires knowledge going far back into the droplet's past. Below is a snapshot of the droplet and its wave field at a single point of time, along with the path the droplet traced in the past.

The droplet does not always move at the same speed. It encounters its wave field as it moves forward, and sometimes this field is sloped behind it, so the droplet speeds up, and sometimes it is sloped ahead, so the droplet collides with it and slows down. The *average* speed of the droplet is about 8.6 millimeters per second, but the speed at any given moment can vary wildly from this value. The droplet can be almost at a standstill, moving only at 2 or 3 millimeters per second, or it could be going as fast as 30 millimeters per second.

Now, here is the showstopper. The speed of the droplet is not random, but it will depend on where the droplet is in the corral. There is a

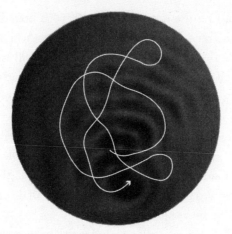

The chaotic trajectory of the walker confined to a circular corral in high memory. Reprinted with permission from Harris et al. 2013. "Wavelike Statistics from
Pilot-Wave Dynamics in a Circular Corral." *Phys Rev E* 88:1.

kind of speed dependence that arises from the system geometry itself.
If the droplet is at the center of the corral, it tends to be going slow.
But at a small radius from the center, it tends to be going fast. A little
farther still and it goes slow again, and this pattern repeats.

Imagine that we divided the corral up into regions, and stared at
the droplet and blinked really fast, and then counted how many times
we found the droplet in a given region. If the droplet was in one region
a lot, that region would have a large number associated with it. If we
hardly ever saw the droplet in a different region, it would have a low
number. We can make a histogram out of these separate regions, which
is essentially a discrete probability distribution of the droplet location.
We have a higher-probability chance of seeing the droplet in those regions with a high number than elsewhere.

The fact that the speed variations of the droplet are associated with
geometry means that the histogram we make will also depend on the
geometry. If the droplet is going slow in a certain region, we are more
likely to see it there, so the probability distribution is more peaked in
the slow regions. Conversely, if the droplet is moving fast in a particular
region, we will see it there less because it will spend less time there.
With these facts in mind, let us examine droplet paths of increasing
length, and color-code the paths with the droplet speed.

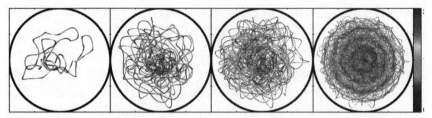

Four different times following the color-coded droplet path. Credit: Daniel M. Harris and John W. M. Bush.

At first, there is no discernible pattern. But eventually, as more and more time passes, the law of large numbers rears its head and the regularity emerges. The droplet, moving randomly, actually obeys a very simple and beautiful probability distribution of its location.

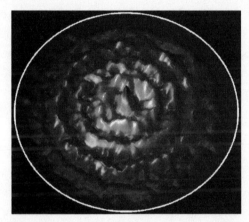

The probability distribution of droplet position formed in the long time limit. This and the above image are reprinted with permission from Harris et al. 2013. "Wave-like Statistics from Pilot-Wave Dynamics in a Circular Corral." *Phys Rev E* 88:1.

The probability distribution given by the Schrödinger equation, the electric field measured by a scanning tunneling microscope in a quantum corral, and the probability density function of a droplet walking in a corral *all have the same shape*. Because of this, the analogy is excellent. The quantum statistics and the droplet statistics have an extremely close mathematical form.

How amazing is this? If I had to pick a single analogy to present the promise of HQA, this would be it. There are many droplet analogies that are significant, but this one has it all. Analogies just don't get better than this. It's beautiful. It's simple. It's so clear what is actually going on. And yet, it comments very clearly on a profound mystery.

This is what Einstein was saying. There is a deeper level beneath the statistics. Bohr disagreed and said complementarity explained why the statistics are final. Einstein's view is very natural, very plausible. Bohr's view, besides being unnatural and implausible, was never formulated clearly. Nobody could ever really understand what he was saying.

And now we have a macroscopic fluid analogy that *directly* supports Einstein's view. As the authors remark, with characteristic academic understatement:

> Whether or not the statistical description provided by quantum mechanics represents a complete description of physical reality was the subject of the celebrated debate between Einstein and Bohr. Whatever the case may be in quantum mechanics, the linear statistics is clearly an incomplete description of our fluid system and is underlaid by a complex, nonlinear, pilot-wave dynamics.[3]

We can go a little further. There is a phenomenon called the *quantum mirage*, most evident in corrals that are not circular, but elliptical. If you remember your geometry, ellipses have two foci. A straight line emerging from one focus will always reflect to the other one.

Because of this, quantum corrals in this shape have some curious features. The quantum probability density is enhanced at the foci, showing traces of the paths that a classical billiard ball would take. In 2000, several researchers at the IBM Almaden research center used cobalt atoms to construct an elliptical quantum corral. When they placed another cobalt atom at one focus, an electronic image of the atom appeared at the other focus. This work was reported in *New Scientist* in an article called "They've Seen a Ghost."

There is also a hydrodynamic analogy of the quantum mirage. The work was done in 2017 by Pedro Sáenz, Tudor Cristea-Platon, and John

Bush, and published in *Nature Physics*. (Sáenz now runs his own HQA lab at the University of North Carolina at Chapel Hill.) As expected in the high-memory regime, the droplet takes a chaotic trajectory through the elliptical bath, driven by a wave field that is extremely complex. However, by averaging the wave field over thirty minutes, one obtains what is called the *mean wave field*. This is highly structured and intimately related to the geometry of the ellipse.

Three different times showing the highly unpredictable form of the wave field. Credit: Pedro Sáenz, UNC Chapel Hill.

Then, running the experiment for a long time, the researchers found the probability distribution for droplet position to be closely related to the mean wave field.

The shape of the mean wave field can be understood by looking at the Faraday waves of the elliptical bath. When the vibration is large enough, the surface of the bath itself vibrates up and down in various shapes. (It's important to remember that in droplet experiments the acceleration is *below* the Faraday threshold, so without the droplet there would be no waves.) These Faraday waves are called the characteristic cavity modes of the ellipse, and are indexed by pairs of whole numbers. The two most dominant modes are (1, 5) and (4, 4). By adding them together, we arrive at a shape that is extremely close to the mean wave field.

So far, this is a simple recreation of the circular corral but with just a different topology, and two dominant modes instead of one. The twist came when Sáenz changed the topology of the bath floor, placing a well beneath one of the foci. It is well-known that wells of this kind attract droplets, so this will skew the statistics, as well as the mean wave field. It was *how* this skew happened that they were able to explain.

The well excited the (1, 5) cavity modes, making its effects on the statistics more dominant. But this is the mode that had maximal values at the two foci. Thus, putting a well at one focus caused the droplet to travel more to the other, empty, focus. A perturbation at one focus "induced a signal at the other focus, a nonlocal statistical projection effect."[4] This is, as they noted, "strongly reminiscent of the quantum mirage."[5]

Part XI

Return to Clarity

I feel that the last word hasn't been said yet on the relationship between waves and particles.

—Paul Dirac[1]

Chapter 39

Bene Respondere!

Bohr believed that what he called the "inevitable dilemma" of particle-wave duality had been solved by his philosophical insight. Complementarity allows us to "harmonize the different views, apparently so divergent."[1] But where, exactly, is this resolution? We explained complementarity in Chapter 26, but was there any place that the connection between particles and waves was actually clarified?

There is a textbook from 1951, indeed we will look closely at it soon, which states that particles and waves are "opposing but complementary modes of realization of the potentialities contained in a given piece of matter."[2] Yes, the wave-particle duality arises from matter, that's the point, that's the *problem*. The utterly weird thing is that, because of complementarity, this statement is supposed to be some kind of solution. As Mara Beller put it:

> To say that microscopic objects exhibit wave and particle properties because they obey "a complementarity between waves and particles"...is not much better than to say that sugar is sweet because it has an essential quality of "sweetness."[3]

This reminds me of a wonderful moment in Molière's *Imaginary Invalid*, about a man named Argan who has hypochondria. The doctors get sick of diagnosing his every malady, and decide to convince him he

297

is a doctor so he can diagnose himself. There follows a hilarious sham examination in pseudo-Latin. The first doctor asks Argan what is the cause and reason why opium makes one sleep, to which Argan replies:

> Because there is in it a *sleepy virtue*, of which the nature is
> to stupefy the senses.

And the chorus proclaims,

> *Bene, bene, bene respondere!*

Given that the particle-wave duality is one of the oldest problems in physics, and it was conceived all along as a *physical* problem about the nature of light, which then became a physical problem about the nature of matter as well, Bohr's philosophical solution is anticlimactic to say the least. It recalls Horace's *parturiunt montes, nascetur ridiculus mus*: the mountains labor, a ridiculous mouse will be born. In order for this flimsy answer to rise to the occasion, we must magnify its size and significance. If we can establish complementarity as a *foundational* principle behind everything, as a *critical* insight into nature itself, the mouse grows in stature and may begin to do some heavy lifting. This is exactly what was claimed.

Bohr thought complementarity arose from the very core of nature, and "was convinced that complementarity was the deepest insight into nature found within the quantum theory."[4] This intuition was elaborated by comparing complementarity to relativity. Pauli said "in analogy to the term relativity theory one could therefore call modern quantum theory complementarity theory."[5] Pais called complementarity "a new kind of relativity."[6] Bohr himself thought along these lines, saying that the introduction of subjectivity into physics was "unavoidable," because the "theory of relativity...was destined to reveal the subjective character of all the concepts of classical physics."[7] The implication is clear: the quantum reality, and therefore all reality, depends crucially on the "frame of reference" with which one looks at it, namely, on the observer. And so, "just as Einstein solved pseudoproblems in classical electrodynamics so Bohr overcame the linguistic paradoxes of atomic physics."[8]

The particle-wave duality is nothing more than a "linguistic paradox." To solve it, we must examine our concepts, not the physical world. Like relativity, complementarity is supposed to revolutionize our entire viewpoint. Bohr's loyal disciple Rosenfeld "considered complementarity the most profound intellectual insight of the twentieth century, the pinnacle of the physical understanding of nature."[9] Wheeler echoed this: "Bohr's principle of complementarity is the most revolutionary scientific concept of this century."[10] This was the twentieth century, so Wheeler evaluates complementarity above even Einstein's relativity. Pais said that Bohr's formulation of complementarity "makes him one of the most important twentieth century philosophers.... Since Bohr, the very definition of what constitutes a phenomenon has wrought changes that, unfortunately, have not yet sunk in sufficiently among professional philosophers."[11]

If we alter fundamentally our conception of what a phenomenon is, if events do not actually have any real existence until they are observed, if every description has two contradictory concepts—position/momentum, energy/time—caught in a tug-of-war over the quantum of action, then we must admit a fundamental unanalyzability has entered into our models. The recognition of this duality at the heart of nature, "represents a thoroughgoing change in the type of concept that is appropriate for the description of matter at the quantum level."[12] In other words, if we drop our misguided ideal of "understanding in the usual sense of the word," then there is no problem anymore. The real problem was trying to solve the problem by situating atoms in space and time, with the "classical" intuition that effects have causes and the world is intelligible.

Complementarity really does restate the problem *as* the answer, with all the dialectical implosions such a move implies. It is as though, being asked a question, we repeat the question in a very profound tone of voice. Why does matter have both particle and wavelike properties? It is because particles and waves are "opposing but complementary modes of realization of the potentialities contained in a given piece of matter." *Bene, bene, bene respondere!*

This is the interchange of "what is foundational and what is derivative, *explanandum* and *explanans*," which Beller has said is so characteristic of Copenhagen philosophy.[13] Because of this, there is no physical explanation in the usual sense of the word; to expect one is "classical." The attempt to find an explanation will lead nowhere; it is "as ridiculous as hoping that $2 \times 2 = 5$."[14] Instead, we must understand that it is our language and our concepts that are generating the situation, because they are inherently limited by complementarity.

If we are being critical, it is hard to escape the sense that all this is a philosophical attempt in the spirit of Comte. Let's solve the hard scientific problems by demonstrating that they arise from the limits of our own minds. This is where Einstein diverged from Bohr. Einstein always believed there was an answer to the question, only it was deeply shrouded in mystery. It would take a long time to finally uncover. Like Comte, Bohr solved the problem with epistemological pronouncements, and invested his energies in what he saw as a profound lesson arising out of the insolubility of the problem. In the process, he became what Einstein called "the Talmudic philosopher [who] doesn't give a hoot about reality, which he regards as a hobgoblin of the naive."[15]

Indeed, the closer we look, the more reasons we find to be suspicious of Bohr's methods. Recall that Chapter 26 briefly mentioned the complementarity of vitalism and mechanism in biology. Vitalism is one of those superseded theories we could have based a parable on. It was supported, in various ways, in the ancient world all the way up through the 1900s. The basic idea is that living beings are *special*, and that the physical laws describing inanimate matter will never be able to explain the life force, the élan vital, the soul. Needless to say, current science has refuted this view entirely.

Bohr, applying complementarity to biology, decided that vitalism was half of the truth. He said a physical or chemical explanation of life was "impossible," and in this sense it was "analogous to the insufficiency" of classical physics to model atoms.

> In every experiment on living organisms there must remain some uncertainty as regards the physical conditions to which they are subjected, and the idea suggests itself that the minimal freedom we must allow the organism will be just large enough to permit it, so to say, to hide its ultimate secrets from us. On this view, the existence of life must be considered an elementary fact that cannot be explained.[16]

This is not because Bohr believed in the soul. In the same paper he says that living organisms are made of inorganic matter. But he comes to the same conclusions as the vitalists because of complementarity, "the most profound intellectual insight of the twentieth century, the pinnacle of the physical understanding of nature."[17] To his credit, Bohr stopped talking this way immediately after the discovery of DNA in 1953, but it's still a really bad look. One wonders whether Bohr, were he around in 1800, would have said that the caloric and vibrational theories of heat were complementary. It actually seems likely, since he applied complementarity liberally to any problematic pair of concepts. For example, "Bohr's claim that his complementarity philosophy provides a resolution of the mind-body problem."[18] This generic applicability far and wide is not a good thing.

> Complementarity is not a rigorous guide to the heart of the quantum mystery. Nor do Bohr's numerous analogies between quantum physics and other domains, such as psychology or biology, withstand close scrutiny. Complementarity does not reveal preexisting similarities; it generates them.[19]

This is what happens when an entire worldview is built up from Fourier analysis. What a surprise, that an algorithm used to compress cat photos for transfer across the internet actually means that, as Heisenberg put it in his uncertainty paper, "quantum mechanics establishes the definitive failure of causality," and that we must renounce the idea of atoms actually existing independently in the external world.[20] Skepticism is warranted. Fourier analysis is just a mathematical tool, like proof by induction, appropriate in a variety of situations. What we should conclude from its applicability to quantum theory is not obvious at all. Saying that Fourier analysis works because it grasps the essence of any situation is quite the leap. Because the uncertainty trade-off applies when sampling audio, should we make profound conclusions about the nature of sound itself? We *could* do this, but why? Such philosophizing is unnecessary and highly fallible.

Imagine a musician-philosopher who does this, claiming that insights gained from sampling audio have redefined the very concept of music. Then he goes further, insisting that this new understanding has profound implications for visual art, politics, and rocketry. Imagine he says these ideas would provide religious comfort and be taught to children in the future. It's possible he could get many followers if he were charismatic enough and understood the guru game, but only among the credulous. Literate and critical people would consider him sadly misguided, or clownish, even perhaps a cynical fraud. Yet, as we have seen, this is uncomfortably close to the way Bohr actually argued.

Complementarity seems simple, even innocuous, expressed by common phrases like "two sides of the same coin." But it is neither simple nor innocuous. It directly implies the final impossibility of a unified description. It is a "principle that in an authoritative, dogmatic way denies the notion of a single truth."[21] In other words, human knowledge will never grasp the nature of things. *That* is the nature of things. "Bohr's doctrine... implies that no direct access to the quantum world is possible, leaving its essence unknowable."[22] All we can do is look at this or that fragment of reality at any one time. But we can never piece together the fragments into a coherent whole because they are mutually contradictory. Quantum physics shows that such a situation emerges from the very nature of our existence and is inescapable. This is the crucial insight that resolves the particle-wave dilemma, for those willing to accept it.

Chapter 40

One Might Say Far-Fetched

If complementarity fails to convince us, what other arguments were made? Turning to Heisenberg's writings, we find no good ones. They are so bad, in fact, that it is hard to believe. Like most people, I grew up admiring Heisenberg and thinking he was a genius. The more I have learned about him, however, the more faded this picture has become.

He almost did not get his PhD because, in the oral exam, he could not describe how a microscope worked. He pushed very hard for strict positivism from his earliest papers. Twenty years later, Heisenberg completely bombed the moral test that was presented by Nazism. Never an experimentalist, he tried, and failed, to build a nuclear bomb for Hitler. When captured by the Allies he pretended his incompetence was sabotage, and yet after the war he said wistfully, "It would have been so beautiful if we had won."[1]

I'm always open to updating my opinion, but these days I think more in terms of the image given by the novelist Benjamín Labatut in *When We Cease to Understand the World*:

> Heisenberg continued working until he had reached the final matrix. When he solved it, he left his bed and ran around his room shouting, "Unobservable! Unimaginable! Unthinkable!" until the entire hotel was awakened. Frau Rosenthal entered in time to see him collapse on the floor and recoiled at the stench of his soiled pyjama bottoms.[2]

"Heisenberg wanted to be the new Kant," but he was not even a second-rate philosopher.[3] His philosophical writing is full of circular arguments and simple assertions. Once he claimed that "when we use concepts unhesitantly, they are exact."[4] But what was his condition for using concepts unhesitantly? He gives us no guidelines "unless we already believe that they are correct and exact—a characteristic circularity of Heisenberg's arguments."[5]

Another time, when Heisenberg's glib analysis got him tangled up in defending both the invalidity and the necessity of classical concepts, he disposed of the difficulty in an equally glib way: "Ingenious Heisenberg had no problem avoiding the impasse: he invented the notions of a 'practical a priori' and 'relative synthetic a priori,' which are, in fact, contradictions in terms."[6]

Here, I will examine just one of Heisenberg's papers. Like anyone who produces more than one work, some of them stand up better than others. I have been rather uncharitable in my selection not to beat a straw man, but to demonstrate the things that were being said when the completeness debate was young. The essay is from a talk Heisenberg gave at Germany's oldest scientific association, the Society of German Natural Scientists and Physicians, in 1934.

Before the members of this august assembly, Heisenberg wondered if abandoning "objective events in time and space independent of observers...represents only a passing crisis," and said that "there seems to be the strongest evidence, that this renunciation will be final."[7] Okay, what is this strongest evidence? First, he gives us an analogy:

> Previous to the beginnings of science in antiquity, the world was conceived as a flat disc, and only the discovery of America and the first circumnavigation of the world destroyed this belief for all time.... The hope that new experiments will yet lead us back to objective events in time and space are about as well founded as the hope of discovering the end of the world somewhere in the unexplored regions of the Antarctic.[8]

The idea is clear enough, but the fact that people used to be wrong about something has no bearing whatsoever on the question of hidden variables. All Heisenberg is doing here is revealing the way *he* thinks about the problem. He makes many florid comparisons to the end of the world, but when he goes beyond the metaphor, his "strongest evidence" boils down to one single assertion:

> The experimental results of Compton, Geiger and Bothe are such clear proof of the necessity of making use of the new lines of thought introduced by the quantum theory, that the loss of concepts of classical physics no longer appears a loss.[9]

That's it. There is no development of the supposed strongest evidence. He could have gone into *what* about the experimental results should establish the final failure of causality, but instead he spent his time implying that Einstein and Schrödinger were flat-earthers.

How the essay develops is also instructive. Heisenberg says that each domain of physics—like mechanics, thermodynamics, electrodynamics, and quantum mechanics—is essentially cut off from the others and relies on its own systems of thought. Thus we should not be surprised that classical thought has no bearing on the atomic world.

It's true that each domain has its own mathematical tools and patterns of thinking that are different than the others. But mechanics, thermodynamics, and electrodynamics have a deeper commonality: they are *classical*; they presume causality and provide pictures that exist in space and time. The real fragmentation Heisenberg is trying to establish, between classical and quantum, is not supported by the existence of multiple classical theories (rather the opposite, really).

It is as though you lived in an apple orchard, and were sitting at a table with three apples of different colors. Heisenberg approaches and gives you a piece of fruit in a bag and asserts confidently that, because the apples on the table are different, you *know* that the fruit in the bag cannot be an apple.

These are the sorts of arguments a con artist would make, though I don't think Heisenberg is trying to deceive anyone. It seems that he is very comfortable and does not feel the need to actually argue at all. He is happy to present his opinions as profound facts, without evidence. Therefore, we have no independent reasons to adopt his opinions. As we read on, in fact, it is rather the opposite.

Heisenberg tips his hand when he predicts that the fragmentation of science should apply also to biology. He says that "according to Bohr, we should expect the laws characteristic of living organisms to be separated from purely physical laws."[10] The picture that Heisenberg gives of fragmented science is inspired by complementarity. And yet, the prediction Heisenberg made about biology based on this point of view was just following Bohr and completely wrong. Heisenberg presented his opinions, without any support, and in the process gave us reasons to *dis*believe them.

Precisely this sort of thing is what led to Mara Beller's condemnation:

> One would expect that the proponents of the Copenhagen interpretation were in possession of some very strong arguments, if not for inevitability, at least for high plausibility. Yet a critical reading reveals that all the far-reaching, or one might say far-fetched, claims of inevitability are built on shaky circular arguments, on intuitively appealing but incorrect statements, on metaphorical allusions.[11]

And, we might add, on rhetoric. Bohr "advanced the Copenhagen interpretation relentlessly" with rhetoric "designed to introduce controversial knowledge in the guise of intuitively appealing propositions."[12] He "preached closure, finality, inevitability, and the impossibility of alternatives."[13] The dialect spoken in Copenhagen was constantly inflected with a rhetoric of "finality and inevitability."[14] Bohr sprinkled his writings with phrases like "it cannot be otherwise," "there is no way around," "the only rational interpretation," "obvious," and "unavoidable."[15] He repeatedly began arguments with phrases like "it must be realized," influencing the listener not to evaluate the following proposition, but to catch what they must admit.[16]

Rhetoric is the attempt to condition a response using techniques that are not fully bathed in the light of rationality. The reason the Copenhagen dogma was so marked by rhetoric was that its arguments were weak. They have always been weak. This combination of authoritative rhetoric and weak argument is a very serious red flag. As physicists adopted Bohr's point of view, the rhetoric became normalized. Others repeatedly spoke so as to *close down alternatives* and to present something merely possible as absolutely necessary.

Heisenberg said that "we can no longer speak of the behavior of the particle independently of the process of observation"[17] and that "the idea of an objective real world...is impossible."[18] Remember that Wheeler claimed "we could not even imagine a universe that did not somewhere and for some stretch of time contain observers"[19] and that "we have to say that we ourselves have an undeniable part in shaping what we have always called the past."[20] Then there is that zinger by David Mermin: "We now know that the moon is demonstrably not there when nobody looks."[21] The antirealist position taken by these physicists is conceived as the *only* position. Note the word choice: no longer, impossible, we could not even imagine, we have to say, undeniable, demonstrably.

This sort of thing is what led Schrödinger to say that the orthodox Copenhagen interpretation was "administered fairly early and authoritatively."[22] Isn't that interesting? There was no examination of alternatives, as we saw from the way de Broglie was treated. Complementarity just took over. It was *administered*—both *early* and *authoritatively*—that quantum mechanics was complete, the laws of physics are random, and objective reality does not exist.

In fact, we can see the truth of Schrödinger's observation in the very essay we just considered. It was written only seven years after the fifth Solvay conference, and two years after von Neumann's proof. That Heisenberg is content with mere assertions and metaphors says everything about the culture of the time. He was on the winning side of a dispute that, in his mind, was already over. There was no need to advance real arguments; it was enough to report on the victory. It is clear that, because we are still arguing about these issues today, Heisenberg's confidence was seriously misplaced.

We could spend a whole book picking through the morass of bad philosophizing and rhetoric that came out of the Copenhagen-Göttingen group. Instead, let us consider what should be the best argument in support of quantum completeness.

David Bohm was an American physicist, born in Pennsylvania. He studied at Caltech but disliked it because he found the physics community there most interested in "competition and getting ahead and mastering techniques."[23] So he transferred to UC Berkeley, where he studied with one of the most influential American physicists of the twentieth century, J. Robert Oppenheimer.

The atmosphere in the research group, as expressed by one of Bohm's fellow PhD students, was that "Bohr was God and Oppie was his prophet."[24] Because Oppenheimer was all in on the Copenhagen interpretation, and the other people in Bohm's group were very intense and convincing about it, Bohm said he got "carried away with it."[25] Also like other members of the group, Bohm became increasingly interested in Communist politics.

World War II upended everything in the Berkeley physics department, as Oppenheimer was chosen to head the special weapons division of the Manhattan Project. Bohm's own doctoral research was considered useful for the war effort and classified. Lacking the security clearance to access his own ideas, Bohm was forbidden from writing up his dissertation. Nevertheless, he got the nod from Oppenheimer and obtained his PhD.

After the war, Bohm was an assistant professor at Princeton. Here he taught physics and began to work on a textbook. This was *Quantum Theory*, "rightly considered the best Bohrian textbook."[26] Everything is expressed as clearly and plainly as possible. Not only that, it was published in 1951, so it had the benefit of absorbing twenty years of arguments for completeness. We should get the strongest possible proof of Bohr's position here.

We find the first mention of the subject in chapter 2, section 5, called "Unlikelihood of Completely Deterministic Laws on a Deeper Level." It is two paragraphs long and Bohm makes the obvious comparison to thermodynamics. Because heat is a probabilistic quantity, to find its "underlying causal laws, we must accept a description in terms of the individual molecules." Bohm then says, "The idea immediately suggests itself that probability in quantum processes arises in a similar way. Perhaps there are hidden variables."[27] However, he says that there are "strong theoretical arguments which make it unlikely that such variables exist." And he promises to discuss them later, in chapter 22, section 19.

This section is what we are looking for; it is titled "Proof That Quantum Theory Is Inconsistent with Hidden Variables." But, surprise! This section is also only two paragraphs long. Bohm says that if there are hidden variables for position and momentum, they must correspond to "simultaneously existing elements of reality." In other words, these variables must describe the real, actual position and momentum, both at the same time.

> To interpret the uncertainty principle, we would then have to assume that we are simply unable to measure the values of the two simultaneously with complete precision. But we saw in Chapt. 6, Sec. 11, that any such assumption would lead to a contradiction with the uncertainty principle.[28]

So now we need to go back to chapter 6, to the section titled "On the Reality of the Wave Properties of Matter." Here, Bohm considers a thought experiment "in which we observe the position of electrons with the aid of a proton microscope"[29] and shows that if we treat the protons as particles only, we obtain ranges for position and momentum that are "much less than the minimum permitted by the uncertainty principle." But if we can do that, then the uncertainty principle breaks down. This means that quantum mechanics itself would break down as a formal theory, which would "make the entire wave-particle duality untenable."

So this is the sum total of the "proof" that quantum mechanics is complete. If it was incomplete, the current way of thinking about waves and particles would be untenable. The theory would require significant further analysis, because it would not be as trustworthy as we thought. But this is exactly what Einstein was saying! If quantum mechanics is not the whole story, then there is a regime in which the quantum mathematics breaks down, just as the heat equation is completely incapable of generating the Maxwell-Boltzmann distribution. In fact, the heat equation is also "inconsistent with hidden variables," in Bohm's sense, since understanding why it works requires replacing its machinery.

When it comes to quantum mechanics, Bohm assumes this possibility to be impossible, and so his argument reduces to this: *If quantum mechanics were incomplete, then quantum mechanics would be incomplete; therefore, quantum mechanics is complete.* He promised strong arguments but did not produce a single idea. The performance is utterly shambolic. That anyone would even publish this mess, not to mention call it a "strong argument," reveals the psychological pressure physicists were under during this time. Conform to the Copenhagen dogma, for it is the light, the truth, and the way.

However, this is far from the end of the story. Bohm eventually became one of the greatest critics of the Copenhagen interpretation and quantum completeness. He made a remarkable volte-face almost immediately after publishing his textbook. He said:

> The conclusion that there is no deeper level of causally determined motion is just a piece of circular reasoning, since it will follow only if we assume beforehand that no such level exists.[30]

And regarding the uncertainty principle specifically, Bohm concluded that it was valid because we have no way of knowing or controlling the position or angles of scatter of particles. This limitation was not fundamental to nature, as Bohr insisted, but rather arose "from the existing quantum theory."[31] In other words, the uncertainty relations are a "deduction from quantum theory in its current form"[32] and, as such,

> attempts to prove their "absolute and final validity" are based on nothing but an a priori, illegitimate insistence on the finality of quantum theory.[33]

Chapter 41

Discarded Diamonds

How Bohm's views evolved is quite dramatic. Shortly after the book appeared, Einstein telephoned Bohm where he was staying with some of Einstein's friends. According to Einstein, the textbook gave "the best presentation he had ever seen of the case against him."[1] They arranged a time to discuss it. Murray Gell-Mann heard about this, and recalled:

> Naturally, when I next saw David I was dying to know how their conversation had gone, and I asked him about it. He looked rather sheepish and said, "He talked me out of it. I'm back where I was before I wrote the book."[2]

The arguments Einstein made are those for incompleteness that we have already encountered repeatedly. "Basically, his objections were that the theory was conceptually incomplete, that this wave function was not a complete description of the reality and there was more to it than that."[3] Einstein's intervention had a powerful effect on Bohm, who now saw that his attempt to justify the Copenhagen interpretation was a failure. As he recalled:

> It seemed that Einstein was right and I already felt dissatisfied, that somehow people were turning a method of calculation into an explanation of reality. They wanted to say that

their method of calculation corresponded to reality. It was part of the current positivist approach.[4]

Bohm began seriously considering Einstein's view on the quantum formalism. "Maybe it doesn't really give the whole picture, it's just a probability calculation."[5] It is quite remarkable, a testament to not only Bohm's skill but also the power of allowing yourself to ask obvious questions, that within a matter of weeks, Bohm had formulated an entirely new perspective on quantum mechanics.

Well, it wasn't entirely new—in fact, it was very close to de Broglie's original pilot-wave theory. Its guiding idea was that the particle was localized at all times in space, it traced out trajectories, but that its motion was completely different than what Newtonian mechanics predicted, because something called the *quantum potential*, derivable from the wave function, pushed the particle around.

Interpreted as a physical thing, the quantum potential is quite strange. It has an influence that does not decay with distance, no matter how far away two interacting particles may be. Bohm believed this idea led to something he called the "implicate order," a hidden level of reality in which all things were really one. He compared the quantum potential to the signal a ship might receive: the signal is a very small electromagnetic wave, incapable of moving a grain of sand, yet after receiving it the entire ship changes course.

The quantum potential in Bohm's theory was something of a mystery, and it has remained so. Entire books have been dedicated to the subject. It seems to transcend our usual notion of a physical guiding force, because changes in this potential are instantaneous. Thus, as Bell put it, "Terrible things happened in the Bohm theory....Particles were instantaneously changed when anyone moved a magnet anywhere in the universe."[6]

Nevertheless, if we change the form of the guidance equation that Bohm presented, we have the particle being pushed around by the gradient of the wave function. Here, again, we have the idea of a particle undergoing gradient-driven motion. These deep similarities are why Bohm and de Broglie's theories, while actually different in several respects, are usually bundled together, and the general picture of physical waves guiding physical particles is usually called de Broglie–Bohm pilot-wave theory.

Only months after his fateful meeting with Einstein, Bohm was forced to flee the country. He had run afoul of the House Un-American Activities Committee, for refusing to name acquaintances who might have communist sympathies. He arrived in Brazil on October 10, 1951, and in January 1952 he published the now famous two-part paper "A Suggested Interpretation of the Quantum Theory in Terms of 'Hidden' Variables."

Unsurprisingly, a theory that stirred the pot in this way was not favorably received. Some responses were downright nasty. Oppenheimer himself said, "If we cannot disprove Bohm, we must agree to ignore him," certainly a very depressing thing to hear from your former supervisor.[7] Bohm was called "a public nuisance"[8] and his theory "juvenile deviationism."[9] The worst was Léon Rosenfeld, the Belgian physicist who was a "particularly fierce and uncompromising defender of the Copenhagen Interpretation."[10]

Rosenfeld refused Bohm the dignity of discussion, writing to Bohm that "I certainly shall not enter into any controversy with you or anybody else on the subject of complementarity, for the simple reason that there is not the slightest controversial point about it."[11] Afterward, he spent "remarkable effort to preventing the spread of Bohm's ideas."[12] He went on a crusade of sorts, stopping Bohm from publishing a paper in *Nature*, and stopping Bohm's fellow travelers from publishing translations of their own books.

Pauli engaged with Bohm far more constructively, writing back to revive the same criticisms he had leveled against de Broglie at the Solvay conference. This time, Bohm answered them. They discussed the issue of priority and who should get credit for this new theory, and Bohm wrote to Pauli:

> If one man finds a diamond and then throws it away because he falsely concludes it is a valueless stone, and if this stone is later found by another man who recognizes its true value, would you not say that the stone belongs to the second man?[13]

There followed a pointed exchange of letters between the two. As an adversary, Pauli was sarcastic, witty, and incredibly withering. But Bohm had inherent intellectual and moral strength. He gave as well as he got, answering all of Pauli's criticism and even criticizing in turn. In the ensuing engagement with the "Wrath of God," Bohm won. Pauli eventually admitted that Bohm's theory was internally consistent. Nevertheless, it moved no needles for him, and he called the theory "a check which cannot be cashed."[14]

Bohr, on the other hand, did not write back at all. The best we have is a somewhat dubious secondhand account that, upon hearing of Bohm's hidden-variable theory, the reincarnation of Socrates considered it to be "very foolish."[15] Schrödinger did not write back either. He had his secretary do it, informing Bohm that "His Eminence," as Bohm referred to Schrödinger after this, was not interested in the work. Unfortunately, Einstein did not approve of the theory either. He was intent on unifying all of physics, of sneaking a look at God's cards. To do this he pursued a nonlinear theory that modeled the geometric structure of space-time, hoping that quantum mechanics would emerge as a statistical theory of that deeper reality. Thus, he thought Bohm's work too incremental. The phrase he used was "too cheap."

The most interesting was de Broglie's reaction, which was immediate. Other than some letters and brief notes presented at the Academy of Sciences, his first extended work came in 1952. In "Will Quantum Physics Remain Indeterministic?" de Broglie discussed the theory of "a young American physicist," which "took up again in their entirety, more or less under the same forms that I had given them, my conceptions of 1927."[16] The great de Broglie had been teaching complementarity for twenty-five years, he had fully incorporated the Copenhagen interpretation in his thinking, and so he was understandably cautious. He said he would not want to assert that a deterministic quantum theory was possible. Yet, still, he thought the question was worth examining. What followed was a miniature historical summary, hitting many of the same beats that we have done in this book. De Broglie said that "the great drama of contemporary microphysics has been, as you know, the discovery of the duality of waves and particles."[17] He recounts his own contribution, and the ideas developed from it. The double solution, the pilot-wave theory, the ill-fated trip to Brussels where he gained only slender support from one person (Einstein) for his ideas. He described the quantum theory of Bohr and Heisenberg, how profoundly the wave function in that theory differed in interpretation from his double solution. How Einstein and Schrödinger never accepted complementarity and "opposed it with troubling objections."[18] He called Bohr the "Rembrandt of contemporary physics," because of a penchant for the "clear obscure" (this was scrubbed from the essay's English translation).[19] And he remarked how strange it was for the orthodox quantum theory to create its own type of probability that did not emerge from our ignorance or inability to cope with complexity. "In the current accepted interpretation of quantum physics," he wrote, "we are dealing with a *pure probability* that does not result from a hidden determinism."[20] Also, de Broglie gave expression to his own intuitions of realism. Smoldering for decades as an almost-spent coal, here it finally burst into flame. He said,

> The return to clear, Cartesian conceptions respecting the validity of the frame of space and time, would certainly satisfy a great many minds and permit us not only to remove the troubling objections of Einstein and Schrödinger, but also to avoid certain strange consequences of the current interpretation.[21]

And de Broglie went on to echo Einstein, who said something very similar to this in his eulogy for Ernst Mach: that the "history of the sciences shows that the progress of knowledge has constantly been hindered by the tyrannical influence of certain conceptions."[22] His restraint and caution have not disappeared, but within their confines, de Broglie argues forcefully for something that is seemingly irreproachable: the freedom to question and inquire.

> We can ask if it is not rather towards a return to the clarity of spatio-temporal representations that we ought to orient ourselves. In any case, it is certainly useful to take up again the very difficult problem of the interpretation of wave mechanics, in order to see if today's orthodox view is truly the only one we might adopt.[23]

The very last line of this paper was repressed in the English translation, but it is most telling of all. De Broglie finished with flair, quoting Boileau: *Vingt fois sur le métier, remettez votre ouvrage*: Twenty times on the loom redo your work.[24] In this way, de Broglie rejects the very attitude that Einstein said, multiple times, was behind the popularity of the Copenhagen orthodoxy: "Scarcely anyone is inclined to abandon an enterprise in which he has invested a great deal of work."[25] This is, of course, the sunk cost fallacy. It reminds me of that poignant Turkish proverb: *No matter how far down the wrong road you have gone, turn back now.*

This paper is a valuable historical document, and a subtle argument for the necessity of intellectual freedom. One can feel de Broglie's strength growing as he renews his commitment to the search for an explanation of the quantum world that does not abandon the traditional standards of science. No matter how far down the wrong road we have gone, let us turn back now. Twenty times, let us redo the work, until we have something that actually makes sense out of the "great drama" of particles and waves.

It was 1952, de Broglie was sixty years old, and he was back.

Chapter 42

It Falls Apart in Your Hands

Bohm's work led to a resurgence in discussion about hidden variables, and von Neumann's proof that quantum mechanics could never be supplemented by them. An essay appearing in a small German periodical in 1959 exemplifies this. It was titled "A Question for Mathematicians and Physicists—Was the Parameter Proof of J. v. Neumann Refuted?" and it described the divergent views of Pauli, March, Bopp, Rosenfeld, de Broglie, Destouches, Feyerabend, Fényes, Zinnes, and others. It ended with a request: "Which assessment of Neumann's proof is actually correct?...The sender would be grateful if this question, to which there is apparently a clear answer, would be answered by an expert."[1]

This question took some time to answer. It was not until the mid-1960s that a transformation occurred. As Max Jammer put it:

> The cogency of von Neumann's "proof," once regarded as
> incontestable, began to be called into question and it seemed
> no longer unreasonable, in spite of von Neumann, to search
> for a deterministic refinement of quantum mechanics.[2]

However, the same transformation occurred much more quickly within de Broglie's mind. Before Bohm's papers appeared, de Broglie believed that

> J. von Neumann has proven that the probability laws of the
> new mechanics are incompatible with the existence of a hidden
> determinism, which makes it most improbable that determin-
> ism in atomic physics will be re-established in the future.[3]

And yet, after Bohm published, de Broglie was led "to a reexamina-
tion of the demonstration of this theorem," and soon he saw a problem:
von Neumann had assumed that hidden-variable theories need to obey
the postulates of quantum mechanics, which simply eliminated hidden
variables from consideration.

> Hence there is a sort of vicious circle, and the theorem of
> von Neumann does not seem to have any longer the import
> which even I had attributed to it in recent years.[4]

He wrote this in 1953. In fact, the same criticism of circularity had
been leveled against von Neumann's theorem almost two decades earlier by the German mathematician and philosopher Grete Hermann.
She studied mathematics with Emmy Noether, the only female professor at the University of Göttingen and indeed the first female professor
in all of Germany. Yet after getting her PhD, Hermann immediately
switched to philosophy. Noether responded with shock, saying of Hermann: "She studies mathematics for four years, and suddenly she dis-
covers her philosophical heart!"[5]

Hermann is a remarkable figure because she was not only a woman
in a field completely dominated by men, but also an exceptionally clear
and prescient thinker, far more so than most of the people around her.
She also had a strong moral backbone. When Hitler came to power,
she worked with the underground resistance before fleeing to England.
There, Hermann married a Brit named Edward Henry, changed her
name to Henry-Hermann, and thereby avoided being put in an intern-
ment camp, a decision of the government she anticipated by years.

It is most remarkable that in 1935, very soon after von Neumann
published his flawed proof, Hermann saw its weakest link and criticized
it in a larger paper devoted to the question of causality in quantum mechanics. She called that section "The Circle in Neumann's Proof," and
in her summary she observed that

a detailed assessment shows here, too, that this mathematically otherwise faultless argumentation introduces into its
formal assumptions, without justification, a statement equivalent to the thesis to be proven.[6]

Hermann is referring to what has come to be called von Neumann's
additivity assumption. We don't need the logical specifics to appreciate
the shape of the critique. It goes like this: There is a specific property
that holds of quantum measurements. Everyone agrees that it holds,
so it is not unreasonable to include it as an axiom. However, when reasoning about hidden variables, which by definition reside in the murky
darkness of our ignorance, we must be very careful to not smuggle in
assumptions that may be unwarranted. This is, according to Hermann,
exactly what von Neumann did. He *assumed* that the specific properties of quantum mechanics would apply to the hidden variables. In
short, he assumed that any theory that underpinned quantum mechanics would have certain quantum mechanical features. As Jammer put
it, "Hermann's criticism may be summarized as follows: von Neumann
has proved that quantum mechanics is a complete theory, but only as
far as quantum mechanical states are concerned."[7]

And yet, despite the fact that she had worked with Heisenberg and
von Weizsäcker very closely, her paper quickly fell into oblivion. It took
decades for the argument she made to be dimly glimpsed by de Broglie,
and even longer for its validity and author to be recognized. Indeed, as
Léna Soler has observed, "Had Grete Henry-Hermann's refutation not
remained a dead letter, the history of the interpretations of quantum
physics would certainly have been very different."[8]

But even before de Broglie was inspired to examine von Neumann's
theorem, and to repeat the core idea of Hermann's critique, the correct-
ness of von Neumann's logic was being questioned. It is little known
that by around 1938, only a few years after Hermann's critique, Einstein had independently identified the theorem's critical weakness.

Einstein knew about von Neumann's theorem. Peter Berg-
mann told Abner Shimony that he, Valentin Bargmann
(both assistants of Einstein at Princeton), and Einstein once
discussed it in Einstein's office. On that occasion Einstein

took down von Neumann's book, pointed to the additivity assumption, and asked why we should believe in that.[9]

And Bohm said something very similar, even identifying the same circularity as Hermann and de Broglie, and independently of them:

> The conclusion that there is no deeper level of causally determined motion is just a piece of circular reasoning, since it will follow only if we assume beforehand that no such level exists. A rather similar analysis can be made with regard to von Neumann's theorem.[10]

But all of these criticisms were scattered across time and space. They could not dent the respect and awe that von Neumann's proof commanded. For that, it would take a campaign of repeated papers, each drilling deeper into the same subject. Who would risk their career to do this? That would require a forceful and bold intellect, equipped with depths of bravery. The person who did this would have to be a solid physicist and a respected insider, but someone with backbone, not easily swayed by waving hands. Also preferably Irish.

The world did eventually get all of these desirable traits, embodied in the person of the singular and wonderful John Stewart Bell.

John Bell was born in Belfast, on July 28, 1928. He described his parents as "poor but honest." His entire family were "carpenters, blacksmiths, laborers, farm workers, and horse dealers," and his siblings all left school and got jobs by the age of fourteen. John was the first one to ever complete high school. The greatest extent of any education in his family was his half uncle, a village blacksmith who "taught himself something about electricity."[11]

From this background, as unlikely as Fourier's to generate scientific achievement, came one of the most important physicists of the twentieth century. Bell was unassuming, witty, and brilliant, and he advanced our understanding of hidden variables and the status of quantum completeness more than probably anyone else since Bohr and Einstein.

Bell did not display any traits of mathematical precocity, but he was a rather bookish child who was "hostile to the idea of sports."[12] He regretted this because he grew up to be what he described as "a seven stone weakling" (a stone is 14 pounds, about 6.35 kilos). Out of high school he got a job in the physics lab of Queen's University Belfast. When he enrolled the next year his former bosses became his professors. It was in the last two years as an undergraduate that Bell started learning quantum mechanics.

From the beginning, he was interested in foundational questions. He argued with one of his professors, "a Doctor Sloane," about the uncertainty principle. "Poor Sloane—he must have felt that he was being attacked by a red-haired dervish."[13] He wondered about the division line between the quantum and classical worlds. He read Max Born's *Natural Philosophy of Cause and Chance*, which described von Neumann's impossibility proof. Bell was "very impressed" by this result. He read Bohr and Pauli and Heisenberg and tried to make sense of complementarity. And then, he stopped thinking about these things and "rather deliberately" focused on accelerator physics. As he put it: "I had the feeling then that getting involved in these questions so early might be a hole I wouldn't get out of."[14]

After obtaining his degree, Bell did not go to graduate school. Because he felt guilty for living off his parents for so much longer than his siblings had, he went to get a job. He wound up at the Atomic Energy Research Establishment, the UK's main researcher of nuclear power, thirteen miles south of Oxford in the sleepy town of Harwell.

Around this time, Bohm's papers came out. As Bell recalls, "I was enormously impressed with them. I saw that von Neumann must have been just wrong."[15] Because von Neumann's book was not yet translated into English and Bell did not read German, he had to be content with discussing the issue with a German colleague, Franz Mandl. These talks planted the seed. As he said: "I already felt that I saw what von Neumann's unreasonable axiom was."[16]

Within a couple short years, Bell would get tenure (at the age of twenty-one), meet a Scottish lass named Mary Ross (also a physicist), marry her (at twenty-two), and then go off to graduate school to study quantum field theory under Rudolf Peierls at the University of Birmingham. Then, in 1960, the Bells left the security of tenure and moved to Geneva, for a dual appointment at CERN. They resided and worked there, childless and happy, for the next thirty years.

In the early 1960s Bell learned of the work of Josef Maria Jauch, one of his colleagues, on strengthening von Neumann's impossibility theorem. Bell thought von Neumann must be wrong, because the pilot-wave theory existed. So his discussions with Jauch had a stimulating effect. Bell said: "For me, that was like a red light to a bull."[17] He began thinking more about the foundations of quantum mechanics. Soon afterward, the Bells had a year working at Stanford. They arrived in late 1963 at the worst possible time, one day after President Kennedy was assassinated. Mary was quickly integrated into the accelerator group, but John was isolated, doing his own theory. He had time and leisure, and as so often happens in these circumstances, he made a breakthrough.

Bell examined von Neumann's original proof (an English translation had by then appeared, in 1955), as well as Jauch's work, published with a coauthor Piron. He also looked at the latest so-called impossibility proof of Andrew M. Gleason, which took a different approach and came to the same conclusions. So, just to summarize the state of things when Bell staged his intervention in the early 1960s: Nearly everyone believed von Neumann had definitively killed off Einstein's program of dissent. Not only that, but more and more demonstrations to this effect were coming out, even stronger than the original. Who in their right mind would doubt the result? Well, as it turned out, exactly someone in their right mind. Someone clearheaded. Bell showed, in a single paper, that the current understanding was completely backward, the physicists were simply fooling themselves, and that all of these so-called proofs "leave the real question untouched."[18]

In Bell's analysis of von Neumann's famous impossibility theorem, he concludes that "the formal proof of von Neumann does not justify his informal conclusion," because

> it was not the objective measurable predictions of quantum
> mechanics which ruled out hidden variables. It was the arbi-
> trary assumption of a particular (and impossible) relation.[19]

The proof of Jauch and Piron has a critical axiom that commits the same sin. As Bell kindly puts it: "The objection to this is the same as before.... The axiom holds for quantum mechanical states. But it is a quite peculiar property of them, in no way a necessity of thought."[20] And so there is the same circularity in both proofs. Bell also considers another well-known impossibility proof of Gleason, and uncovers a hidden and completely arbitrary assumption about measurement that need not be true. What Bell writes about Gleason could well be applied to all three of them:

> That so much follows from such apparently innocent assumptions leads us to question their innocence. Are the requirements imposed, which are satisfied by quantum mechanical states, reasonable requirements on the [hypothetical deterministic] states? Indeed they are not.[21]

This is quite something. The great von Neumann assumed a particular and *impossible* relationship to be true. Bell wrote and spoke often about these issues, often elaborating on them with color. Here is something he said in an interview with *Omni*:

> The physicists didn't want to be bothered with the idea that maybe quantum theory is only provisional. A horn of plenty had been spilled before them, and every physicist could find something to apply quantum mechanics to. They were pleased to think that this great mathematician had shown it was so.

> Yet the von Neumann proof, if you actually come to grips with it, falls apart in your hands! There is nothing to it. It's not just flawed, it's *silly*.... You may quote me on that: The proof of von Neumann is not merely false but foolish.[22]

Part XII

Interlude

Chapter 43

Unintuitive Droplets

Toward the end of Einstein's life, he wrote to his lifelong friend Michele Besso:

> The whole fifty years of continuous brooding have not brought me nearer to the answer to the question "What are light quanta?" Nowadays every Tom, Dick and Harry thinks he knows it, but he is mistaken.[1]

There is another story from about the same time that rings true because it expresses the same sentiment. I could not find a reliable source, so I repeat it here with the caveat that it may be apocryphal.

In the 1950s, physics was exploding with new particles. There was a whole menagerie of new types of matter, and the field had an embarrassment of uncategorized riches. Some physicists speaking with Einstein once asked him what he thought of these "huge numbers of short-lived heavy particles, kaons, pions, quarks, mesons, etc. found using high-energy accelerators and enormous amounts of time and money." Einstein said to them: "I would just like to know what an electron is."[2]

If he were alive today, Einstein might still be asking this question. One popular undergraduate textbook states: "The electron (as far as we know) is a structureless point, and its spin angular momentum cannot be decomposed into orbital angular momenta of constituent parts."[3] A structureless point with internal angular momentum is a contradiction in terms. But that is where we are.

Over the years many physicists have wondered about this extra property of the electron, proposing various physical explanations for it. For example, Hans Ohanian wrote a paper called "What Is Spin?" in 1986, the first line of which is, "According to the prevailing belief, the spin of the electron or of some other particle is a mysterious internal angular momentum for which no concrete physical picture is available."[4] He went on to reference a calculation of Belinfante, from 1939, that tried to rationalize spin as generated by a circulating flow of energy in the wave field.

In 1930 Schrödinger pointed out that the quantum theory implied the existence of *zitterbewegung*, a "trembling motion" within the core of the electron. This is usually interpreted as a curiosity of the theory, even an artifact, but in 1990 the physicist David Hestenes went so far as to claim that this trembling motion was "a ubiquitous phenomenon with manifestations in every application of quantum mechanics."[5] Hestenes began his paper, titled "The Zitterbewegung Interpretation of Quantum Mechanics," like this: "The idea that the electron spin and magnetic moment are generated by a localized circulatory motion of the electron has been proposed independently by many physicists."[6] Hestenes attempts to develop this intuition, noting that "the essential feature of the zitterbewegung idea is the association of the spin with a local circulatory motion."[7]

For those who think along similar lines, and for anyone who is hesitant to accept blatantly contradictory physical concepts like a "structureless point with angular momentum," the following droplet experiment should be very interesting indeed.

We have already seen how fundamentally memory can affect a system. Memory can turn a droplet into a walker; it can cause two droplets to orbit each other; it can even create quantization of orbital radius. One thing that researchers wondered was whether memory could induce highly non-Newtonian behavior. For example, if a single droplet was moving in a circle because of a rotation of the bath, what happens if we suddenly turn off the rotation? The droplet does something very suggestive: it continues moving in a circle, all by itself, a handful of times, before it destabilizes into the walking state and shoots off in a straight line.

The obvious question is now: Can a droplet ever just zip around in a circle all by itself? What sorts of conditions would be required for this to happen? For almost a decade, this was an open question. It was addressed in several papers by mathematicians Anand Oza, Matt Durey, and Ruben Rosales, all of whom worked with John Bush at MIT. Using a simplified model, they found that droplets *could* theoretically act this way. In the field of hydrodynamic quantum analogues, this is called a *spin state*. The name invokes the mysterious property of electron spin.

It was thought for some time that these spin states were inherently unstable. While researchers could set them up in clever ways, "initializing" the wave field into the right shape, the droplets would not stay in a circle for more than five or six revolutions, no matter what they did. It is a testament to the richness of the droplet system that it took over a decade after the spin states were first observed for a stable regime to be discovered.

Bernard-Bernardet, Fleury, and Fort used small droplets around a half millimeter in diameter, and a high memory time with a wave field constructed from between seventy and one hundred previous bounces. They also surrounded the bath with a tapering beach where the fluid depth got increasingly shallow. This was to attenuate the outgoing waves, and thus minimize the reflection from the walls. By doing this, the size of the reflected waves was truly tiny, on the order of one micron, whereas the wave field in the vicinity of the drop has a maximum amplitude of about one hundred microns (there are ten thousand microns in one centimeter). With this setup, the walking droplet can *spontaneously* enter the spin state, and then stay there.[8]

This picture is the macroscopic version of Hestenes's electron, in which the spin magnetic moment is created by a charge zipping around in a circle. The fact that we may have a hydrodynamic quantum analog of the electron is intriguing, and it tells us something: the ideas about the nature of spin that have been proposed for almost a century are *physically* possible. That something analogous to these proposals has now been realized in a system that links particles and waves, and exhibits many other quantum analogs, provides further encouragement that this is the right direction to go.

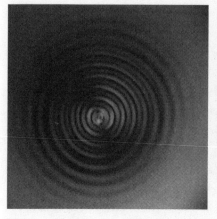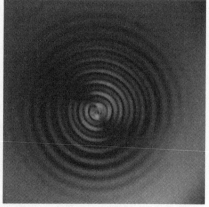

The wave field at two different times, about 100 seconds apart, of a self-orbiting droplet in the spin state. Credit: Samuel Bernardet.

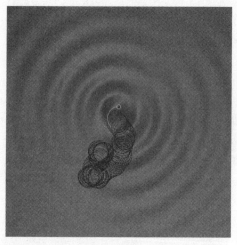

The path of a spinning droplet trapped in its own wave field. The rotations are quite slow, about once every five seconds, so the path pictured takes about seven minutes. The drift is caused by small effects from wall-reflected waves, and likely some extremely weak air currents. Credit: Samuel Bernardet.

Regardless of whether we are seeing an analogy of an electron or not, there is a larger lesson we can draw from the spin states. The quantum world is often described as "a non-classical world."[9] It is common to remark that classical physics thought of the world as some kind of clock, totally determined in its operation. This view is often associated with Laplace, whom we have encountered several times now.

In *A Philosophical Essay on Probabilities*, published in 1814, Laplace imagined "an intelligence" that could comprehend, for example, the positions and velocities of all molecules in a cubic centimeter of air. We have already seen that this is absolutely impossible, but never mind. *If* such an intelligence existed, Laplace says, it would be able to know the state of the air at any value of time in the future or past. It's a short step to expand to a cubic meter of air, and since we're just playing games with words now, the whole universe. This intelligence is today known as Laplace's demon.

The appeal of Laplace's demon was based on the undeniable power of Newton's differential equations. When Maxwell considered molecules of gas to be perfectly hard spheres that collided and recoiled from each other, he used Newton's laws as the foundation. Assuming a deterministic interaction, he was able to derive the statistical laws we use today. In fact, most any differential equation from classical physics proves the same point. In the heat equation, we saw how heat changed in time according to the spatial curvature. The math is deterministic. If you take a single snapshot of an iron bar, and from this read off the heat at each point, you can describe with ironclad certainty what the heat will look like at any other time. You can predict the future and even retrodict the past. It's just like fast-forwarding and rewinding a movie.[10]

The conceptual assumption made here is that you can specify the variables you need to know as closely as required. In quantum theory, however, this assumption fails. There is an irreducible grain of imprecision, so that if you are getting increasingly clear about, say, the position of a particle, its momentum becomes increasingly blurry, and vice versa. You can have one or the other, but not both. This is Heisenberg's famous uncertainty principle, which, as we have seen, inspired him to proclaim the death of classical physics ("quantum mechanics establishes the final failure of causality").

For a century people have been saying that causality is dead, alongside many other classical things, like pictures that make sense in space and time. It is therefore quite surprising to come across the bouncing droplets and especially beautiful and surprising dynamics like the spin state. Nothing like this has ever been imagined in classical physics. A *self-orbiting* particle would have been unthinkable to Newton, and yet the physics of these spinning droplets is essentially just Newton's laws, applied to fluids with memory. The connection of a particle and a wave brings us suddenly beyond the old clockwork universe of particles traveling in straight lines.

There is still another droplet analogy that brings out this point beautifully.

In 1992, four physicists criticized Bohmian mechanics because the particle "trajectories, which David Bohm invented in his attempt at a *realistic* interpretation of quantum mechanics, are in fact *surrealistic*."[11] The Bohmian trajectories, they argued, are surreal because they will interact with features of the environment they do not even approach, thus reflecting off of nothing at all. They decided that "one cannot attribute reality to the Bohm trajectories." One of the authors, picking up the work five years later, said that "particles do not follow the Bohm trajectories as we would expect from a classical type model," and finally declared "the Bohmian picture to be at variance with common sense."[12]

Making judgments about what is or is not "classical" based on "common sense" is, as the droplets have shown us, a dangerous strategy. Matt Durey, a mathematician at the University of Glasgow who has worked for almost a decade on the droplet system, put it this way: "I have long ago stopped trusting my intuition about what will happen in these systems. You just have to do the math and see."

In 2022, Valeri Frumkin and David Darrow, both in John Bush's research group, alongside Ward Struyve in Belgium, built a droplet experiment inspired by the so-called surreal trajectories. In it, a particle can be reflected by obstacles in two paths. The particle takes one path and is reflected, but the wave field takes both paths. Because the particle interacts with the wave field reflecting from the path not taken, it reflects again. If we were unaware of the pilot wave, we would say it reflected from nothing at all. In other words, these researchers immediately observed exactly what Bohmian mechanics predicts. This led to a paper with the kind of sweet title one rarely sees in academia: "Real Surreal Trajectories in Pilot-Wave Hydrodynamics."

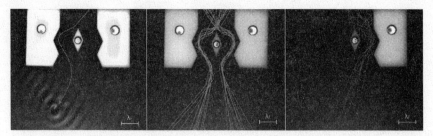

Left: A single trajectory when both barriers are present. *Middle*: A series of trajectories. Note that after emerging from the apparatus their paths do not cross, a phenomenon predicted by Bohmian mechanics. *Right*: Classical "intuitive" trajectories in the absence of the other reflecting barrier. Reprinted with permission from Frumkin et al. 2022. "Real Surreal Trajectories in Pilot-Wave Hydrodynamics," *Phys Rev A* 106:1.

This result led the authors to make a pointed observation: "In the absence of experimental measurements of actual particle paths, the designation of Bohmian trajectories as real or surreal necessarily depends only on one's preconceptions as to how quantum particles should behave."[13]

The moral of both these analogies is something like this: before categorically declaring the "final failure" of a conceptual scheme, we should carefully examine all of the assumptions required for that scheme to work. It may be that, by relaxing one of these assumptions, we can escape the scheme's limitations, and yet retain its core aspects. This is precisely what happens here. Spinning droplets and real surreal trajectories are *classical*, in the sense that they use known fluid dynamics and assume strict causality. And yet, by requiring more than a single past time of the droplet to specify the wave field, and linking the droplet dynamics to this wave field, we have violated a very common assumption of classical physics.

If we fail to properly identify all instances of some thing, then our judgments regarding that thing will be untrustworthy. Heisenberg's proclamation of the "final failure of causality" was dealing with a truncated subset of the causal mechanisms available to classical physics. The old ideas are not exhausted. They are still surprising us.

Chapter 44

Parable of Plate Tectonics

Here's a good one.

Abraham Ortelius was born in 1527, in the Duchy of Brabant, a region of Europe that is today part of the Netherlands. A geographer and cartographer, he created the first modern atlas. In 1587, Ortelius's work with maps led him to publish a dictionary of place names called *Thesaurus Geographicus* (Geographical Treasury). In it, he remarked on "the matching coastlines of the Old and New Worlds."[1] So far as we know, he was the first to mention this puzzling coincidence.

It's obvious to us that the continents move, and that a long, long time ago, they were in direct contact. Any child with a globe and imagination could have this idea, because the continents fit together like literal puzzle pieces. What is more, this childish idea could not be more important. Continental drift, and its modern incarnation as plate tectonics, is today axiomatic to geophysics and fundamental to all of modern geology. It has rightly been called "one of the greatest unifying theories of all time."[2]

Yet this striking congruity of coastlines, especially between South America and Africa, meant something only to a handful of people through the centuries. There are only about a dozen people that we know of perspicacious enough to have had this obvious idea. They include Thomas Young (there the good doctor is again, on the right side of history...), Jean-Baptiste Lamarck, and Alexander von Humboldt. In 1858 the naturalist Antonio Snider-Pellegrini explicitly suggested the idea, as we can see from the maps he drew depicting the world before and after what he called "the separation."

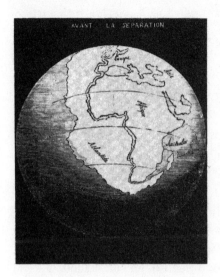

But these people were ignored. Nobody believed them or built anything on their ideas. Why not? It was because most everyone had a style of thought based on fixity. Their models assumed that the continents were essentially stationary, and they tried to answer all questions in their field from this viewpoint.

And there *were* questions, which began piling up. For example, paleontologists found the same kinds of fossils in Brazil and South Africa. They were found right where the puzzle pieces fit together. This pattern was repeated on multiple continents, not just between the Old and New Worlds.

Then there were several geological similarities exactly where you would expect to find them if the continents were connected once upon a time. For example, the coastline location of mountain ranges, and the existence of layered formations that dropped off on one side of the ocean and reappeared on the other side, "as if someone had torn a newspaper page in two and yet you could read across the tear."[3]

On Christmas Day, in 1910, a German meteorologist named Alfred Wegener was at a friend's house. Probably sated on food and drink, he was idly flipping through a new atlas. He suddenly had the same thought that has surely occurred to countless children, and anyone whose minds were not besotted with fixity.

The difference was that Wegener really did something about it. He started by cutting out the continents and fitting them together (not using his friend's atlas, one hopes). He was able to do this with surprising ease, and he called the result Pangea (All-land). He then collected extensive evidence, reading recent books that drew from hundreds of academic papers.

Then, almost exactly a year later, he gave a presentation on his ideas to the German Geological Society. His argument depended on three main elements: the continents fit together like puzzle pieces; we see the same flora and fauna in fossil records exactly where we would expect, if the continents were once connected; and the geological features like mountains and strata also line up. We don't know what anyone thought because the meeting minutes say there was "no discussion due to the advanced hour." Wegener published his ideas in *The Origin of Continents and Oceans*, a German book that appeared in 1915. Then, seven years later, the book was translated into English, and all hell broke loose.

The geoscientists of the age called Wegener's ideas "delirious ravings."[4] They said he had "moving crust disease and wandering pole plague."[5] Lingering anti-German sentiment from the war no doubt fueled the accusations of "Germanic pseudo-science," concocted by a moron in "a state of auto-intoxication."[6] At one meeting of the Royal Geographical Society, a speaker criticized Wegener's theory and afterward someone in the audience stood up and thanked the speaker for exploding Wegener, and then thanked the absent "Professor Wegener for offering himself for the explosion." Anything critics could take a shot at, they did. For example, since Wegener was not a geoscientist, but a meteorologist who studied air currents in Greenland, whose PhD was actually in astronomy, his points were invalid. It was a matter of moral principle, because it was "wrong for a stranger to the facts he handles to generalize from them."[7] What's your PhD in, bro? What college you go to? Ha ha ha ha! Your ideas are bad!

Perhaps this came down to habits of thought. Those who believed in mobility seemed more mentally flexible as well. Wegener, when he was advised to be more theoretically cautious by his academic mentor and future father-in-law, replied: "Why should we hesitate to toss the old views overboard?"[8] On the other hand, those who believed in fixity said things like this: "If we are to believe Wegener's hypothesis we must forget everything which has been learned in the last 70 years and start all over again."[9]

Though Wegener was mocked, he never backed down. "That was always his response: Just assert it again, even more strongly."[10]

> By the time Wegener published the final version of his theory, in 1929, he was certain it would sweep other theories aside and pull together all the accumulating evidence into a unifying vision of the earth's history.[11]

And so began a debate between the mobilists, also called the drifters, and the fixists. It lasted five decades. Continental drift could explain so many things with ease, but it had one big problem: What was the cause? Without a mechanism, hardheaded scientists simply would not consider the idea that continents moved. Cherry Lewis called the establishment view a "dogma [that] formed obstructions to continental drift," and said that "only a few enlightened individuals recognised early on that it was the only way so many geological phenomena could be explained."[12]

The dogma was everywhere. In the latter 1940s Sir David Attenborough, the beloved naturalist and maestro of British broadcasting, was at Clare College, Cambridge (the best college, by the way). He recounted this experience:

> I once asked one of my lecturers why he was not talking to us about continental drift and I was told, sneeringly, that if I could prove there was a force that could move continents, then he might think about it. The idea was moonshine, I was informed.[13]

The fixists explained the coincidence of fossils on separate continents by land bridges. Land bridges actually exist; Panama is an example. Three million years ago it connected North and South America, allowing for what is called the Great American Biotic Interchange. This isthmus is about 480 miles (772 kilometers) long. The proposal of the fixists was that there had once been a land bridge ten times this length connecting Brazil and South Africa. They gave it a name: Archhelenis.

There was another one in the North Atlantic called Archiboreis, and one connecting Central America to Hawaii and thence to Asia (roughly 10,000 miles, or 16,093 kilometers), called Archigalenis.

To us moderns, who benefit from all the struggles of the past, the idea of these land bridges seems rather desperate. But back then, when nobody knew the answer, intuitions and their associated styles of thought were strongly at play. If your mind was favorable to mobilism, this idea of land bridges was ridiculous. To a fixist, the idea of the continents moving was the ridiculous thing.

The drifters believed that the continents were emerging out of the sea floor, from mid-ocean ridges, enormous fissures that extend down the very center of oceans. During the Second World War, those searching for submarines with delicate magnetic instruments discovered what is called magnetic striping. These were stripes of magnetized rock on the seafloor that could be hundreds of miles wide. If one stripe was polarized in one direction, the stripes on either side would have the opposite magnetic polarization. The stripes were also *symmetric* on either side of the ocean ridge. This was a mystery until it was discovered that the magnetic field of the Earth itself switched back and forth, once every half million years on average. This is called *geomagnetic reversal*, and it explained the magnetic stripes, assuming that there was new Earth crust emerging from the ocean floor and spreading outward.

> Either side of the ridge, stripes of exactly the same age could be matched up with one another. So the oceans really were opening, and the continents really were drifting apart. Wild miracle confirmed! Continental drift at long last became a reality.[14]

These explanations happened in the 1960s, and within a few years, "a revolution had occurred in the earth sciences."[15] Even after most geologists accepted it, there were people so mentally fixed they could not ever change. The British mathematician and geophysicist Harold Jeffreys was one of these. He said that continental drift was "out of the question."[16] He said that "the earth probably became solid within 15,000 years of its ejection from the sun." And he refused to accept plate tectonics until the day he died, in 1989.

We like to think we have a lot of knowledge, and we do. But it's healthy to remember that the entire science of geology was a shambles only a handful of decades ago. As the geologist Finn Surlyk said:

> When I started my geology studies in 1962 what we learned above the level of minerals and fossils was absolutely nonsense. The poor teachers did not understand what they were lecturing, but hid their ignorance behind an enormous terminology. All this changed with the theory of plate tectonics.[17]

An idea that is so obvious that it occurs to children was criticized by experts as delirious raving and pseudoscience. But it turned out to be right. Not only that, it today forms the bedrock of the entire field.

Part XIII

Hidden Variables Today

The double slit experiment with electrons was first predicted in 1923 by Louis de Broglie. I repeat. At that time, nobody thought an electron would behave like a wave. De Broglie in his first papers in 1923 predicted electron diffraction and interference, on the basis of a theory where the electron is a particle that is guided by a wave. This is historically most bizarre. In 1915 Einstein predicted the deflection of light by the sun, and in 1919 people observed it. Imagine if that observation had been taken as proof *against* general relativity. Ridiculous. Somehow, de Broglie predicted [electron diffraction] on the basis of a theory with trajectories. When it was observed by Davisson and Germer in 1927 it was hailed by Heisenberg and others as a *proof* of the non-existence of trajectories.

—Antony Valentini[1]

Chapter 45

A Bolt from the Blue

In 1932, the same year that von Neumann's impossibility proof was published, Einstein left Berlin for a tour of the United States. He had a little summer house in Caputh, Brandenburg, and upon departing he said to his wife, Elsa, "Take a very good look at it. You will never see it again."[1] This premonition turned out to be true. The next year Hitler was named chancellor, and Einstein's apartment in Berlin was raided repeatedly in February and March. In April, Jews were banned from holding any public post in Germany. There was hardly a peep of protest. Einstein surrendered his German passport, renouncing his citizenship, and in 1934 took up a post at Princeton's Institute of Advanced Study. He was one of the first appointees to the new institute, along with John von Neumann, Kurt Gödel, and Hermann Weyl. Einstein's Caputh home was converted into a Hitler Youth camp.

Soon after Einstein's appointment, he took on an institute fellow as his assistant. This was Nathan Rosen, a kid from Brooklyn who had studied at MIT during the Great Depression. The Russian physicist Boris Podolsky was also a recent fellow at the institute. Einstein had met him earlier during a visit to Caltech, and knew that he had collaborated with Fock, Dirac, and Landau. Podolsky's appointment as a fellow was supported by Einstein, who wrote to the director of the institute that Podolsky was "one of the most brilliant of the younger men who has worked and published with Dirac."[2]

During 1934, immediately after settling in at Princeton, Einstein, Podolsky, and Rosen wrote a paper together that has become a foundation of modern physics. It is now Einstein's most-cited paper of all time (with at least 22,350 citations at the time of writing), and is so well-known and often-discussed that it is abbreviated by its authors' names as EPR. The paper was titled "Can Quantum-Mechanical Description of Reality Be Considered Complete?" and it came together like this: "Einstein contributed the general point of view and its implications," Rosen did the math, and Podolsky wrote it up. It was four pages long and appeared in *Physical Review* on May 15, 1935.

The paper caused quite a stir, "like a pike in a goldfish pond," Schrödinger said, and he was quite pleased that Einstein "caught dogmatic Copenhagen by the coattails." In Cambridge, Dirac said: "Now we have to start all over again, because Einstein proved that it does not work."[3] Pauli was in Zürich, and he wrote to Heisenberg in Leipzig: "Einstein has once again made a public statement about quantum mechanics....As is well known, that is a disaster whenever it happens."[4] He called Podolsky and Rosen "no good company" and gave the paper a barbed compliment: if a freshman had raised such objections, Pauli would consider the student "quite intelligent and promising." Pauli considered "squandering paper and ink" to explain the "facts" that "cause Einstein special mental troubles," but he did not stoop to the task.[5] (Landsman, probably thinking about Pauli, said that "the immediate response to EPR by the Bohr camp reveals their breathtaking arrogance towards Einstein's critique of quantum theory."[6]) Rosenfeld, Bohr's assistant at the time, called EPR an "onslaught," and said it "came down on us as a bolt from the blue."[7] Bohr immediately dropped everything and worked full-time on a reply, which he finished in record time (six weeks).

What stirred up so much fuss? Because Einstein complained later that "the essential thing was, so to speak, smothered by the [mathematics]" let us explain it in the simpler terms he used in a letter to Schrödinger.[8]

Einstein imagined that a ball was placed in one of two boxes, which are then closed and moved far apart from each other. An observer approaches one of the boxes. Because she does not know which box holds the ball, she would say that there is a fifty-fifty chance that the ball is in either box. If she wrote down a ball-in-the-box theory, she would use probabilities. Now, Einstein says, we have a choice. It is the same choice we have encountered multiple times. He wrote:

> *The probability is 1/2 that the ball is in the first box.* Is this a complete description?
>
> NO: A complete statement is: The ball is (or is not) in the first box. That is how the characterization of the state of affairs must appear in a complete description.
>
> YES: Before I open them, the ball is by no means in one of the two boxes. Being in a definite box only comes about when I lift the covers...
>
> We face similar alternatives when we want to explain the relation of quantum mechanics to reality. With regard to the ball-system, naturally, the [YES answer] is absurd, and the man on the street would only take the first [NO answer] seriously. But the Talmudic philosopher dismisses "reality" as a frightening hobgoblin of the naive mind, and declares that the two conceptions differ only in their mode of expression.

If we insist that the statistical theory is a *complete* description of the system, then we are forced to conclude that before opening the box, the ball was *both* there and not there, in *both* boxes. It was in a "superposition" of thereness, in a "world of potentialities or possibilities rather than one of things or facts,"[9] to use Heisenberg's locution.

Simplifying things like this makes quantum completeness appear blatantly absurd. The formula of "the man on the street" is here an appeal to a generic person who is not blinded by a theory. We saw how belief in miasma led people to do obviously stupid things like fill their drinking water with sewage. Einstein is saying that quantum completeness has had a similar effect.

The takeaway is twofold. First, science should deal with reality, what Einstein called the "real factual situation that exists independent of our observations."[10] Rejecting reality as a frightening hobgoblin on the back of some obscure philosophizing is unacceptable. Second, if we assume completeness, we have a problem with the other box. Something must have happened to the other box the instant the observer opened the first one. If she opened the nearby box and observed it was empty, the ball must have materialized in the distant box, because the probability is now 100 percent that it is there. It suddenly stopped being a "potentiality and possibility" (50 percent there and 50 percent not there) and became a hard fact of reality (100 percent there).

Worse, this needs to occur the moment our observer opens the first box *no matter how far away the second box is*. It could be on the moon, orbiting Alpha Centauri, or lost on the other side of the universe. This sudden "collapse" would still need to happen instantly. Remember, this was the point of Einstein's sole comment during the fifth Solvay conference, in 1927. In a letter to Max Born in 1947, Einstein found the phrase that stuck: *spukhafte Fernwirkung*, spooky action at a distance.

Chapter 46

Entanglement

There is a simple and beautiful experiment with atoms which has become the standard way of approaching these issues. It is due to David Bohm, who recast the original EPR thought experiment in a very useful and suggestive way, while maintaining the essence of Einstein's point.

Remember when we discussed spin, that mysterious quality of atoms and subatomic particles that involves, somehow, an idea of rotation? In 1921, Otto Stern proposed a way to observe this spin: if an atom that is otherwise electrically neutral but has a nonzero spin is passed through a strong, spatially varying magnetic field, it will get deflected in one of two directions. In 1922, Walther Gerlach conducted the experiment with a beam of silver atoms. They struck a glass slide in one of two possible locations, proving that the theoretical prediction of spin was correct. This is called a Stern-Gerlach experiment.

If we arranged the magnet so it deflected the particles either to the left or the right, and then sent two hundred particles through the magnet, we would see something like this:

Bohm imagined that "we have a molecule containing two atoms in a state in which the total spin is zero."[1] This means that no matter what the spin of one atom, the other atom must have equal and opposite spin. The total spin will remain zero even if the molecule breaks apart, assuming there was no torque acting on it, and so if the two atoms fly apart they will still have equal and opposite spin. If we assign a value of +1 to one, we must give the other the value −1.

So now imagine that Alice and Bob are on two different sides of a room, each with their own Stern-Gerlach magnets, catching the two atoms of the broken molecule. We know that each one of these atoms has opposite spin, and with their magnets Alice and Bob can measure it. The orientation of these magnets matters, because each has north and south poles. First, let us assume that Alice and Bob have their magnets oriented in the same way.

In this case, if Alice sees an atom appear on the left (which we will say is −1), Bob will see its twin appear on the right (+1). If Alice sees one on the right, Bob will see one on the left. Without fail. This happens because the two atoms are described by the same wave function, and there is no way of separating the wave function into its constituent parts of the two atoms. In this case, it is said that the two atoms are *entangled*, a term Schrödinger introduced in the same paper that debuted his cat.

Forgetting experimental artifacts created by the messy real world, the entangled measurements are *always* opposite, and this pattern is repeated without exception.

$$+ + - + - - + - - + + + - - -$$
$$- - + - + + - + + - - - + + +$$

And their two experimental setups will look like this, where I have written ☆ to indicate the location of the last observed particle, and ▷ to indicate the orientation of the magnet:

Alice ▷   Bob ▷

We know what will happen if we change the orientation of one of the magnets. Let's say Bob begins rotating his magnet so that it is no longer perfectly aligned with Alice's. At first, very little will change. The measurements will be almost perfectly anticorrelated, as before, but not quite perfect. Very, very rarely, maybe once in a thousand trials, Alice and Bob will measure the same thing.

As Bob changes his angle more, the frequency of $++$ and $--$ events starts going up. Eventually, when Bob's magnet is at a right angle to Alice's there will be an equal number of same and different events. There is no discernible connection between what Alice sees and what Bob sees; it really is random, like two people flipping different coins in the same room.

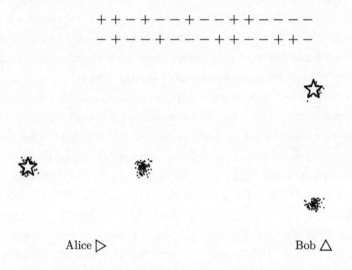

$$+ + - + - - + - - + + - - - - -$$
$$- + - - + - - - + + - - + + -$$

Alice ▷   Bob △

If Bob keeps rotating in the same direction, past a right angle, the number of same events starts to increase. We will still get some cases where the outcomes are opposite, $+ -$ and $- +$ still happen, but they are getting rarer and rarer. Eventually, when Bob's magnet has switched orientations completely by 180 degrees, the two sequences are identical. We now *never* see opposite things happening.

$$+ + - + - - + - - + + - + - -$$
$$+ + - + - - + - - + + - + - -$$

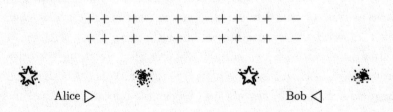

Alice ▷   Bob ◁

Let us stop for a moment and just think about this. The molecule is a physical thing; the atoms that make it up are physical, as is the magnetic field. When the atom passes through the field, its spin generates motion that is manifested as a physical distance between the left and right. So clearly the spin is physical too.

The science of physics is about understanding exactly this sort of thing. The two atoms came from the same place, and they have opposite spins. We can prove this. So something physical must be going on. Quantum theory only tells us that we have a fifty-fifty chance for any single atom to appear either left or right, and quantum completeness insists that there is nothing more to say, *ever*, because the universe is fundamentally unpredictable. But actually, it is not so unpredictable, is it? Because when their magnets are perfectly (anti-)aligned, knowing Alice's observation means we also know Bob's. As Bell said, these sorts of correlations "almost seem to cry out for a hidden variable interpretation."[2]

It is now the smallest step to think that there *really is* a dynamical fact of the matter, and that if we knew it, we would be able to predict exactly where each atom would wind up, right or left, well before Alice made her observation. Indeed, we may even be able to make the right prediction before the atom split apart. This is the classical intuition, favored by those who believe that quantum mechanics is incomplete. It's just like a coin flip. We know that a coin lands heads or tails because of how it was thrown, how it hit the air currents, and how it landed. We can't predict heads or tails, but that is just because we don't know all the variables. Why should we say that this entangled spin experiment is any different?

As we can see, this setup has the same logic as Einstein's box. Alice has a fifty-fifty chance of seeing the electron go in one direction or the other. We can ask: Is this a complete description of the situation? If we answer NO, we are allowed to search for a deeper explanation for the link between Alice's and Bob's observations. We can say the atom has a complex structure that is changing in time in some way; its spin is likewise a complex structure. Maybe there are other complex structures as well, or concepts we have yet to even imagine are at play. We don't know what they are, but we know that they must be there.

If we answer YES, we cannot say any of these things because the probability statement is a complete description. When Alice's and Bob's magnets are aligned, they always observe the opposite results. But how can two purely random events be so perfectly in sync? One must know about the other one *instantaneously* upon measurement, just as one box must know instantaneously about whether the ball is in the other one. It's spooky action at a distance.

For Einstein, this is a strong argument in favor of incompleteness because physics simply has no place for such magic. The theories of gravitational and electric forces were originally thought to operate in this way, but then scientists discovered fields, and understood that all gravitational and electromagnetic influences are passed progressively through space. Wave phenomena are a perfect example of this idea. When you throw a rock in a pond, the ripples spread out from where the rock fell. As they propagate, they move out from where they are now, and proceed forward into the space they are touching. This is because physical events are happening that propagate the energy from one tiny portion of space to the neighboring ones. The old-timers used to call this a contact process. The term today is *locality*.

This was the point that was "smothered" in the mathematics of the paper; it emerged only in the penultimate paragraph. If the reality of a particle's momentum far away depends on a measurement we make nearby, then there is spooky action at a distance. As EPR put it: "No reasonable definition of reality could be expected to permit this."[3] Einstein was forcing everyone onto the horns of a dilemma. Take your pick: admit that either quantum mechanics is incomplete or physical reality is nonlocal.

Bell was always on the lookout for examples that could illuminate the driving idea behind hidden variables. He imagined that the faces of a coin were separated and each half slipped into the pockets of two different people. No matter how far away the two people are, he said, "The first to look, finding that he has head or tail, will know immediately what the other will subsequently find. Are the quantum mechanical correlations any different? Indeed they are not, according to Einstein."[4]

Another of his favorite examples was a case that led to the foundation for the Institute for the Study of Twins (in the Twin Cities, of all places). Two identical twins, James Arthur Springer and James Edward Lewis, were separated as babies. They lived thirty-nine years ignorant of each other, and reunited at forty; they were astounded by a list of coincidences. As children, they both had dogs named Toy. At school they both loved math and hated spelling. At the age of eighteen, they both started getting muscle-tension headaches which they described with identical language. They both liked block lettering and mechanical drawing, and both worked with wood—one made little tables and the other made little rocking chairs. Both had worked part-time as deputy sheriffs. They both bit their nails and smoked the same brand of cigarette. They both bought the same model of car in the same color. They vacationed at the same three-block-long beach near St. Petersburg, Florida. They had both married on the same day and both had ex-wives named Linda, and current wives named Betty. One had a son named James Alan, the other had a son named James Allen.[5]

In his talks, Bell showed a picture of these twins. The correlations in this case, while marvelous, are not mystical. They can be explained in a hand-waving sort of way by appeal to the identical genetic material. Should not the entangled atoms that go flying toward Alice's and Bob's magnets not have identical "genes" as well? Who is Bohr, or anyone, to say they don't? This is how Bell conceived of the matter:

> The observed perfect quantum correlations seem to demand something like the "genetic" hypothesis. For me, it is so reasonable to assume that the [particles] in those experiments carry with them programs, which have been correlated in advance, telling them how to behave. This is so rational that I think that when Einstein saw that, and the others refused to see it, he was the rational man....I feel that Einstein's intellectual superiority over Bohr, in this instance, was enormous; a vast gulf between the man who saw clearly what was needed, and the obscurantist.[6]

Now, the way this quote ends is surprising. Bell says: "So for me, it is a pity that Einstein's idea doesn't work. The reasonable thing just doesn't work." He said this because of a result that he obtained and which bears his name. It is called Bell's theorem. The philosopher Tim Maudlin called it "the most astonishing thing in the history of physics,"[7] and a recent paper in *Reviews of Modern Physics* said that it "ranks among the most profound scientific discoveries ever made."[8]

Chapter 47

Bell's Theorem

The primary conceptual element of Bell's theorem is a statistical correlation, so let us spend a moment understanding what correlations are. Imagine that you have a sequence of numbers, for example, a coin flip where every heads represents +1 and every tails is −1. Alice is standing there flipping a coin every five seconds, and marking the number down on a piece of paper. Bob is also in the room with a coin, doing the exact same thing.

A correlation is a number that expresses the degree of relatedness or sameness between the two lists that Alice and Bob create. Assume that, for some odd reason, Alice and Bob always got the same result when they flipped their two coins. Alice: heads, Bob: heads. Alice: tails, Bob: tails. When we brought these two sequences of numbers together and compared them, their correlation would be 1. They are 100 percent correlated; they are the same sequences.

The way we compute this number from the actual sequences is to multiply each pair of results, and then take the average. Because Alice and Bob always get the same thing, their results are both either +1 or −1. Multiplying the two results together always results in +1. If they got different results, we would get −1 when we multiplied them together. Now, let's say Alice and Bob flip their coins N times. We will let S be the number of times their results are the same, and D be the number of times their results are different. Note that always $S + D = N$. The correlation is defined like this:

$$\text{Correlation} = \frac{S - D}{N}$$

If Alice and Bob always get the same result, then $D = 0$. That means $S = N$, and so the correlation is $N/N = 1$. If Bob got the exact opposite of Alice every single time, their sequences would be the reverse of each other. Everywhere Alice wrote $+1$, Bob wrote -1, and vice versa. In this case, the situation is reversed: $S = 0$ and $D = N$. Then we have $-N/N = -1$. The two sequences are perfectly anticorrelated.

In reality, if we actually did this experiment and Alice and Bob both have fair coins, the correlation will be close to zero. This is because the two coin flips have no relationship to each other. Sometimes they are the same, sometimes different. The more trials we do, the more the same and different results wash out. In other words, both S and D will tend to $N/2$. They might not be exactly this number, so maybe $S - D$ is not exactly zero, but it will be small. Because N is large, we are dividing a small number by a large number, which makes the small number even smaller. The larger N gets, the closer to 0 the correlation will be.

Beyond correlations, Bell's theorem also assumes the existence of hidden variables, but it does not assume anything specific about them. The hidden variables must be mathematical magnitudes of some sort, so we can do calculus with them, but they could be complicated continuous functions, or a single fixed number, or anything in between. These variables could be something like position and momentum, or "something more exotic, perhaps some state space as yet unthought of."[1]

Beyond this, Bell uses three main assumptions. They go by various names, but we will call them *locality*, *reality*, and *freedom*. These terms are actually just convenient handles for precise mathematical statements, and we will explain each one in turn.

Imagine randomly selecting a single human being out of the roughly eight billion people in the world today. What is the chance that they have blue eyes? I'm going to make up some okay-sounding numbers. Say it is about 9 percent. What is the chance they have green eyes? About 2 percent. Brown eyes? Roughly 75 percent or so. We can do the same thing with hair color. What is the chance our random human has dark brown or black hair? It's high, maybe 84 percent or so. Medium brown hair? About 11 percent. Blond and red hair only make up about 3 percent and 2 percent, respectively.

After picking our random human, we can ask: What is the probability that they have *both* brown hair and brown eyes, or brown hair and blue eyes, or...? You can go through all the combinations. You can combine these with any other random trait you can think of. For example, what is the probability they wear glasses? Say this is roughly 66 percent.

If you select someone and you notice that they have black hair, what is the probability that they wear glasses? Well, it is still 66 percent because hair color has no influence on glasses wearing. The two of them are utterly indifferent to each other. When this happens for any two factors we might want to make probabilities out of, we call those factors *statistically independent.*

It's even more informative to consider cases that are not statistically independent. Noticing our random human has black hair, we can ask: What is the probability they have blue eyes? Sure, picking someone at random from the entire population of Earth gives us a 9 percent chance of getting blue eyes, but if we first notice the person we picked has black hair, it is ridiculous to say that they still have a 9 percent chance of having blue eyes. This is even easier to see considering skin color. Do 9 percent of people with black or very dark skin have blue eyes? This does happen, but much, much less than 1 percent of the time; it is extremely rare.

Skin, hair, and eye color are all connected, because all three depend on the amount of melanin that a person has in their body. Therefore, these three factors are *not* statistically independent. They tend to go together.

This observation allows us to express Bell's first assumption of locality. We imagine the Stern-Gerlach experiment with entangled atoms, as before. Alice is on one side of the room (or the planet), and Bob is on the other. Each one of them has their own magnet orientation which they are free to choose any way they like. Finally, we have some hidden variables. We don't know what they are, but we imagine that Einstein was right and so the atom is specified by a rich, hidden, deterministic system. If we knew the hidden variables beforehand, we would be able to predict with accuracy what Alice and Bob will measure, for any one of their angles.

The assumption of locality says this: The probability of Alice seeing +1 depends on the hidden variables in the atom, and Alice's magnet angle. The probability of Bob seeing +1 depends on the hidden variables in the atom, and Bob's magnet angle. What Alice sees does *not* depend on Bob's measurement setting or the hidden variables in his atom, and vice versa for Bob. What Bob sees has nothing to do with Alice's magnet or her atom.

In other words, these two probabilities are statistically independent. They are like hair color and glasses wearing. One does not depend on the other. At first, both atoms are in contact, but after the molecule separates and they fly toward their respective magnets, they are on their own. They might change their state in time, and the magnet might alter that state again, but whatever happens to one atom has nothing to do with the other atom, which is far away. This is the expression for locality in statistical terms. It is the first assumption of Bell's theorem.

Next comes reality, which is also called counterfactual definiteness. Imagine we have a pair of identical twins, like James Springer and James Lewis. We allow Alice to ask one question of Springer, and Bob to ask one of Lewis. Let's say that Alice asks about hair color and Bob asks about eye color. Alice's answer is "brown," and Bob's answer is "blue." Alice did not ask about eye color, and she is not allowed to ask another question. But these are identical twins, so we have a strong intuition that *if* Alice had asked Springer about eye color, the answer would have been blue.

A counterfactual is a hypothetical way things could have gone. When we assume counterfactual definiteness, we are assuming that the state of reality supports clear answers to a whole set of questions, even if we are only allowed to ask one, and only one, question. We *could have* asked a different question, and the answer would have reflected the real properties of the thing we are querying.

This matters because when we send a single pair of entangled atoms toward Alice's and Bob's magnets, the experiment needs to be set up in a particular way. The atom's path will depend on the magnet angle, and we cannot rewind time to test what would have happened to the atom if one of the angles had been different. Nor can we set up exactly the same microstate in the next molecule, to essentially run the same experiment with a different magnet angle.

Finally, there is the assumption of freedom. It has also been called no-conspiracy. When Alice and Bob are asking questions of a stream of identical twins, we assume that there is no hidden connection between the questions they actually do ask, and the genetic code of the twins. Alice and Bob are assumed to have the free ability to choose their own questions, regardless of anything else. Equivalently, their choices of what to measure are independent of the things they are measuring.

Without this assumption, you can get some weird things going on. For example, imagine that whenever any twins have blue eyes, Alice and Bob ask about eye color. This in itself will skew the statistics; they will think that blue eyes are more common than they actually are. But if Alice and Bob are influenced to *never* ask about eye color unless the eyes are blue, they will think that every twin has blue eyes.

There is some debate back and forth about this. Do Alice and Bob have free will, really? The view that tries to escape Bell's results by denying this assumption of "freedom" is called *superdeterminism*, which is determinism with some extra assumptions. But invoking superdeterminism to explain statistics is desperate and strange, not least because it's something we could do for every single scientific result we did not like. Imagine someone who said that the negative effects of vaccines were drastically underreported, because seemingly random events yesterday, twenty years ago, or at the beginning of the universe were not actually random—they conspired to assure only people resistant to the side effects of vaccines reported for the vaccine trials. As Shimony, Horne, and Clauser put it: "Unless we proceed under the assumption that hidden conspiracies of this sort do not occur, we have abandoned in advance the whole enterprise of discovering the laws of nature by experimentation."[2]

All three of these assumptions appear to be very commonsensical. Denying any one of them is rather desperate and involves us in some extravagant ideas about the universe. Bell's great contribution was to show how the correlations of entangled spins must behave under these assumptions. Crucially, the quantum theory predicts something different.

This is not the place for the actual mathematical derivation, of course. But I will write down the inequality itself and explain it, because it will come up again. We consider three different measurement angles, and assume Alice and Bob can choose them freely. These angles will be x, y, and z.

If Alice chooses x and Bob chooses z, and then we run the experiment for a long time, they will see some correlation in their measurements. As we know, it depends on the angle between x and z and will vary between −1 and 1. Similarly if Alice chooses y and Bob chooses z, and so on. We will write $C(x, y)$ for the correlations observed, given choices of two measurement angles x and y.

So, here is Bell's inequality, which follows as a matter of logic from the assumptions above:

$$|C(x, y) - C(y, z)| \leq 1 - C(x, z)$$

The symbol $|x|$ means the "absolute value of x," which makes x positive if it is negative, and does nothing otherwise. So for example, $|-2| = |+2| = +2$. The math is saying that correlations generated by local realistic variables must be of a certain type; they must never break the bounds set by this inequality.

Quantum mechanics predicts the correlations will break the inequality. So quantum mechanics is in direct opposition to at least one of the three commonsense assumptions we made:

Locality

Reality

Freedom

Chapter 48

It's as If Reality ... Didn't Exist

Bell's original version of his theorem was published in 1964 in a new and little-known journal called *Physics, Physique, Fizika*. A few years later, in the library of the Goddard Institute for Space Studies, a part of Columbia University, this unusual title caught the eye of a young PhD student named John Clauser. He was interested in quantum foundations even though, as we have already seen, "nearly every physicist of the time learned to avoid such questions as part of their training."[1] He read Bell's paper, and recognized its importance immediately:

> At first I didn't believe it. I went through a cycle of trying to find a counter-example, trying to prove it, trying to refute it. Then one day I realized there was nothing wrong. Jesus Christ, I thought, this is a very important paper. Why had no one else picked up on it?[2]

Clauser was sympathetic to the hidden variables program. He found Einstein's arguments "much more persuasive than Bohr's ... a perfectly logical solution to the problem. By holding that opinion," he said, "I was certainly branded as a heretic by many, and undoubtedly as a quack by others."[3] Against the recommendation of his PhD supervisor,

359

Clauser wrote to Bell asking if anyone had undertaken the experiment suggested by his theorem.

An *experimentum crucis* is an experiment that settles a dispute between two options. We have already seen some of these in our survey of the past. Newton's experiment with the prism showed colors resided in the very properties of light, and Arago's experiment of the speed of light in dense media, done by Fizeau and Foucault, settled the particle-wave debate definitively against Newton and for Fresnel (at least until 1905). The experiment suggested by Bell's theorem seems to be of this kind. There is no wonder that its importance has only grown over the last sixty years.

The history of the experiments to test Bell's inequalities is well-known, and well told in several places.[4] Compressed into a single paragraph, it goes like this:

Clauser teamed up with several others who were working on the project at the same time, and in 1969 Clauser, Horne, Shimony, and Holt published a proposal to test Bell's inequalities. In 1972, this experiment was actually done and the work published using light and polarization analyses. Quantum mechanics was confirmed by six standard deviations. This means there was about a two in a billion chance of the result being a random fluke. The experiments were not fully convincing, however, because a variety of loopholes existed which could allow defenders of local realism to explain the results. For example, the number of detection events, as compared to the total number of events, was low, possibly skewing the statistics. The detector settings could have influenced each other in an unknown way, invalidating Bell's assumption of locality, and so on. A crucial advance was made by Alain Aspect and his colleagues, who did an experiment in 1982 where the analyzer was randomly selected faster than any signal could get from Alice to Bob. (When Aspect approached Bell to get his opinion on this experiment, Bell first asked if Aspect had a permanent position because this kind of research was still unfashionable and risky.) Following on for almost twenty years, an increasing number of sophisticated experiments were made, closing one loophole after another. Measuring entanglement correlations became standard fare, and in the process, the quantum information age was born. This has been called the second quantum revolution, largely associated with John Bell. This work today is at the cutting edge of quantum computing, and entanglement is now recognized as a physical resource, like wind or fire, which can be exploited to accomplish a variety of tasks.

So what are we to conclude from the experimental violation, repeatedly demonstrated in progressively improved experiments to the general satisfaction of nearly everybody? Bell listed four possibilities:

1. The inefficiencies of the counter, and so on, are essential. Quantum mechanics will fail in sufficiently critical experiments.
2. There are influences going faster than light, even if we cannot control them for practical telegraphy. Einstein's local causality fails, and we must live with this.
3. Einstein's local causality can survive. But apparently separate parts of the world become deeply entangled, and our apparent free will is entangled with them.
4. The whole analysis can be ignored. The lesson of quantum mechanics is not to look behind the predictions of the formalism. As for the correlations, well, that's quantum mechanics...

As the experiments improved, option 1 became less and less tenable. It has now been fifty years of Bell tests, with an increasing number of "loophole-free" tests accomplished in the past decade. At some point it just becomes irrational to insist on experimental error. Furthermore, the actual ways out provided by the loopholes (conspiratorial skewing of statistics, weird backward signaling) were never particularly compelling anyway.

The physics establishment has, in fact, chosen option 4. The common view is that "Bell disproved hidden variables." They don't say we now know that the deeper theory is nonlocal; or we now know that free will doesn't exist and what we chose to measure and what we are measuring depend on each other. They say there are no hidden variables. Of course, this is just the Copenhagen interpretation, easily adopted because it was already the dominant view in 1927. I will give a few examples.

In October 2022 the Nobel Prize was awarded in the field of quantum foundations. It was shared between Alain Aspect, John F. Clauser, and Anton Zeilinger, "for experiments with entangled photons, establishing the violation of Bell inequalities and pioneering quantum information science."[5]

The citation went into further detail about why we should care about the Bell inequalities. According to the Nobel committee:

> In the 1960s, John Stewart Bell developed the mathematical inequality that is named after him. This states that if there are hidden variables, the correlation between the results of a large number of measurements will never exceed a certain value.[6]

Because quantum theory predicts the correlation exceeds this certain value, and the many experiments have vindicated the quantum prediction, the conclusion is inescapable: hidden variables do not exist.

The award was predictably followed by a blizzard of popular science explainers, both articles and videos, informing us that "the universe is not real," and asking, "Has quantum mechanics proved that reality does not exist?" Mermin, in his own explainer, preempted these comments by forty years and flung out the zinger I have quoted multiple times already: "We now know that the moon is demonstrably not there when nobody looks."[7] All this is very much in line with several popular science books published recently. In one notable example, a pair of grown men, not teenagers into solipsism, sit on a beach and breathlessly wonder: "Can we believe this? It's as if reality...didn't exist."[8]

Another example. A blog post by a former Harvard physics professor makes similar assumptions. It has since been taken down, but I copy-pasted it years ago just in case that happened. The relevant passages are popcorn-worthy:

> There's no language "deeper than quantum mechanics" that could be used to interpret quantum mechanics. Unfortunately, what the "interpretation of quantum mechanics" ends up with is an attempt to find a hypothetical "deeper classical description" underneath the basic wheels and gears of quantum mechanics. But there's demonstrably none. . . .
>
> So the people who are constantly distracted by some instinctive ideas that the world has to be realist or deterministic or

> classical...are politely asked to stop reading this article and
> return to their everyday activities involving their garden or
> lunch....
>
> Now, when the obnoxious medieval bigots are gone, we may
> finally talk about some serious physics.[9]

There is "demonstrably" no language deeper than quantum physics,
that is, no hidden variables, because of Bell's theorem. Anyone who
would say otherwise is an "obnoxious medieval bigot."

Finally, here is an example pitched between the Olympian heights
of the Swedish Academy of Sciences and the street cant of scientific
cutpurses: Allan Adams, in his very first lecture teaching quantum mechanics to first-year undergrads at MIT, says this:

> The first natural move is: Oh, look, surely there's some addi-
> tional property of the electron that we just haven't measured
> yet to determine whether an electron comes out [spin up or
> down]....And we will see later on, using Bell's inequality, we
> can pretty much nail that such things don't exist. But this
> tells us something really disturbing....There is something
> intrinsically unpredictable, non-deterministic, and random
> about physical processes.[10]

As we can see, the scientific establishment has chosen Bell's option 4:
The lesson of quantum mechanics is not to look behind the predictions
of the formalism. Yet the majority view is only one of (at least) four options. As Maudlin remarks, the common chant "Sorry, Einstein, God
plays dice, Bell disproved hidden variables," displays "complete incom-
prehension of what Bell did, what Einstein thought, what the situation
with respect to deterministic 'hidden variables' theories is."[11]

The Nobel announcement is a bit puzzling for anyone even moder-
ately conversant with the quantum foundations literature. What about
Bell's three other possibilities? However, a remark by Howard Wiseman
frames the issue in a way that explains somewhat why so many just
blithely state that hidden variables do not exist:

As Bell proved in 1964, this leaves two options for the nature of reality. The first is that reality is irreducibly random, meaning that there are no hidden variables that "determine the results of individual measurements." The second option is that reality is "non-local," meaning that "the setting of one measuring device can influence the reading of another instrument, however remote."

Most physicists are localists: they recognize the two options but choose the first, because hidden variables are, by definition, empirically inaccessible.[12]

It is certainly wrong to claim hidden variables are "by definition" empirically inaccessible. This is a misunderstanding I addressed earlier in the book. But the comment is still revealing: the majority of physicists opt for "no hidden variables" because they find nonlocal interactions even worse, or they believe the hidden variables can never be observed and so might as well not exist.

Those with a deeper understanding say Bell's theorem disproves "local realism." Then, we can choose which of these we discard. The Nobel committee, and most physicists today, reject realism. They follow Bohr and the old Copenhagen view. But one could in fact choose to reject locality. Indeed, nonlocality is constantly discussed in the foundation literature, and "a wide range of concepts and technical tools have been developed for describing and studying the nonlocality of quantum theory."[13]

Those who go this way choose Bell's option 2, and say that Bell's theorem proves the existence of "spooky action at a distance." This view is supported by Bohmians because, as we saw, Bohm's 1952 pilot-wave theory is explicitly nonlocal. Bell tests are very frequently seen as support of Bohmian mechanics. Putnam, Beller, Bricmont, and many others, including Bell himself, have considered quantum nonlocality to be a prediction of Bohm's theory.

Either way you slice it, though, the universe must be weird. Either Heisenberg and Bohr were right about objective reality not existing, or objective reality does exist but it operates by "telepathic methods" that reach across arbitrary regions of space, across the solar system, the galaxy, the entire universe, instantaneously. And so everywhere you look, you read sentences like this one:

> The evidence is now overwhelming that Einstein's program to "complete" the quantum theory with a local deterministic theory was misguided. Local realism simply does not work.[14]

And yet, leaving the story here would be to fundamentally misrepresent the situation.

Chapter 49

The Fourth Assumption

Yes, the majority believes that there is "demonstrably" no deeper theory than quantum mechanics. A smaller group believes that nonlocal interactions are proven, and that we should all be Bohmians. But there are many other smaller and smaller viewpoints splintering off from here, and no solid consensus exists. Take, for example, the following comment:

> Bell's inequality really beats all records on publications, citations, discussions and controversies. As an organizer of a few international conferences on foundations of quantum theory, I was really disappointed by the stormy debates on Bell's inequality. I could not find any problem in quantum foundations which can be compared with Bell's inequality by intensiveness of discussions and strength of reactions to opponent's views.[1]

I have never seen a book of this kind delve into the controversy, probably because no popular science writer wants to meddle with the freak show of the literature on this subject. It is so much safer and easier to take the current majority view and report that to the public. But this book is about dissent, about outsiders, and about growth in human knowledge, sometimes even at the expense of the majority. So, damn the torpedoes, full speed ahead!

Some of the back-and-forth between the majority and minority resembles a street fight, only not with fists and knives but with Clifford algebra and Bayesian inference. For example, Richard Gill, an emeritus professor of statistics who got interested in this area, laments the ability of Joy Christian to publish his work in Royal Society journals:

> Editors of the serious journals to which [Christian] submits such papers should immediately realise that they are either dealing with an almost inconceivably revolutionary genius, or with a (pardon the expression) ordinary crackpot....Does [the Royal Society] want to encourage serious debate on foundational issues in science, or does it just want to get internet clicks by publishing outrageous claims, despite the down-side that it thereby clouds serious debate by publishing superfluous noise? *O tempora, o mores!*[2]

This is an eerie echo of Henry Brougham's critique of Thomas Young. It has all the same features: the majority talking down to the minority in a haughty tone, directly addressing the Royal Society and upbraiding it for low standards, complete with Latin exclamation at the end. Recall Brougham's was *Proh, pudor!* For shame!

Of course, there is little similarity between Young and Christian, and the fact that Gill sounds like Brougham and Brougham was wrong does not make Gill wrong. I am not siding with either Gill or Christian here. The point is that there is a very real debate going on. While some of those criticizing Bell's theorem surely are crackpots, it won't do to discard all criticisms as the merest crackpottery. We know that science enters eddies from time to time where it gets trapped in its assumptions. It could be that the majority view is just wrong.

New ideas need to come along from somewhere, after all, and the truth does not care about what university conferred your degree or even if you have a degree. Like a magic bird, it will select the person who listens to its song, even if they are a total outsider whose job is building roads (I'm looking at you, Fresnel). We should pay close attention to all of the critiques of Bell's theorem, at least once, because picking between reality and locality feels like a trick question and a lose-lose proposition. As one physicist put it, letting go of either one results in

a major barrier to simple views on the reality underlying
quantum theory and a rather hopeless starting point for
attempts to improve on quantum theory by formulating a
subquantum theory....Before giving [local realism] up, we
should really have not any other option.[3]

In short, if science has prematurely closed the door on hidden variables *again*, this is absolutely crucial to figure out. I say again because,
as we saw, this already happened before with von Neumann's theorem.
Now we have another theorem that is the subject of considerable debate. There are maybe a hundred papers that say Bell's theorem has a
flaw. They have titles like these:

"Bell's 'Theorem,' Is It the Biggest Mistake in the History of
Science?"
"Bell's Theorem Does Not Eliminate Fully Causal Hidden
Variables"
"Where Bell Went Wrong"
"Disproof of Bell's Theorem by Clifford Algebra Valued Lo-
cal Variables"

Because of the technical nature of the subject, I cannot cover even
a small selection of these critiques in any respectable way. Instead, I
have chosen just one, based on the following criteria: (1) maintenance
of Einstein's program; (2) generation of more clarity; (3) support by a
reasonably large community of professional scientists; (4) plausible from
a classical viewpoint.

Any one of these is easy to break. For example, we can get out
of Bell's theorem by assuming that there are faster-than-light signals,
but it breaks my last two criteria. Superdeterminism can save us from
Bell's theorem, but it also breaks criteria two and three. We can dic-
tate that all probabilities encode knowledge and none encode causal
facts, but this breaks criterion 2—it does not generate more clarity into
quantum mechanics. Finally, we can find dense mathematical examples
that are short on physical reasoning. These are disqualified because
they generate anti-clarity, and regarding such papers I wish to quote an
observation:

Harold Jeffreys likened this to trying to strengthen a building by anchoring steel beams into plaster. An argument which makes it clear intuitively *why* a result is correct is actually more trustworthy, and more likely of a permanent place in science, than is one that makes a great overt show of mathematical rigor unaccompanied by understanding.[4]

So far as I can tell, the choice I have made is actually the only one that satisfies all four of the above criteria. Having been transparent about why I selected this one, let us now turn to the critique of Bell that centers on something called *compatibility*.

Alice and Bob really deserve their own romantic comedy. They've been doing experiments so long together, I'm sure the air between them is practically crackling.

For anyone who doesn't know, Alice and Bob are, of course, colleagues in the same physics department. They have worked together for a long time, and something seems to have changed recently between them. Ask the head of the department, the postdocs—even the undergrads have noticed it.

Bob is a creature of habit, and there is a fifty-fifty chance that he will be in the café, or in the lab. Whenever someone goes looking for him, it's assured that he will be in one of these two places, with equal probability. But these days everyone is talking about how Bob and Alice are always together. If Bob is in the café, then Alice is too. And if Bob is in the lab, sure enough, Alice is there working beside him. Their locations are perfectly correlated. Oh là là!

This goes on just long enough for the rumors and giggles to start, but as everyone knows, in a good romantic comedy love does not come easy. Therefore, something must throw a spanner in Alice's and Bob's happiness. This something is a some*one*, another professor, Carol, who is very appealing. She can invert 9×9 matrices in her head, and the heat flux through a perfect sphere enclosing her is practically infinite. Bob—can we forgive him?—changes his mind. He doesn't change his habits, of course. He is either in the café or the lab, but now whenever we see him in the café, Carol is sipping her latte too. And whenever Bob is working in the lab, sure enough, Carol is by his side.

Nobody is giggling anymore; everyone is watching these developments with rapt seriousness. It might not surprise you to hear that, these days, if we find Carol in the café then Alice will definitely be in the lab, and whenever Carol is working in the lab, Alice is frowning at a book in the café. Their locations are perfectly anti-correlated. Storm clouds are on the horizon!

This little drama would be understood by crows, but it reveals something surprising. Imagine that one of the graduate students has decided to examine the probability structure of Alice's, Bob's, and Carol's locations. Seeing a person in a place will be an event, and the graduate student is interested in how often two events occur together. Writing P(Alice café, Bob café) for the probability that Bob is in the café and Alice is in the café, the graduate student creates the following table:

$$P(\text{Alice café, Bob café}) = P(\text{Alice lab, Bob lab}) = 50\%$$

$$P(\text{Bob café, Carol café}) = P(\text{Bob lab, Carol lab}) = 50\%$$

$$P(\text{Alice café, Carol lab}) = P(\text{Alice lab, Carol café}) = 50\%$$

But now, the graduate student is scratching his head. The table makes no sense. Assume that Bob is to be found in the café. Well, by the first line of probabilities, we know that Alice is also in the café. By the second line of probabilities, we know that Carol is also in the café. But the third line tells us that if Alice is in the café, then Carol is in the lab and vice versa. So there is a contradiction. What has happened?

Anybody on the street can solve this problem. A child could solve it. The graduate student has foolishly assumed that the three pairs of probabilities all happened at the same time. In fact, for this drama to even make sense, we know that at least the first two lines need to have happened at *different* times.

This is an intuitive way of explaining an important idea in probability theory. Recall how, in Chapter 13, we defined a sample space as the collection of all events that can happen. Once this is defined, we count all the different ways all the events can happen, and we get a probability distribution. However, when we compare probabilities, or collect things into a distribution, we are assuming that the things we collect do not vary in some crucial way that messes up the statistics. Usually this is a safe assumption.

Going the *other* way is more problematic, as these examples show us. Given a collection of probabilities with dependency relations between the random events, it is not always assured that they can be stitched together into a single sample space, from which we can make an all-encompassing distribution. When we can do it, the individual probabilities are said to be *compatible*. Otherwise they are incompatible. The mistake the graduate student made was trying to combine three incompatible probabilities into a single space.

This is an abstract idea, but it is also simple once you come to grips with it. Imagine that you have three random variables, x, y, z, each obeying a bell curve distribution, and you make a three-dimensional probability distribution out of them. You can imagine the distribution of all three together as a spherical cloud that is dense at the center and transparent at the fringes. If you fix a single variable, say $x = 0$, this is like selecting a two-dimensional slice of the y-z distribution from the cloud. This slice of the spherical cloud will be a circular cloud, dense in the center and transparent at the perimeter. Slices like this are called *marginal probability distributions*. There are many such slices you could select, and the appearance of each one will depend on the original shape of the cloud, and where you actually decided to slice. Now, forget about this cloud, and imagine that I gave you a box with a bunch of slices in it, all jumbled up. There are x-y, y-z, and x-z slices. I tell you: put these slices together into a cloud, or even multiple clouds. With enough time you might be able to do this, but it also might be an impossible task. You can see this with only two slices that intersect: What if they do not have the same value at the intersection point? In other words, if the slices do not satisfy the right conditions of mutual dependence, they can never be assembled into a cloud that does not contradict itself. They are incompatible.

The first person to notice this feature of probabilities was one of the patron saints of the modern age, a man named George Boole.

Boole's influence has been felt by anyone who has interacted with an information-processing device. In other words, we live today in the world Boole made possible. He was English, born in Lincolnshire in 1815, and like Faraday, Green, and only a few others in the annals of science, he was largely self-taught. He became a teacher at the age of sixteen, supporting both his parents and three younger siblings. At the age of thirty-four he was named professor of mathematics at Queen's College (now University College Cork), in Ireland. Boole was interested in fundamental mathematical and logical questions, and he is best known for his book *The Laws of Thought*, published in 1854. This book lays the foundation of the entire treatment of modern logic, which gives the rules for building up statements out of connectives like AND, OR, NOT, and so on, and for evaluating the truth of logical expressions by using 1 and 0 to mean true and false. As we know today, these ideas allow infinite extension. They form the basis of logic gates, which are the building blocks of computers.

Boole was also interested in probability, and the latter half of *The Laws of Thought* was devoted to this subject. Thinking through any issue from first principles reveals surprising problems that cannot be anticipated at the outset, and Boole encountered one of these. It turns out that if you are considering multiple events together, their probabilities need to have certain relationships in order to even be possible.

This is the way Boole put this question before the Royal Society in 1862 (I have changed his variables x to "Bob in the café" and so forth):

> *Problem.* Given that the probability of the concurrence of Bob and Alice being in the café is $C(\text{Bob, Alice})$, of Bob and Carol being in the café, $C(\text{Bob, Carol})$, and of Alice and Carol being in the café, $C(\text{Alice, Carol})$. Required the conditions to which [the concurrences] must be subject in order that the above data may be consistent with a possible experience.[5]

Boole talked about "conditions of possible experience" because, as he saw it, if the events did not satisfy these conditions, they could not exist in the real world and therefore could not be experienced. By the "probability of the concurrence," Boole meant the probability that when one happens, the other happens. This is our notion of correlation, and under suitable conditions we can translate back and forth between our probabilities and correlations. We wrote P(Bob, Alice) and so on before, but it's easy to translate to correlations. We already know that the professor locations are perfectly (anti)correlated:

$$C(\text{Bob, Alice}) = C(\text{Bob, Carol}) = +1$$

$$C(\text{Alice, Carol}) = -1$$

Now, immediately after posing his problem, Boole solved it. He found that three correlations were needed to satisfy three inequalities, called Boole's inequalities.

One of them is

$$C(\text{Alice, Carol}) \geq C(\text{Bob, Alice}) + C(\text{Bob, Carol}) - 1.$$

Plugging in the specific values for our problem, we get $-1 \geq +1$, which is false. Boole's inequality is violated. Thus he would say that the joint probabilities violate the conditions of a possible experience. The example he gave the Royal Society, slightly modified, goes like this.

Imagine that John Snow was investigating the Soho cholera outbreak, and he found that three symptoms appeared together in pairs. Over many observations, Snow generates correlations, deciding that dehydration and vomiting have a correlation of 40 percent, vomiting and cramps are correlated at 66 percent, and cramps and dehydration have a correlation of 80 percent. Well, there is no way of this happening. These correlations are actually impossible. If Snow came to us with this data, we would say he *must* be wrong. Either he collected the correlations at different times, or made some other mistake. We know this because these three numbers violate Boole's inequality.

Regarding such violations, Boole said "the evidence is contradictory." Incidentally, this is also what we would say to the graduate student who came to us with the table of correlated professor locations.

But now here is a surprise. We will do a little algebra on the inequality above. From both sides, subtract C(Bob, Alice) and multiply by -1.

So far we haven't changed the equation at all. Now take the absolute value of the left-hand side (this does change the equation but in the current case does not change the result). We will get

$$|C(\text{Bob, Alice}) - C(\text{Alice, Carol})| \le 1 - C(\text{Bob, Carol}).$$

This is Bell's inequality. Go ahead and check it against what we wrote down at the end of Chapter 47. Plugging in our values for $C(\text{Alice, Carol})$ and so forth, this results in the false statement $+2 \le 0$. Our thought experiment with amorous professors also violates Bell's inequalities.

As you may suspect, we are dealing with more than just algebraic tricks here. There are intimate conceptual connections, so close that some authors have started referring to the Boole-Bell inequality.[6]

Boole's problem has attracted the attention of mathematicians for 150 years. In 1936, the Italian mathematician Carlo Bonferroni worked on it, and managed to generalize the conditions to an arbitrary number of random variables (professors, in our example). These are called the Bonferroni inequalities. In the 1940s the French mathematician René Fréchet wrote two books on the subject. In 1955, Schell considered an economic problem ("a hypothetical soap manufacturer...equipped to produce m types of soap: bars, chips, granulated, liquid, etc.") and proved consistency conditions for three random variables in a way similar to Boole. In the same year, another French mathematician, Jean Bass, proved some necessary conditions for the existence of a probability space of three random variables.[7] Then in 1959, five years before Bell published his epochal paper, the Russian mathematician Nikolai Vorob'ev generalized the result of Bass to any number of random variables.[8] Others who have contributed more recently include Dall'Aglio, Kellerer, Joe, Schweizer, and Sklar. Work in the field continues to this day.

Here, the theorems of Bass and Vorob'ev are the most relevant for our purposes. Given several probability distributions with dependency relationships between their random variables, under what conditions are they compatible? Bass and Vorob'ev showed that these conditions are equivalent to certain inequalities being satisfied. If even one inequality is violated, the distributions are incompatible. Intuitively, they cannot have a consistent dependence structure and so we cannot talk about all of the random variables as though they are similar enough to occur under the same conditions. As Boole would say, the evidence is contradictory and cannot form a possible experience.

Our example with Alice, Bob, and Carol was in fact translated directly from the introduction of Vorob'ev's paper. The joint distributions, P(Bob, Alice) and so on, were given. The random variables Alice, Bob, Carol then had a certain dependence structure. However, because they do not satisfy Boole's inequality, the theorem of Bass tells us that there is no way to construct a single probability distribution P(Alice, Bob, Carol) of all three variables that would also produce the marginal distributions, that is, the slices P(Bob, Alice), P(Bob, Carol), and P(Alice, Carol).

As you may have guessed, Bell's inequalities just are a particularly simple case of the inequalities that have been studied by mathematicians for over 150 years.

> Abstracted from any physical assumption, Bell type inequalities have a long history, beginning in mid-19th century. They have important applications in various branches of mathematics and mathematical physics: Combinatorial theory, probability theory, propositional logic, the theory of computational complexity, the Ising spin model and neural networks, and of course, the foundations of quantum mechanics.[9]

The Bell inequalities are a result of classical probability theory that was first explored before Bell was even born. This is very remarkable, and the more we investigate, the more connections we find. For example, Boole and Bell both made the same crucial logical step. In order to define his inequalities, the very first thing Boole did was gather all three of his random variables x, y, and z into a single probability space, which he wrote xyz. Because his inequalities are violated, and his reasoning contains no errors, we could say that his assumption that xyz exists is invalid.

Bell does the exact same thing when he imagines the probability distribution of the hidden variables, which he writes $\rho(\lambda)$. "Bell started his considerations by assuming that such a single probability measure exists. He represented all correlations [in terms of] the same probability measure ρ."[10] As we saw, these hidden variables are supposed to work with the magnet angles locally to determine the measurement outcomes. In defining his problem, therefore, Bell simply assumed that there was a single probability space able to describe the dependency relations of the three different correlations. From here, just as in Boole's case, Bell used reasoning that contains no errors to derive his inequality.

This inequality is violated by the predictions of quantum theory and by the results of decades of painstaking experiments. As we have seen, this fact has been interpreted in the most mind-blowing way possible. But according to the mathematical results in this area going back 150 years, what it actually means is that "the random variables A, B and C, that are supposed to form the basis for the model of this idealized experiment, can not be defined on one common probability space."[11] Not only that, this actually makes intuitive sense: "Physically this is not a complete surprise, because anyhow the relevant experiments could not be carried out simultaneously."[12] This is the Achilles' heel of all Bell tests and extensions based on his work: the inequality must be expressed using correlations that cannot all be observed simultaneously.

Ana María Cetto is a physicist at the University of Mexico, who has been thinking about quantum foundations for a very long time. Her first publication on Bell's inequalities was in 1972, and Bell's reply to it is actually included in his classic book *Speakable and Unspeakable in Quantum Mechanics*. I called her to talk about that paper, and she explained the issue in no uncertain terms.

> Reality is there. We don't create it. Locality, also, is a basic principle of physics that we cannot sacrifice just because we don't understand what is happening. We first have to understand, and then we can draw conclusions, not jump to conclusions as the community has been doing. I think a more modern way to say what we said fifty years ago is that you can't mix the outcomes of different settings.

> In a Bell test, you have a certain instrument setting in the laboratory, you have your state prepared, you do many measurements. You get a correlation. Okay. Then, you change

the setting, and measure a new correlation. You cannot mix the outcomes of the new setting with the old setting. In Bell's presentation, and all the presentations that have followed, CHSH, you name it, they mix them. They put the outcomes from different measurement settings under the same integral. That is not legitimate. That is a mistake.[13]

This conversation was what put me on the trail of the compatibility literature. According to Cetto and quite a few others, there is, in fact, a sneaky fourth assumption required to derive Bell's theorem. We need locality, reality, freedom, *and* compatibility. This gives us another way out, because "Bell inequalities are valid if and only if the three random variables involved [in a Bell test] can actually be defined on a common probability space."[14] Because the inequalities are not valid, we can simply conclude that there is no common probability space, and this "has nothing to do with nonlocality or death of reality."[15]

The reason we got into such a mess with Bell's theorem is that "Boole's results were totally forgotten.... Vorob'ev's results were also practically forgotten."[16] Even though a rather sizable community has rediscovered these results, knowledge has not spread widely among physicists because of communication problems between the physical and the mathematical communities. Nevertheless, they are encouraged by the fact that, as Khrennikov notes, "so many people came to the same conclusion practically independently."[17] Kupczynski cites forty-one different papers, and then says:

> Several authors arrived, often independently, to similar conclusions and explained rationally why Bell inequalities might be violated. Strangely enough these explanations have been neglected by the majority of the quantum information community and remain unknown to the general public.[18]

Let's give these researchers the benefit of the doubt, and assume they are right. What does it all actually mean? As usual, we can get a rough idea by considering examples.

We often do mix the results generated in different conditions, and we are used to doing so. The SAT test mentioned in Chapter 13 is an example. It is taken every year by over a million high school students, in thousands of different locations, on dozens of different days. But the data is all brought together and connected into a single probability distribution, and of course this makes sense. It's normal to believe that the different administrators, the different weather, the different clocks on the different colored walls, do not matter for the test scores. We need to forget about irrelevant context to gather any statistics at all.

But if we forget about relevant context, we will mess up. Would we want to mix the results if we learned that, for half of the students, a fire alarm kept going off and they were evacuated no less than six times before they finished? Or that all students whose first names included the letters A, C, or E were mailed the test they would be taking a week beforehand? If we wanted to gather anything useful from data under these circumstances, we would separate the relevant sample spaces from each other.

This idea of context is not vague; it has actually been used to develop an extension of classical probability theory, one in which interference of probability is observed not just for quantum events, but for any kind of statistical data. In this "contextual probabilistic model," also called the Växjö model, the Born rule is no longer a mysterious postulate.[19] The development of a theory that has interference for more than just particles is very encouraging, because such a generalization is likely a condition for truly understanding the quantum mystery. Remember how the heat equation was finally understood: as one member of a wide class of diffusion processes. It would be reasonable to hope for the same thing when it comes to the Schrödinger equation.

The explanation of Bell's theorem, according to this contextual probability, is that the different magnet angles used in different runs of the experiment could be inducing different contexts, to the extent that we are not permitted to connect the data. In other words, the three different contexts act to make the three probability distributions of hidden variables incompatible. There are several independent reasons to believe this is the case.

First of all, the statistics of any Bell test rely centrally on collecting data over multiple runs by sending many atoms through different magnet angles. Alice and Bob have to pick one angle for their magnet, and then shoot many atoms through the magnets until they get a statistically robust correlation. (This is called a *static test*. There is another kind called a dynamic test, but my point here applies to both.) If, somehow, Alice and Bob could see all the relevant correlations at the same time, then clearly they would satisfy the conditions of a possible experience. But they cannot ever do this, by the very design of Bell's theorem. It is therefore entirely possible that the correlations might break Boole's condition of a possible experience.

Second, several physicists have used insights from statistical compatibility to produce local, realistic models of Bell tests.[20] The central feature in work of this kind is a separation of the three different contexts. The idea is that "states of measurement devices play an important role in determining the experimental context."[21] In this case, the measurement apparatus has an active role in the measurement result. Both Kupczynski and Cetto have argued that this is the correct meaning of the term "measurement dependence." According to these physicists, it has nothing to do with the free will of experimenters, or the state of the atom as it leaves the source depending on both measurement angles; instead "it refers to the dependence of the partitioning of the probability space on the measurement setting."[22]

Third and finally, statistical incompatibility is not some far-fetched idea. It is down-to-earth, a known phenomenon that often occurs.

> The situation when pairwise probability distributions exist, but a single probability measure P could not be constructed is rather standard. What would be a reason for the existence of P in the case when the simultaneous measurement of three projections of [spin] is impossible?[23]

The idea that quantum statistics depend very strongly on context is appealing because it allows us to not only explain Bell's theorem but also clarify the quantum mystery. According to Nieuwenhuizen, "Bell went wrong...because of the *contextuality loophole*, that cannot be closed."[24] It is perhaps interesting that context is something Bohr discussed frequently, without making it precise in this way. Part of his redefinition of the concept "phenomenon" included the necessity of thinking about how the phenomenon was registered and recorded. If this contextuality loophole really was the answer to Bell's theorem, in a certain sense it would render both Bohr and Einstein correct, an interesting outcome that I don't think anyone has ever considered before.

Contextuality subsumes Bell's theorem into more general (and therefore more powerful) mathematical theorems. "In this situation, strong arguments are needed to [justify the assumption] that these three probability distributions could be obtained from a single probability measure."[25] The burden of proof is actually on anyone who wants to deny local realism, because that is a much more extravagant position than denying compatibility. Bell's inequality might just imply that there is no single probability space for the hidden variables, and that the variable written as $\rho(\lambda)$ does not exist.

If the scientists pursuing this direction are correct, there is a very straightforward way to maintain Einstein's program. Quantum mechanics is vindicated experimentally, of course it is, and that means one of the four assumptions behind Bell's theorem must be wrong. It is easy to see which one we should discard.

Locality

Reality

Freedom

~~*Compatibility*~~

Part XIV

Interlude

Chapter 50

Nature Is Whispering

In the summer of 2023, I went camping out on the plains of Colorado. I stayed on a ranch, and to get there had driven past huge fields of produce being watered by center-pivot sprinklers, those green circles and semicircles you see from airplanes.

There was a stately row of cottonwoods by my little tepee, then a stream, and beyond that another field where a tractor came and went and came again. That evening, a fierce thunderstorm appeared and I sat in the little bed watching the canvas of the tepee snap. I thought it would last forever; then it was gone. I went outside. As the day's light faded, the air was damp and fresh. Brooding clouds flashed on the horizon.

I had to leave the next day, but I lingered in the morning. Somehow, I wound up down by the stream, just looking at the trees and the thick green plants on the other side. It wasn't much of a stream, more like a stagnant strand. I was standing there for some time before I noticed the motion.

The reflection of the greenery in the water was moving. It was moving everywhere. The plants themselves were motionless, I checked. There was no wind.

I came to the bank and crouched in tall grasses. The entire surface of the water was vibrating. Then, a different motion caught my eye and I saw them, the water striders. The surface was actually full of them. Each time they moved, they sent vibrating waves shooting out in concentric rings. The waves were so small, they hardly did anything to the water's surface. There was something about the angle of the leaves, the patchwork pattern of their reflection, that made the bending water visible. Three or four big crests were easiest to see as they shot out, but smaller ones followed. The waves went very far, surprisingly far away from the little bugs that had made them. When two sets of rings collided, which was happening all the time, they just passed through each other. Their overlapping made this pattern, a convolving cross-corrugation that was gone the instant I saw it.

It was so completely calm and still. The only sound was my breathing, and the occasional trill of a bird. I stared at the water striders, at the living water and its vibrating reflections of leaves, for a long time. I was thinking about how I had almost missed this. It was a secret I had stumbled on by chance.

Something was here, something important and deep. I was lucky to see it, even if its meaning escaped me. I knew with an immediacy beyond language that this world, this mysterious world, was one seamless unity. That it was not strange for nature to whisper profundities to someone who stopped to listen. Silent whispers in the movement of water striders.

Eventually, I had to leave the stream and pack my car. Back to everyday reality. And yet not quite. Staring at the water had done something to my mind. When I closed my eyes, I saw concentric rings spreading out from many central points.

Chapter 51

Parable of Icicles

The closer you look, the more wonderful and mysterious everything reveals itself to be. The world is full of surprises.

Here's one: icicles have waves in them. Have you ever noticed this? They are never perfectly straight, but always have ripples that are one centimeter long. This pattern is repeated everywhere, and does not depend on properties of the icicle itself. Regardless of the icicle's length, thickness, internal temperature, and growth time, the waves in its surface are one centimeter.

This is an outstanding riddle of hydrodynamics. As it turns out, nobody really knows why icicles have waves in them. There have been some proposals. One article, called "Waves on Icicles" by Naohisa Ogawa and Yoshinori Furukawa, of the Institute of Low Temperature Science at Hokkaido University, tried to "explain this remarkable phenomenon by introducing a new instability theory."[1]

This surprise is one reason why Stephen Morris has spent ten years growing and photographing icicles. It is one reason why Antony Szu-Han Chen did his PhD with Morris on the growth of icicles. Together they have compiled the Icicle Atlas, a website hosted by the University of Toronto. It has over 230,000 images of icicles, as well as hundreds of time-lapse videos and 3D printing files.

A piece by the *New York Times* titled "Why Icicles Look the Way They Do" discusses the Icicle Atlas, and its authors. It says that Morris "likes to think that his pursuits capture the spirit of 19th-century natural philosophers, who did not separate beauty, form, mathematics and science into different domains."[2]

It also mentions that the proposal of Ogawa and Furukawa, that the ripples emerge from surface tension and instability, was disproven by Chen and Morris. They tried growing icicles with distilled water, and the waves and ribs disappeared. So, the waves depend on impurities in the water. That's progress, but the physics are still not yet understood. For example, the waves did not depend at all on the amount of impurity. As Morris said: "The ripples have a universal wavelength of exactly one centimeter, no matter what you do. That's the real mystery."[3]

The mystery has so far resisted an answer. Scientists can have an almost emotional reaction to these kinds of simple, beautiful, intractable problems. I heard one mathematician describe the ribbing of icicles as "so frustrating."

Is it not astounding that something so common as frozen water hanging from your eaves contains a deep mystery? We walk and breathe and live in a world full of fascinating problems.

So we don't even understand icicles. But sure, quantum mechanics is complete.

Part XV

The Ancient Strength of Rational Thinking

There is no need to ask if organic and inorganic bodies are, or are not, of the same nature—that is an insoluble question.

—Auguste Comte[1]

How is the cholera generated?—how spread?...These problems are, and will probably ever remain, among the inscrutable secrets of nature. They belong to a class of questions radically inaccessible to the human intelligence.

—*Times* (London), September 1849

Indeed, it may be doubted whether we shall ever properly understand the realities ultimately involved in atoms; they may well be so fundamental as to be beyond the grasp of the human mind.

—James Jeans[2]

Chapter 52

Positivism, Again?

The majority of physicists have decided that there are no hidden variables. They believe a simple line of reasoning demonstrates this as a fact. Perhaps they would prefer a physical explanation situated in space and time, but it is a weak and watered-down preference, like wanting truffle fries when the restaurant only has regular fries. What can you do? They shrug and say, "That's quantum mechanics." They can even pat themselves on the back for being members of the "enlightened cognoscenti."[1] Those who believe in hidden variables are even sometimes despised, seen as "obnoxious medieval bigots"[2] and "deplorable scientific reactionaries."[3]

When someone declares that "Bell disproved hidden variables," they are asserting that the perfect correlations observed between the measurements of Alice and Bob when their magnets are at 0 or 180 degrees *have no explanation*. Similarly, as Bob alters his magnet orientation, the correlation will change smoothly in a predictable, regular way, from −1 to 0 to +1 and back again.

The amount of correlation depends entirely on the amount of spatial overlap between the two magnets. This is encoded in the mathematics of the quantum formalism as a dot product, and that is what a dot product is. It projects one line onto another and tells you how big the result is. If the two lines are parallel, they overlap fully. If they are perpendicular, they have zero overlap. That's why the observed spin

389

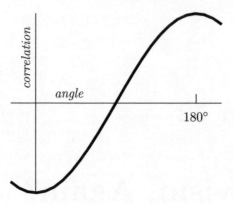

correlation varies with magnet angle—the amount of spatial overlap between the magnets is changing. But, supposedly, the fact that the correlations depend in a fully regular way on the spatial similarity of the magnets *has no explanation*. The majority of physicists just shrug.

This critical point is not stressed enough. This is where we are today, in science and culture. We can see the same thing repeated endlessly at every level, from the citation for the Nobel Prize, to blogs by Harvard professors, to popular science explainers on YouTube. And yet, it seems likely that those who comfortably proclaim that Bell disproved hidden variables would begin to squirm if forced to acknowledge the immediate corollary: certain perfect correlations in nature simply have no explanation. I had the opportunity to press a well-known defender of Bell's Theorem on this issue recently, and he admitted to having worried about the same thing for forty years, and had concluded that "the universe is absurd."

The modern view that there are no explanations for the quantum correlations strikes me as the weakest point of today's consensus view. Correlations, as Bell said, "cry out for explanation." This is a deep intuition which most people would not want to deny. When there is complete correlation between things, there *must* be a cause forcing these two things to always appear together. I'm not saying correlation implies causation. This is incorrect. Just because the number of pirates has been going down and the global temperature has been going up does not mean we need more pirates to combat global warming. I am talking about the converse statement: causation absolutely does imply correlation. Indeed, the search for causes is why we even care about correlation to begin with. Drop an apple and it falls. There is a perfect correlation between dropping the apple and it falling because gravity causes things to fall. Gravity always works. All of physics, all of science, the very existence of knowledge itself, is built on such perfect correlations. Things that *always* go together must have some causal relationship that partakes of physical law.

It is simply irrational to deny this. And yet the majority view is precisely such a denial when it comes to the atomic world. There are no hidden variables that make the correlations behave the way they do. Thus, there is no explanation. Everyone who cares about this subject needs to be clear-eyed about this. If hidden variables do not exist, then the perfect correlations between entangled particles, and the perfectly causal way these correlations change in a cosine curve, *have no explanation.*

And so here we are again, breathing the air of a shameless positivism. One very forceful expression of the current state of things was given by Edwin Jaynes, who always drops gems:

> In biology or medicine, if we note that an effect does not occur unless a condition is present, it seems natural to infer that the condition is a necessary causative agent for the effect....But suppose that condition does not always lead to the effect; what further inferences should a scientist draw? At this point, the reasoning formats of biology and quantum theory diverge sharply.

> In the biological sciences, one takes it for granted that in addition to the condition, there must be some other causative factor, not yet identified. One searches for it, tracking down the assumed cause by a process of elimination of possibilities that is sometimes extremely tedious. But persistence pays off; over and over again....Most enzymes, vitamins, viruses, and other biologically active substances owe their discovery to this reasoning process.

> What is done in quantum theory today is just the opposite; when no cause is apparent one simply postulates that no cause exists—ergo, the laws of physics are indeterministic and can be expressed only in probability form.

Biologists have a mechanistic picture of the world because, being trained to believe in causes, they continue to use the full power of their brains to search for them—and so they find them. Quantum physicists have only probability laws because for two generations we have been indoctrinated not to believe in causes—and so we have stopped looking for them. Indeed, any attempt to search for the causes of micro-phenomena is met with scorn and a charge of professional incompetence and "obsolete mechanistic materialism."[4]

Today's quantum positivism is a sinister mutation of Comte's view because it has stopped even believing in mechanism, causation, and intelligibility. Even Comte was not so depraved as to deny the very existence of causes. He believed causes were real, only that Fourier was right to say, "Primary causes are not at all known to us," because, as Comte said, they are "necessarily insoluble to our intelligence." For him, it was epistemologically naive to attempt to explain the heat equation using concepts alone.

The culture of quantum completeness that we are drowning in today makes the same move. The lesson is don't look behind the predictions of the formalism, because there is nothing there. Functionally, these two types of positivism are identical; it is only their justification that differs. For Comte and Mach, careful analysis revealed the limitations of metaphysical thinking. For Heisenberg and Bohr, the Fourier un-certainty trade-off, along with some vague philosophizing, proved that probabilities were inescapable. For most physicists today, Bell's theo-rem is the nail in the coffin of Einstein's program.

This brings out the most suspicious thing about the current inter-pretation of Bell's theorem. We are told that forty years of advanced scientific experiments has disproved causal mechanism. But when those experiments started, causal mechanism was already considered dis-proved. Remember John Nash, the beautiful mind who was stricken with schizophrenia shortly after attempting to clear up the quantum muddle? In 1957 he apologized to Oppenheimer in a letter about how aggressive he had been in an argument, but voiced his frustration at "most physicists (also some mathematicians who have studied Quan-tum Theory)" whom he found "quite too dogmatic in their attitudes."

They treat "anyone with any sort of questioning attitude or a belief in 'hidden parameters'...as stupid or at best a quite ignorant person."[5]

John Clauser, who started the craze for Bell tests, grew up in this era:

> As part of the "common wisdom" taught in typical under-graduate and graduate curricula, students were told simply that Bohr was right and Einstein was wrong. That was the end of the story, and the end of the discussion. Any student who questioned the theory's foundations, or, God forbid, considered studying the associated problems as a legitimate pursuit in physics was sternly advised that he would ruin his career by doing so.[6]

The Bell tests that have supposedly killed off Einstein's program started in an era when Einstein's program was already supposedly dead. It was career suicide to work on hidden variables because everybody already knew that there was no causal mechanism. The lesson we have learned from Bell's theorem is the one everybody already knew, the one that reinforces Bohr's ideas. As we saw, the original justification of completeness was based in complementarity and astonishingly weak. It could not explain the particle-wave duality, and it boiled down to assertions and rhetoric. Almost nobody talks about these reasons anymore or works on them. Complementarity is now almost completely forgotten. It seems likely that our new realization that hidden variables don't exist is just our old realization in disguise, and that nothing, actually, has changed. This is just the Copenhagen interpretation again. Plain old quantum completeness. Just because the wine is in new bottles does not mean the drunk emperor is wearing clothes.

The majority viewpoint is so unbelievably unnatural that I want to make the point over and over. Quantum theory predicts that if a pair of entangled particles with anti-aligned spins pass through a pair of Stern-Gerlach magnets, they will *always* yield the opposite result when the magnets are aligned. If you flip one of the magnets around 180 degrees, the particles will *always* yield the same result. When Bell said that these sorts of correlations "almost seem to cry out for a hidden variable interpretation" we must say the "almost" is entirely too weak and tentative. Hidden variables just mean a deeper story, that is, an explanation. Clearly, these correlations *absolutely* cry out for a hidden variable interpretation.

But on the back of an abstract probabilistic argument, and the fact that careful experiments confirm every prediction of quantum theory, the conclusion is that Bell's theorem disproved hidden variables. Thus, the majority believes that *perfect correlations have no explanation.* Honestly, what has happened to physics? Since when do perfect correlations have no explanation? It's so absolutely perverse that the ghost of Comte is standing over us like some kind of god. History has amply proved that positivism is unsound. More, because positivism has the power to impede and even halt progress in science, it is actively dangerous. We really should know better by now.

If we were to bet on the future of science, it would hardly be possible to find a safer place to put our money than the idea that, one day, these correlations will be explained. It is just so unnatural to say otherwise, and so hard to reconcile with what we might call the beating heart of science, which is the faith, hitherto brilliantly confirmed, that effects have causes. Why should we make an exception for atoms? It seems so much more likely that, as Einstein repeatedly said, we don't understand them yet. The claim that these correlations will never be understood seems destined for the same trash can overflowing with identical positivistic proclamations.

In fact, Bell himself seemed to believe this. He referred to his own theorem as an impossibility proof, and yet, perhaps inspired by his experience with von Neumann, Bell took a dim view of impossibility proofs in general. He wrote, in an article reviewing pilot-wave theory almost twenty years after developing his theorem:

> Long may Louis de Broglie continue to inspire those who suspect that what is proved by impossibility proofs is lack of imagination.[7]

Chapter 53

The Lion's Share

When engineers want to test the pressure of wind on a building, or how water will flow around a submarine, they build toy models first. So long as these models have the exact same shape, they can be one hundred, even one thousand times smaller than the real thing. But why should we trust what a wind tunnel says about a seven-foot-tall skyscraper? The answer lies in something called *dynamic similarity*.

In the Navier-Stokes equation (or any equation), when you divide two different terms you get a dimensionless number that tells you the relative magnitude of those terms. One of the most important of these is called the Reynolds number. It describes the ratio of inertia to viscous forces at a given point in a given situation. At low Reynolds numbers, viscosity dominates and flow is laminar, in layered sheets. At high numbers, inertia dominates and the fluid becomes turbulent.

Dynamic similarity occurs when fluid systems of different scales have the same Reynolds number. It is this number, and the shape of the body, that determines the shape of the flowlines, where turbulence occurs, the properties of the vorticity and wake, the stress on the bodies, and so on. When a toy model is buffeted with high speeds in a wind tunnel, its Reynolds number is the same as the real thing in normal weather. Thus, the behavior of the wind around the skyscraper will also be the same. These two systems, so different in scale, are actually identical.

Similarly, in labs that model geophysical flows, they think about the Rossby number, which is the relative magnitude of the inertial and rotational terms, and also the Ekman number, the ratio of rotational to viscous terms. A lab can easily make a one-meter fluid tank rotate once every five seconds, more than 17,000 times faster than the Earth. And yet, because the dimensionless ratios are the same in both cases, you achieve dynamic similarity with the fluid flows of the Earth. Then, the vortices that arise in the tank are literal weather patterns.

It is somewhat a case of definition whether or not we want to call a toy skyscraper in a wind tunnel an analogy of the real skyscraper in the wind. For dramatic purposes, let's agree to call dynamic similarity the best possible analogy. In this case, reasoning by analogy *is* physical reasoning.

Analogies come in a spectrum. Some are perfect, some are okay, some are downright terrible. False analogies occur when the similarities between the unknown and the known are only superficial. Reasoning by analogy in this case gets us into hot water, because now we are operating with a false theory rather than no theory at all.

One example of a bad analogy comes from geology's problem of orogenesis. Where do mountains come from? Why isn't the Earth just flat? Before Wegener, the explanation was that the Earth is contracting. The main proponent of this idea was Eduard Suess, who said, "What we are witnessing is the collapse of the world."[1] He said the Earth is cooling and drying out, causing mountains to form, "like the wrinkled skin of a drying apple."[2] The wrinkled surfaces of dried spherical fruits *do* look like mountains, kind of, but in this case the analogy is only skin deep.

The study of analogical reasoning in science has a rich literature, and it intersects rather marvelously with the main themes of this book. In one of her several important studies on analogy, Mary Hesse examined the opposing ideas of two physicists, Pierre Duhem and Norman Robert Campbell. Duhem thought that the French and English had fundamentally different approaches to science, and that the French one was better (Duhem was French). As he explained it, the French preference was for pure logic and mathematical rigor with the fewest possible assumptions, without distracting pictures of what might "really" be going on. The English approach was short on logical connections but long on pictures and analogies, which were actually just distracting.

Campbell disagreed (he was English). He said,

> Analogies are not "aids" to the establishment of theories; they are an utterly essential part of theories, without which theories would be completely valueless and unworthy of the name. It is often suggested that the analogy leads to the formulation of the theory, but that once the theory is formulated the analogy has served its purpose and may be removed or forgotten. Such a suggestion is absolutely false and perniciously misleading.[3]

Campbell argued, "perceptively" in Hesse's opinion, that the picture of atoms as billiard balls was not just a visual crutch for Boltzmann and Maxwell, but a central and generative assumption that suggested how the theory should be developed step by step. Without such a mental model, each step would have been simply arbitrary. Not only that (I am adding this), each step would have been much less likely to have been taken in the first place, since without pictures of hidden reality, the only guidance is mathematical rigor. This makes the search space explode, because then all steps that do not lead immediately to contradiction are on equal footing.

At a certain scale, which turned out to be the right one for the problem of heat and pressure in gases, the logical structure of the interactions of atoms *is* the structure of billiard balls flying around and clacking into each other. Here, we achieve the next best thing to dynamic similarity, which fully justifies the application of the analogy. But only in the correct regime. If we were to zoom in and try to resolve the full details of a single collision, our assumptions of Newton's laws would no longer be valid at all. As we zoom in, the picture of billiard balls slides down the spectrum of analogies from good to bad.

All of this is to clear the ground for some rather obvious questions. How are we to think about the droplet analogies? How good are they? What are they telling us? Can we really trust experiments that are literally *billions* of times larger than the phenomena they are supposedly revealing?

The short answer is that nobody knows yet, and that only time and work will tell.

So far, most professionals working on modern physics are not impressed by the analogy. Yes, the droplets might be beautiful fluid mechanics, but they think that's about it. Gerard 't Hooft, a very well-known particle physicist at Utrecht University, said, "Personally, I think it has little to do with quantum mechanics."[4] Frank Wilczek, another famous physicist at MIT, said, "I think the experiments are very clever and mind-expanding, but they take you only a few steps along what would have to be a very long road."[5]

Part of the problem is that HQA is so intimately tied to pilot-wave theory. As we saw, this theory was dropped for decades immediately after it was formulated, then it was reformulated and actively suppressed. This has resulted in the theory being in an extremely immature and incomplete state, compared to modern physics. Side by side, they appear like a mouse and a lion.

Not only that, from the viewpoint of modern physics, pilot waves are actively defective because they need particles. In quantum field theory, most people think that the particles are actually manifestations of fields. Sean Carroll, whom we met in Chapter 35, expresses the issue like this:

> De Broglie–Bohm has a really tough time fitting that into the formalism. Real fields. It was founded in a point particle, non-relativistic context. You have particles. That's what is there in your ontology. Of course you'd like to adapt it, and people have. It never smells right.[6]

To a professional physicist, one of the biggest smells of all will be the fact that HQA is classical. All of physics has developed in the exact opposite direction. Those at the heights of their discipline have very little interest in the old ways. We can thus expect most of them to find *negative* value in the droplet analogies. Like thinking dried apples support your theory of global contraction, these droplets may be misleading as a quantum model.

This may be. But maybe always cuts both ways. Maybe the relation between pilot waves and quantum field theory really will be like that between a mouse and a lion. Aesop tells of a lion who disregards the mouse's significance, and soon after is caught in a tangle of vines. The mouse comes and gnaws through the vines, freeing the lion. Stranger things have happened in science, and the first part of the story has already come to pass.

While we are entertaining what may be, maybe physicists' icy reception of the droplets is actually a good thing. Maybe Carver Mead was right when he said we can't look to the physicists for the future of physics. And maybe John Bush is right to point out that the common idea that classical physics has been transcended is not quite as meaningful as it appears, since classical physics is hardly taught anymore. As he put it,

> If you are to declare that something cannot be understood from a classical perspective, you had better be a master of classical mechanics. Picasso was a great master, and at a certain stage of life he adopted a cubist style. That spawned an entire generation of cubists, but some cubists couldn't paint a realistic bowl of fruit if their lives depended on it. Some physicists I have spoken to think in terms of mathematical abstraction rather than physical picture. They have not been trained to think in terms of mechanism.[7]

Moreover, it is good for us to remember that classical physics is not "done." It is surprisingly rich and holds many mysteries. The droplets exemplified this, and here is another fun example.

In 2014, Terry Tao, one of the most accomplished, skilled, and knowledgeable mathematicians alive today, appeared on the popular talk show *Colbert Report*. He and Stephen Colbert mostly discussed prime numbers. At the very end of the encounter, Colbert asked:

"What's the next thing that's happening in math?"

"I myself am working on fluid equations," Tao said. "I'm trying to work out whether water can spontaneously blow up."

Colbert leaned forward with a worried look: "That would be good to know."[8]

Just as an exercise, let us agree with Campbell that analogies are crucial in theory formation, and with George Pólya, who said, "Analogy has a share in every discovery, but in some it has the lion's share." Let us assume that the droplets really do have the lion's share in a great discovery to come and that, in the future, they will be talked about the same way we talk about Young's ripple tank. What are they telling us?

As we have seen, quantization in HQA arises out of a combination of three features: resonance between the particle and wave (in de Broglie's terms, the harmony of phases), a regular wavelength (also a feature of de Broglie's picture), and persistent wave memory (not mentioned by de Broglie). When the waves laid down by the droplet last for a long time, they all add up and create a topology for the droplet to explore. The wave reflections from distant objects a long time ago can affect the droplet here and now, and the droplet's own past persists as a dynamic influence. If other droplets are present, their past trajectories also contribute.

The underlying question here is: What are the particle and the wave made out of? The particle question is even less clear than the wave one, so let's focus on fields, which are very common in physics. Are we talking about a new kind of field, previously unimagined? Or a field we already know, like the gravitational or electromagnetic one?

We might look for connections in a theory called stochastic electrodynamics, or SED, which grew out of work in the early 1960s, and has been developed in the 1970s to the present day by many researchers, most notably Trevor Marshall, Timothy Boyer, Emilio Santos, Luis de la Peña, and Ana María Cetto.

The idea of SED is to build on the zero-point radiation field, described by Planck in 1911. This is an electromagnetic background that is full of natural activity so that particles are constantly being influenced by it. The influence is chaotic at one scale, but at larger scales gives rise to dynamics described by the Schrödinger equation. This is still largely dismissed by the physics community, just like all attempts to get beneath quantum mechanics. Nevertheless, SED is more than just another interpretation; it actually tries to provide a picture of subquantum reality and explain things. The phenomena SED rationalizes include atomic stability, spin, and entanglement.[9] It may also shed light on current mysteries like the problem of neutrino mass.[10] It also seems to make some predictions different than quantum mechanics on very small scales that are not yet accessible, so there are real possibilities for experimental progress.

One other possibility that has been called "the most beautiful" by Bush is the idea that the wave field is gravitational, and that the particle is undergoing mass oscillations. It would be nice if this were the case, since this would, presumably, create a connection between general relativity and quantum mechanics. But so far this idea is like all proposals for the subquantum field in HQA: extremely speculative.

Another very important feature of this system has to do with characteristic timescales. The droplets have at least three, each very different than the other. This is best seen in the long-running experiments like the quantum corral analogy.

> The walker corral is marked by three distinct timescales, those of droplet bouncing (0.01 sec), droplet translation (2 sec) and statistical convergence (1 hour). Given the vast difference in scales between this experiment and its quantum counterpart (e.g. the corral diameter is 3 cm rather than 75 Angstrom), the ability to resolve all three timescales in the laboratory is quite remarkable.[11]

The bouncing is very rapid when compared to the walking timescale. They differ by a factor of 200. But the statistical timescale is much longer than both. The statistics appear on a scale 3,600 times the characteristic walking timescale, and 180,000 times the bouncing timescale. Here we have a system with dynamics that span five orders of magnitude, from one-hundredth of a second to an hour. As Bush remarked, "It is no small miracle that the timescale for statistical convergence is an hour, which is about the timescale for a reliable experiment. And also the timescale for any experiment you can ask a graduate student to do."[12]

Most interestingly, it is the slowest statistical droplet timescale that corresponds to the Schrödinger-Born description in the corral. As we have seen, the standard view of quantum reality is that this description has no explanation in terms of finer structure. Such a view has entailed the denial of trajectories. As a well-known textbook put it: "In quantum mechanics there is no such concept as the path of a particle."[13] And so we can express quantum completeness in yet another way: the thesis that there is no faster timescale than that of the Schrödinger equation.

If quantum mechanics has a rational explanation in terms of more detailed dynamics, then the characteristic timescale of the Schrödinger equation, over which the statistics appear, would be the longest one. Below that would be the trajectory timescale. Below that again, assuming de Broglie's picture is correct, would be the oscillation frequency of the particle. If the droplet analogy holds, it is directly implying the existence of an extremely fast, hidden dynamical regime.

How fast? It's inconceivable, but let's try for fun. You know those movies where a superhero or alien hedgehog speeds up, and we see from their point of view as the whole rest of the world slows to a crawl? Let's imagine something like that. The hedgehog speeds up so that every second for it is one cycle of a red photon. It would observe everything on Earth to be completely frozen solid. The birds and planes would be hanging in the sky, the fish motionless in water as still as glass, every napping house cat fixed in an eternal catnap.

Think of all recorded history, roughly 6,000 years or so. It turns out this is only 2 percent of everything that has happened to us. *Homo sapiens* evolved something like 300,000 years ago. Multiplying by the number of seconds in a year, and rounding up, we can say there were about 10 trillion seconds in the entire history of our species. Ten trillion seconds from hunting with sticks on the plains of Africa to sipping cappuccinos and flicking screens.

Now, for this hedgehog moving as fast as a red photon, time only appears frozen. Things *are* moving. A single second on Earth, where the birds and fish and cats have actually moved just a little, would take the hedgehog a very, *very* long time to see. It's a good thing the hedgehog is an alien and immortal because to see one Earth second, to see the bird turn its head, the fish flick its tail, the cat breathe once, the superfast alien would have to wait the entire length of human history—all 300,000 years of it—740 times.

A red photon has a frequency of 740 ten trillion cycles per second. In every second that passes on Earth, red light wavers back and forth 740 ten trillion times. Moreover, I feel compelled to point out that, by the standards of the subatomic world, this is *slow*. The oscillation frequency that de Broglie imagined is called the Compton frequency, a property of particles that was discovered by Compton a few years before de Broglie wrote his thesis. For an electron, this frequency is almost 10 billion times faster than the oscillation of a red photon.

Clearly, we are just playing with words here. There is no comprehending these speeds, any more than we can comprehend the number of molecules in a cubic centimeter of air, or the vastness of the space between two galaxies. The point is that, assuming these oscillations are actually real, whatever is oscillating can live through millions of epic stories before we can even blink once. Who knows what kinds of motions these tiny objects are actually undergoing? It is eminently reasonable that the first way we would find to talk about them would be statistical. Quantum mechanics was invented in the 1920s, remember, before television. People were still sending telegraphs.

Finally, let us tie the ideas of timescale and wave field together. One of the guiding ideas that has come out of HQA is that of the *mean wave field*. To get the mean wave field, you simply take all the wave fields over a period of time, and average them. In the elliptical corral, for example, the wave field was obtained by averaging over thirty minutes. When researchers looked at the mean wave, they found it reflected structural components of the environment, what are called "cavity modes." The waves generated by the droplet explore the wider environment and then, because they persist, have a long-lived dynamical influence. When we look at the mean wave field, we can see the fingerprint of that environment and reason about how it is affecting the particle.

There is a nice theoretical result about the mean wave field, obtained by Durey, Milewski, and Bush and called Durey's theorem by Matt Durey's friends. It allows us to compute the mean wave field easily on a computer, and even by hand analytically in some cases. Bush and Oza have considered possible connections with the mysterious quantum potential of Bohmian mechanics. The connection is motivated by the fact that the "mean pilot-wave plays the role of a self-induced nonlocal potential that the walker navigates."[14]

This nonlocal quantum potential is an outstanding mystery of the pilot-wave approach. It is, as Jean Bricmont said, "slightly mysterious to put it mildly."[15] It is also a major difference between de Broglie's and Bohm's theories. While de Broglie–Bohm is usually mentioned in the same breath, they *are* different from each other. The double solution derives the quantum probabilities as emergent effects from the interaction of physical particles with physical waves, situated in space and time. Bohm's theory, on the other hand, considers the wave function to be something physically real, which de Broglie criticized. It's true that the version of de Broglie's theory that he presented at Solvay was very close to Bohmian mechanics, but only because the double solution was mathematically difficult. As he said, "I did not perceive at the time that I had greatly weakened my position."[16]

It is quite dramatic that, inspired by Bohm, de Broglie picked up his old ideas and pursued them until the end of his life. In 1956 he published a book, *Attempt at a Causal and Nonlinear Interpretation of Wave Mechanics*. With several young researchers, de Broglie developed the double solution through the 1950s, '60s, and '70s. In one of his last papers, published in 1972, when he was eighty years old, de Broglie summarized his work, drew connections between quantum and thermodynamic theories, and concluded with these thoughts:

> I think that when this interpretation is further elaborated, extended, and eventually modified in some of its aspects, it will lead to a better understanding of the true coexistence of waves and particles about which actual Quantum mechanics only gives statistical information, often correct, but in my opinion incomplete.[17]

John Bell always admired de Broglie, and it comes across strongly in many of his papers. His last paper remains unpublished, but it was made available to scholars recently by Mary Bell, and it includes this paragraph:

> The courage and vitality of de Broglie, in his late maturity, in breaking from a heavily prevailing orthodoxy, remains for me admirable and inspiring....I was indignant that I have not been told sooner of the pilot wave picture. And I am indignant that even now most students are not told of it at all. For it is a powerful antidote to the complacent notion that the ambiguity and subjectivity of contemporary physical theory are dictated by experimental facts.[18]

The story of pilot waves is intricately tied up with the deepest story of quantum mechanics itself. We saw how, in 1927, physicists could have chosen to pursue either de Broglie's or Bohr's ideas. We saw how de Broglie was defeated, and how von Neumann's proof convinced everyone that the story was over. Twenty-five years later, de Broglie's ideas were rediscovered by David Bohm, and this led to a rekindling of interest in hidden variables, and then the defeat of von Neumann's supposed proof. Shortly afterward, Bell's theorem convinced everyone again that the story was over. Now, pilot waves have been discovered *again* in the form of the bouncing droplets and have *again* rekindled interest in hidden variables.

Pilot-wave theory was discovered, independently, *three times*. In 1925, 1952, and 2005. This is a very curious fact. You have to ask yourself, what are the chances? Even if this is the strangest coincidence, nothing more than a fluke, these little bouncing balls of oil are breathing new life into Einstein's program. Researchers skeptical of completeness have existed in an unbroken line as a scientific subculture for a hundred years. We form a true tradition. Our intuitions are not planted by professors but arise spontaneously and so are difficult, maybe even impossible, to extinguish.

Louis de Broglie died in 1987, one month before his ninety-fifth birthday. He was buried in the ancient cemetery of Neuilly-sur-Seine, on the outskirts of Paris. The droplets were discovered eighteen years later, completely by accident. They could have been discovered fifty years earlier. Some HQA researchers have speculated about what the founding fathers of quantum theory would have thought if they had seen the droplet experiments. When it comes to de Broglie, the answer seems clear. He would have thought they were marvelous.

Chapter 54

Parable of Extramission

Many pages ago, we saw how Euclid and Ptolemy believed that the eye sent out a visual fire that melded with light and resulted in vision. This was called the *extramission* (sending out) theory. We wise moderns know that the intromission theory is the correct one, because light is coming into our eyes. But extramission is not completely gone. We talk about someone's "piercing" eyes, and of the "evil eye," and how else can you explain the sense of being stared at? Indeed, there is something uncanny about extramission—it is psychologically appealing, and it sticks in the mind. A series of papers by Gerald Winer and coauthors reported that roughly 50 percent of college students believed some form of emission from their eyes was responsible for visual perception, a belief that persisted even after vision was explained to them.[1] Extramission seems, somehow, natural. And when I learned that it is a common belief among children, I made my own little experiment. I asked my son, who was five at the time, how he thought his eyes worked.

"I don't know," he said.

(Fair enough.)

"Well," I asked, "do you think something comes into your eyes and you see, or do you think your eyes send something out?"

He replied at once: "Send something out."

A year later, I repeated this experiment with my niece, who was also five at the time. She replied the exact same way.

There was a period in the history of optics, after the Greeks had their day, in which Arabic thinkers made the most progress. The Islamic Golden Age of science started around 800 AD, and it gave rise to Ibn al-Haytham, who lived from 965 to 1040. His name was latinized as Alhazen in Europe, and he is considered by many to be the father of modern optics. His book *De Aspectibus* swept aside the older theories of Euclid and Ptolemy.

Al-Haytham believed that light comes into our eyes, causing us to see. He did a simple experiment in a dark room with a hole in one wall. Outside the room he hung two lanterns at different heights, so that inside the room, he could see two glowing patches of light on the wall opposite the hole. When he covered one lantern (say, the high one), he saw the low light disappear. It was clear that lantern light traveled in straight lines. Was it then necessary to also believe in visual rays traveling in straight lines? "He thus concluded that light does not emanate from the human eye, but is emitted by objects such as lanterns and travels from these objects in straight lines."[2] In other words, there was only one kind of light, not two, as Plato had said.

The troubling thing is that anyone could have done this experiment. Ptolemy, Euclid, Plato, Empedocles. All the way back to the first proponent of extramission, Pythagoras. It requires no complex technology. Lanterns, rooms, walls, and holes were available to everyone. And yet, nobody thought of it. It took over a thousand years for a simple refutation of the extramission theory to be carried out. And, perhaps even more troubling, this refutation was not accepted as such. More centuries would have to pass before visual rays shooting from the eyes were abandoned.

It seems, indeed, that this pattern characterizes discovery in general. Arthur Koestler, known for his novel *Darkness at Noon*, also wrote a book called *The Sleepwalkers* in which he asked why it took so long to rediscover the heliocentric universe. After all, Aristarchus of Samos had the right idea almost two thousand years before Copernicus. Koestler was interested in the psychological process that blinds people to truths which, when they are finally perceived after hundreds, maybe thousands of years, become "heartbreakingly obvious." He likened the cause of the blindness to a blackout shutter, and thought that the shutter operates in the minds of everyone, even great scientists like Ptolemy and Galileo. He said, "It looks as if, while part of their spirit was asking for more light, another part had been crying out for more darkness."[3] Kevin O'Regan has called this being "blinded by the obvious."[4]

Advances in knowledge follow a strange, nonlinear process that can even go backward. The same thing happens with technology. Why did it take two millennia for the telescope to be invented, when the Greeks had burning lenses? They also had railways and steam power, so why did they not invent the steam engine? Why did a bicycle not exist for Sappho, Jesus, Joan of Arc, or even Mozart to ride? Why did it take until Leonardo da Vinci in the 1500s to imagine a submarine? Why does discovery proceed the way it does, in bizarre fits and starts, with long periods of stasis lasting hundreds, even thousands of years, punctuated by flashes of progress?

When we look, we see this sort of thing everywhere. Often, the correct idea could have been had at any time. It did not depend on any particular knowledge. Rather, it depended on taking the right viewpoint. It is viewpoints that blind us or let us see. It can take a hundred years for a person with the right viewpoint to come along and think heartbreakingly obvious thoughts.

Chapter 55

Wisdom and Folly

It may be that the human mind creates physical reality, the moon is not there when nobody looks, observation causes the past to crystallize retrospectively, and we humans are the center of the universe. Or, maybe endless universes are branching away from this one right now, each one containing a different way things could have gone. Yes, maybe we need to completely reexamine what we thought an event and a phenomenon is. We need to realize that very small things don't exist the way stones exist, and in light of this accept final limits on our knowledge. Or, hey, here's a *really* crazy idea: quantum mechanics is incomplete.

The thing that always blew my mind is that the vast majority of quantum weirdness, the supposedly revolutionary import of modern science with its particles in two places at once, its mind-body connections, its multiverses, its vindication of the Indian sage Nagarjuna, blah blah blah, all of it just disappears the moment we are willing to admit that quantum mechanics is statistical because it is like every other statistical theory: it has a deeper explanation.

It is obvious that quantum mechanics is incomplete. It is a statistical theory, which means that it is incomplete as a matter of logic. This is what Einstein meant when he said, "It seems certain to me that the fundamentally statistical character of the theory is simply a consequence of the incompleteness of the description."[1] That is to say, once you understand what probability distributions actually *are*, emergent patterns arising from underlying structure, it follows immediately that they are not fundamental. The only way of escaping this conclusion is to change the meaning of probability distributions, to create a new type of probability that only applies to quantum phenomena, and which emerges from no deeper structures. You could do this. In fact, most people have done this. But why?

Statistical theories are common, after all. The world is unimaginably detailed, and so we should not be surprised to find that most real-world things we want to know about are too complicated to model with full precision. Examples are easy to multiply: stock prices, insurance risks, galaxy classification, employee performance, weather and climate patterns, psychology experiments, medical diagnoses, the effectiveness of vaccines and other medicines, rates of criminal recidivism, the spread of epidemics, consumer demographics, voting and polling, casino edges, marketing conversion rates, intracellular protein interactions, machine learning, the effect of exercise on sleep, natural language sentiment analysis, and on and on and on.

In every one of these examples—*every single one!*—the probabilities emerge as higher-order patterns generated by physical structures that are so complex (or so mysterious) that we cannot reason about them at their causal level. We are told to believe that there is one—*only one!*—area of probability theory in which the concepts grown by Pascal and Fermat apply, but the normal way of thinking about those concepts is wrong. This area is quantum physics. Here, we have a probability distribution, but it supposedly has no deeper explanation. There are supposedly no physical structures underlying it. Rather, we are told to believe that the physical structures themselves *do not exist*. Put into a formula: every statistical theory comes from mechanism, except for quantum theory, which comes from magic.

If this is wrong, it is not just a little wrong. It is a magnificent philosophical blunder where the derived fact is taken to be fundamental, and the fundamental fact is eliminated. We might as well try to make a house of cards without any cards. We might as well climb a tree, cut off the branch we are sitting on, and expect the tree to fall over.

The majority of quantum interpretations were created by people speculating about what the theory of atoms might be telling us, *under the assumption of completeness.* If quantum mechanics is complete, then yes, we must revolutionize the entire project of human knowledge. But if it is not complete, we are allowed to look at the behavior of atoms as nothing more than science's current riddle. Things are allowed to be mysterious; it's not a philosophical emergency that needs to be papered over yesterday. The experimental results of quantum mechanics are strange, but they are far less strange than many of the ideas proffered to explain them. Seen in this light, quantum weirdness is actually just an unnatural theoretical superaddition, the result of attempts to tie definitive meaning to a confusing and provisional formalism. This was Dirac's opinion:

> He thought that the puzzles were associated with the pro-
> visional state of the theory, and that looking back from the
> vantage point of a more developed theory we will regret the
> time spent trying to interpret an underdeveloped one.[2]

It has become commonplace in popular science books on this subject for authors to denounce the recent trend of wallowing in strangeness, and then promise to set the matter straight with reverse causality, woo-woo about the mind, quadrillions of copies of the universe popping out of you every second, or whatever. It's all in quite bad taste. We don't understand why the Schrödinger equation works. Big deal. We didn't understand the heat equation two hundred years ago. Scientists figured out the heat equation in a few decades once they took the right viewpoint. Let's admit that there is something to understand, and start thinking seriously about the problem.

When you hear things like "the universe splits," or "matter can be in two places at once," or "there is no objective world independent of observation," or "consciousness collapses the wave function," you are fully warranted in being deeply skeptical. This book has been geared toward giving you the ammunition to evaluate these intense claims, to be able to defend yourself against the speculative, probably-wrong ideas of authorities at famous institutes, and to be confident in the line that goes: *Sure, maybe. Or, quantum mechanics is incomplete.* And if you prefer to defend yourself with a quote by an authority at a famous institute, just memorize this double-barreled sentence:

The walker system reminds us that much of the inscrutabil-
ity of quantum mechanics and its paradoxes are forced upon
us only if we insist that the statistical description of quantum particles is complete.... The development of a rational
quantum dynamics would allow us to dispense with the very
notion of quantum interpretation, and restore a rational view
of the microscopic world.[3]

The parables we have explored along the way show that scientific
disputes are often two-sided, and in the end, one side wins decisively.
This happened in the debate between the chemists and phlogistonists,
the calorists and vibrationists, the miasmatists and contagionists, the
driftists and fixists. When science is confused, that is usually because
the wrong hypothesis was dominant for some time.

It may seem that the last century of quantum debate has been over
different interpretations, but that is not really the heart of the matter.
The debate has been about completeness. It is about whether we will
one day do better than probability distributions. The fight is between
two different *families* of interpretations—those that assume hidden
variables are impossible, and those that assume they are necessary.
There are precious few interpretations of the latter kind.

The quantum debate is dramatized and encapsulated in the clash between Bohr and Einstein, which is why their encounter is so galvaniz-
ing. In my view, though, their encounter is never given the emphasis it
deserves. It is treated as a side attraction, a juicy story to spice things
up with human interest in the midst of all those collapsing wave functions. But the long duel between the two great men is not a side attraction; it is the main event and it is still going on.

It can be hard, even impossible, to see this in other accounts, maybe
because of the common idea that Bohr won the debate, that Einstein
went senile and "had difficulty" understanding quantum mechanics,
that Bell disproved hidden variables and thus killed Einstein's program.
It makes sense that, in this cultural atmosphere, it would be hard to get
a clear view.

Yes, many people think the debate is over and Bohr won, but that is actually a social phenomenon, one of many that interact to produce the nuance and textured surfaces of the debate as it is actually lived and prosecuted. The debate is not over. No debate is ever really over until the losing side gives up debating, or gives up rationality. As we have seen, Einstein's torch of reason is still carried by many scientists today. In the first lines of their papers they say things like,

> The main question debated since nearly 100 years is: are these probabilities irreducible or do they emerge from some more detailed description of physical reality?[4]

After all, when people tell you something is impossible, they are a little bit exposed. All it takes is a single counterexample, even the smallest one, to disprove them. All the people who believe in completeness have a supposed impossibility proof, but the proof itself is under attack. Just because it has stood up so far does not mean it is invulnerable. Who knows what will happen eventually? Writing to Bohr in 1926, Schrödinger expressed this idea:

> Even if a hundred attempts have failed, one ought not to give up hope of arriving at the goal [of modeling atoms]... through logically consistent conceptions of the true nature of the space-time events. It is extremely likely that this is possible.[5]

In the late 1940s there was a Festschrift for Einstein. This is a celebratory book put together to honor someone, usually an academic, while they are still alive. It was called *Albert Einstein: Philosopher-Scientist*, and was published in 1949 when Einstein was sixty-nine years old. It has proven invaluable to historians over the years.

Bohr's contribution discussed the long duel with his friend. It so happened that when he was composing his chapter, he was in Princeton. Einstein, who preferred to work in smaller spaces, lent his large office to Bohr. This is how Pais and Bohr wound up working together in Einstein's office. Pais recounts:

After we had entered, Bohr asked me to sit down ("I always need an origin for the coordinate system") and soon started to pace furiously around the oblong table in the center of the room. He then asked me if I could note down a few sentences as they emerged during his pacing. It should be explained that, at such sessions, Bohr never had a full sentence ready. He would often dwell on one word, coax it, implore it, to find the continuation.

This could go on for several minutes. At that moment the word was "Einstein." There was Bohr, almost running around the table and repeating: "Einstein...Einstein..." It would have been a curious sight for someone not familiar with him. After a little while he walked to the window, gazed out, repeating every now and then: "Einstein...Einstein..."

At that moment the door opened very softly and Einstein tiptoed in. He indicated to me with a finger on his lips to be very quiet, an urchin smile on his face. He was to explain a few minutes later the reason for his behavior. Einstein was not allowed by his doctor to buy any tobacco. However, the doctor had not forbidden him to steal tobacco, and this was precisely what he set out to do now. Always on tiptoe he made a beeline for Bohr's tobacco pot, which stood on the table at which I was sitting. Meanwhile Bohr, unaware, was standing at the window, muttering "Einstein...Einstein..." I was at a loss what to do, especially because I had at that moment not the faintest idea what Einstein was up to.

Then Bohr, with a firm "Einstein" turned around. There they were, face to face, as if Bohr had summoned him forth. It is an understatement to say that for a moment Bohr was speechless. I, myself, who had seen it coming, had felt distinctly uncanny for a moment, so I could well understand Bohr's own reaction. A moment later the spell was broken when Einstein explained his mission and soon we were all bursting with laughter.[6]

I repeat this anecdote in full not only for its loveliness, but because we may choose to see it as an allegory. I think a marvelous surprise awaits us in microphysics. I think the time will come when the deeper dynamical facts beneath the quantum world, the facts of light and atoms and subatomic particles, are spelled out in astonishing detail. We will be staring out the window, wondering about the riddle, and truth will suddenly appear. Einstein's logical spirit will rise seemingly out of nowhere, and will leave us speechless, shaken awake with the force of an ice bath.

For so many reasons, some of which we have explored in this book, this seems assured to happen. Sometimes it takes nothing more than the correct viewpoint to think heartbreakingly obvious thoughts. Other times, a daunting number of different advances need to come together before a mystery can be solved. Perhaps we cannot imagine the future theory any more than the ancient Romans gazing up at comets could imagine telescopes, heliocentrism, electromagnetism, logarithms, spectroscopy, micrometers, hydrogen, calculus, and the solar wind. They had no chance with comets, yet it was possible to think in the right direction, and some of them did.

Just like folly, after all, wisdom is timeless. We can look back to those who got things right, and we can learn from them. People like Young and Wegener, who had heartbreakingly obvious thoughts, and the courage to develop them within a hostile culture. Like de Broglie, who rejected the sunk cost fallacy even though it meant turning his back on twenty-five years. Like John Snow, who insisted on causes even though he could not imagine them; or Boltzmann, who could imagine but not see them. Finally, there is Seneca, who almost two thousand years ago could neither see nor imagine any causes, and yet took the correct attitude toward the scientific mysteries of his time:

> The day will yet come when the progress of research through
> long ages will reveal to sight the mysteries of nature that are
> now concealed.[7]

Until this day comes for quantum mechanics, we should be humble. We should have the historical perspective and presence of mind to recognize ignorance when we see it. Ignorance is the rule, pervasive all the time. The multiplying beacons of light and truth that we have managed to establish over the millennia are infinitely precious and impressive. But we cannot let them lull us into thinking that ignorance is not the rule. Especially when we venture into the shadows, the semi-light, in the attempt to extend our lighthouses, we should expect ignorance to writhe around us and take many of us down. That is the way it has always been.

One popular textbook on quantum theory has this for the final sentence: "It is entirely possible that future generations will look back from the vantage point of a more sophisticated theory, and wonder how we could have been so gullible."[8] As residents of the breaking crest of time, we have the feeling that we are wise. We are not wise. We are just like the fools who believed that phlogiston, caloric, and miasma were real. We are no different than those who looked up at comets and proclaimed them to be windy exhalations. We poured the foundations of geology and biology so recently that the concrete is still drying. We think we are modern, but we are not. *We* are the ancients. When future generations look back at us, what spectacle will we make of ourselves? What feasts will we be for their historians? It seems certain that our century-long assumption that quantum mechanics was a complete theory will be, to them, lunacy on a civilizational scale. Our descendants will shake their heads at us. They will look at themselves and say: "We know better now, we moderns. *They* were foolish. *We* are wise."

Acknowledgments

I must thank three people above all.

To my dad, the strongest man I know, your support is a force of nature and has made all the difference in my life. My agent Caroline Dawnay, also a force of nature, working with you has been truly grand. And of course John Bush, your good taste and careful attention have made this book much better than it had any right to be.

Without any of you this book would not exist.

Without the help of many other people, this book would have been much diminished, and I would have been a nervous wreck.

On the production front, thank you, T. J. Kelleher, for your long-suffering patience, your fortitude, and encouragement. Also thanks to Ed Lake for the same, and especially for the conversation giving me the courage to spend fifteen thousand words on thermodynamics. Thanks is due to my US publisher Lara Heimert, who stepped in more than once to get this project over the finish line, when no doubt she had a hundred other books clamoring for her attention. Also Kelly Lenkevich, Kristen Kim, Kat Aitken, Melissa Veronesi, and Angela Maurer have helped in many practical and important ways. Finally, endless gratitude to Susie Pitzen, your punctilious eye corrected many (many) embarrassing slips. Sadly nobody can save me from myself, but you really did heroically try.

To the many scientists and philosophers who have indulged my questions, suggested avenues of investigation, and encouraged me along my way: thank you all. In particular, Emmanuel Fort, it was great talking with you about the beginnings of HQA and your experiments. Jean Bricmont, your excellent books provided invaluable scaffolding, and even inspiration for the title. Discussing quantum mechanics with you was illuminating and extremely helpful. Sean Carroll, thanks again for taking time out of your day to entertain the questions of an animadversor. And many thanks to Ana María Cetto, another scientist with excellent taste, for patiently explaining statistical compatibility to me. Last but not least, to Carver Mead; your questing mind and sharpness have been an inspiration for me from the first time we went out to lunch.

Special thanks to my colleagues, current and past, in and around the MIT math lab. Discussions with you have always been interesting and informative. Quentin Louis—for devouring the first long draft months before my editors read the shorter draft and suggesting I actually make it longer—bless you. Valeri Frumkin, for reading and improving chapters in the middle of a literal war. Davis Evans, for summer bicycle rides into Boston gossiping about Ciceronianism and electromagnetism. And Matt Durey, you were the first to really explain the droplets to me in mathematical detail. One day I hope to murder problems the way you do.

I owe more than thanks to my friends but thanks is all you get right now. Owen Bondurant, for your interest, your long friendship, and for reading 140,000 words on a phone. Jeffrey Kirkwood, for trying to keep me honest about Ernst Mach (lol), for enthusing about the project and suggesting scientists to read, and for that text thread discussing the anti–charm bracelet mini quark-a-matron and the true meaning of the double-slit experiment. You're the OG. Julia King, for all the times I came over and ate your food. Julia Qin, the only person who can make me dress to the nines—you deserve everything to the nines. Lan Chen, Anantdeep Parihar, Mert Kaya, Robbie Birrell, and Tomas Imach, all of you have been very patient with my extended diversions into literary land. And of course Gilbert Ramsay, for reading and seriously improving the early draft, especially with your suggestion that I cut the line "what the kids these days call sick burns," which now has made it into the book anyway. See what happens when I don't have your steady hand on my shoulder?

I have had an unusual amount of help from my family. Erin, Wesley, Mom, Julia, Thalia, you are all due more than thanks for support of every imaginable kind, far beyond the call of duty, whatever that is. You are my fascia. And to Matt, for asking that one time about how it was going with the book, let me be the first to express my ever-dying love. Good luck on your test.

Brightest for last: Rowan. Thank you for participating in my survey about extramission, for tolerating the endless times I had to work instead of play with you, and for being the sun that lights my life. I hope one day you read this book and are proud of me.

Notes

Epigraph

1. Seneca, *Questiones Naturales*, 7 (25, 5).

Chapter 1: Weird Science

1. Gribbin 2019, x.
2. Chopra 1989.
3. Tzu 1990, 3.
4. Feynman 1964.
5. Holland 1993, xvii.
6. Jean Bricmont, conversation with the author, February 2, 2022.
7. Bell 1987, 170.
8. Dyson 2007.
9. Nasar 1998, 221.
10. Weinberg 2017.
11. In Bricmont 2016, 10.
12. Putnam 2005, 624.
13. Einstein 1952.
14. Einstein 2011.
15. Einstein 1953b.
16. Einstein 2011.
17. See, for example, de Broglie's book on quantum mechanics, *The Revolution in Physics*.
18. Feynman 2010, vol. 3, ch. 1, sec. 1.
19. Feynman 2010, vol. 1, ch. 37, sec. 1.
20. Davies and Brown 1993, 51.
21. Cox and Forshaw 2011, 24.
22. Al-Khalili 2003, 21.
23. Haldane 1928, 298–299.
24. Baggott 2020, 152.
25. Aristotle 1933, 1.1.
26. Vickers 1984, 149.
27. Pólya 1954, 17.

28. His last name is famously hard for English speakers to pronounce. The right way is

    b   like in *about,*

    r   this is the hard one; it's guttural, like the *ch* in the Scottish *loch*, but recognizably an *r,*

    o   like in *off,*

    y   like in *yet,*

    l   and if you want to be really fancy, you can barely pronounce a silent *l* at the end.

29. In Becker 2018, 90.
30. Bell 1987, 160.

Part I: Parable of Waves and Particles

1. Redhead 1977, 65.

Chapter 2: A Visual Fire

1. Also known as "the Savior," Ptolemy was one of Alexander the Great's generals and one of the four Diadochi, or "successors."
2. Byrne 2010, 3.
3. Plato 1929, 46b.
4. This image of the blind "seeing" with a cane was introduced by Descartes in his theory of light as an instantaneous pressure operating through a medium.
5. Burton 1943, 357.
6. Burton 1943, 357.
7. It would seem that Euclid also tacitly assumes that in order to be seen, a thing must be struck by two rays simultaneously.
8. Zajonc 1995, 25.
9. The last one was Ptolemy XV Caesarion, the son of Julius Caesar and Cleopatra VII.
10. The name in Greek was *Mathematike Syntaxis,* but the work was later titled *Megale Syntaxis* or Great Treatise. Later this was abbreviated to *Megiste, Greatest,* and then translated into Arabic, where it exerted an influence during the Islamic Golden Age of science (many important texts were only preserved in Arabic, while Europe was plunged into the Dark Ages). The modern name of the treatise comes to us from *Al Majisti,* meaning "the greatest" in Arabic.
11. The admiral Eugene of Sicily translated Ptolemy's *Optics* from two different Arabic copies. No trace of any Arabic copy, or the Greek from which they originated, remains.
12. Zubairy 2016, 6.
13. (II.50)

Chapter 3: Baroque Animadversors

1. This is from *De rerum natura*. The famous statue was described by Wordsworth in his poem "Residence at Cambridge." Lying in his bed in a "nook obscure" in St. John's College, he said,

> And from my pillow, looking forth by light
> Of moon or favouring stars, I could behold
> The antechapel where the statue stood
> Of Newton with his prism and silent face,
> The marble index of a mind for ever
> Voyaging through strange seas of Thought, alone.

2. Newton 1958, 27.
3. Newton probably obtained his prisms at the Stourbridge Fair at Midsummer Common, a large field in Cambridge which is still the place for gatherings like the Strawberry Fair and Bonfire Night. Prisms were not scientific instruments at the time, and at the fair they would have been on sale as curiosities and toys. Mills 1981, 14.
4. "And so the true case of the length of that Image was detected to be no other, than that *Light* consists of *Rays differently refrangible*, which without any respect to a difference in their incidence, were, according to their degrees of refrangibility, transmitted towards divers parts of the wall." Newton 1958, 50.
5. Newton 1958, 53.
6. Newton 1958, 43.
7. Hall 1948, 248.
8. Newton 1958, 184.
9. It is disputed whether Huygens got there first.
10. Drake 2006, 135.
11. Newton 1958, 112.
12. Newton 1958, 114.
13. Newton 1958, 38.
14. Newton 1958, 121.
15. Newton 1958, 184.
16. An exception to this is sound waves in the ocean, which can enter the "sound channel," an area where the temperature and pressure interact like mirrors and bounce the sound back and forth so that it loses very little energy. Low-frequency waves can travel for fifteen thousand miles along the sound channel. For example, an explosion in the Bahamas can be heard off the western coast of Africa, the sound crossing the entire Atlantic Ocean in the sound channel.
17. Newton 1958, 482.
18. J. C. Squire replied ingeniously:

> It did not last: the Devil howling "Ho!
> Let Einstein be!" restored the status quo.

19. For example, "A couple that prays together, stays together" or "Your vibe attracts your tribe." Sounds true!
20. Newton 1958, 45.

Chapter 4: The Doctor and the Engineer

1. The first quote is from the epitaph carved on a marble tablet in Westminster Abbey; the second is from Peacock 1855, 117.
2. Nichols 1933.
3. This last epithet sounds like a thriller, and in fact is one, directed by Alfred Hitchcock.
4. Young 1855, 78.
5. He was a Syrian slave who was brought to Rome and, because he was so witty, granted freedom and education. He is often credited with the proverb "A rolling stone gathers no moss." The Rolling Stones are named after this saying.
6. Brougham 1803, 452.
7. Though Domenico Argentieri claimed precedence in the study of diffraction for Leonardo da Vinci, which would be beautiful and unsurprising, if true.
8. Grimaldi 1665.
9. Young 1855, 209.
10. Young 1855, 209.
11. Young 1855, 208.
12. Young 1855, 213.
13. Young 1855, 207.
14. Young 1855, 215.
15. Newton has his miraculous year in 1666 during a plague year, also ensconced in his mother's house. Mothers and their houses, it would seem, play a hitherto unappreciated role in the history of science.
16. Born and Wolf 1999, 141.
17. Arago 1858, 111.
18. Young 1855, 409.

Chapter 5: Seemingly Monstrous Assumptions

1. Einstein 1954, 269.
2. Weinberg 1993, 14.
3. Weinberg 1993, 14.
4. During a plague year in England, Newton went home to live with his mother. He was still a student at Cambridge, but during this time he did his first optical experiments with prisms, formulated the universal law of gravitation, and one more thing I think? Oh yeah, invented calculus.
5. The journalist Albrecht Fölsing called this the most "revolutionary" sentence written by a physicist in the twentieth century.
6. Einstein 1990, 391.
7. Einstein 1990, 394.
8. Einstein 1990, 395.

Chapter 6: Lifting a Corner of the Great Veil

1. Abragam 1988, 25.
2. In Bernstein 1991, 30.
3. In Abragam 1988, 26.
4. Abragam 1988, 28.
5. Abragam 1988, 28.
6. de Broglie 1925a.
7. Jean Perrin was responsible for the first characterization of Brownian motion. Inspired by his work, Einstein deduced a beautiful result in fluid mechanics about the effective viscosity of a dilute suspension of spheres.
8. de Broglie 1967.
9. Popper 1992, 48.
10. Abragam 1988, 29.
11. Abragam 1988, 31.
12. Abragam 1988, 30.
13. Abragam 1988, 30.
14. In Abragam 1988, 30.
15. Abragam 1988, 30.

Chapter 7: Nature Is Talking

1. Redhead 1977, 65.
2. Schilpp 1949, 666.
3. de Broglie 1952, 290.
4. Huerre 2019. An example of such an experiment is Couder's explanation of why sunflower seeds grow in a Fibonacci sequence. He examined ferromagnetic drops supported in a water jet flow. These drops mutually repelled because of their magnetic properties, but they also flowed outward radially in all directions from the central jet. The combination of these two forces balances out in the same pattern as sunflower seeds.
5. Emmanuel Fort, conversation with the author, May 18, 2022.
6. *Through the Wormhole* 2011.
7. Walker 1978.
8. *Through the Wormhole* 2011.
9. Emmanuel Fort, conversation with the author, May 18, 2022.
10. Emmanuel Fort, conversation with the author, May 18, 2022.
11. John Bush, conversation with the author, July 4, 2023.
12. John Bush, conversation with the author, July 4, 2023.
13. John Bush, conversation with the author, July 4, 2023.
14. John Bush, conversation with the author, July 4, 2023.

Chapter 8: Parable of Phlogiston

1. Lavoisier 1965, xxiii.
2. Priestley 1796, 9.

Chapter 9: Something That Has Not Been Seen

1. Bloch 1976, 23.
2. Bloch 1976, 23.
3. Bloch 1976, 23–24.
4. "Erwin Schrödinger" 2023.
5. The original German is
   Gar Manches rechnet Erwin schon
   Mit seiner Wellenfunktion.
   Nur wissen möcht' man gerne wohl
   Was man sich dabei vorstell'n soll.

Part III: The Hidden Nature of Heat

1. Davy 1840, 350.

Chapter 10: The Heat Equation

1. Narasimhan 1999, 151.
2. Herivel 1975, 101.
3. Herivel 1975, 103.
4. In Whittaker 1989, 139.
5. "The Remarkable Story" 2022.

Chapter 11: Parable of Caloric

1. Roller 1961, 62.
2. Roller 1961, 43. This citation is from Black; I could not find the relevant passage in *De Forma Calidi*.
3. Often attributed to Yeats, which would make me happy, but the first known thinker of this idea seems to be Benjamin Franklin in a 1782 letter: "And we now find, that it is not only right to strike while the iron is hot, but that it may be very practicable to heat it by continually striking." For someone unfamiliar with ironworking, this surprising process is awesome to behold. The video I watched to check if this was true noted that because the iron is strongly compressed at the tip by this process, it cannot be hammered red hot again after the first time. And the blacksmiths drew a further moral from this: Do not squander the opportunities you create for yourself by striking. Build a fire the moment you can, for those opportunities cannot be created again.
4. Poinier 1877, note 7.
5. Roller 1961, 62.
6. Roller 1961, 48.
7. Roller 1961, 79.
8. Roller 1961, 75.
9. Fox 1971, 116.
10. Truesdell 1980, 3.
11. Roller 1961, 32.

12. Roller 1961, 40.
13. Roller 1961, 23.
14. Roller 1961, 23.
15. Roller 1961, 45.
16. Roller 1961, 44.
17. Roller 1961, 45.
18. Roller 1961, 44.
19. The connection between Rumford and Lavoisier was more than coincidental. After Lavoisier was beheaded during the French Revolution, Rumford married Lavoisier's widow. The union was short-lived, and Rumford's boasts that he was going to overturn Lavoisier's theory of heat probably did not help.
20. Fox 1971, 79.
21. Fox 1971, 79.
22. Fox 1971, 104.
23. Roller 1961, 48.

Chapter 12: Positivism

1. Narasimhan 1999, 155.
2. Comte 1896, 11.
3. Mill 1865, 10.
4. Comte's view on metaphysics is illuminated by his view on hypothesis. He held some hypotheses to be constructive and others destructive. If an explanation directly bears on "the laws of phenomena," and can be empirically checked, it is deemed "admissible." However, any hypothesis about the "hidden nature," "cause, whether primary or final," or "essential mode of production" of phenomena is inadmissible. Such hypothesizing is of "an anti-scientific character," because it is metaphysical. Guillin 2016, 26.
5. Comte 1830, 3.
6. Comte 1896, 29.
7. Comte 1988.
8. Comte 2009, 1.
9. Bryson 1936, 344.
10. Davies 1997, 28–29.
11. Gros 1908, 34.
12. Bromley and Braden 2016.
13. Romero 2016.
14. Nietzsche 1990, "Maxims and Arrows," sec. 26.
15. Buchwald and Fox 2013, 490.
16. Comte 1830, 17.
17. Comte 1831, 436.
18. Comte 1838, 7.
19. Hearnshaw 2010.
20. Comte 1988, 53.
21. Comte 1838, 683.

22. Comte 1896, 261.
23. Comte 1896, 230.
24. Comte 1896, 239.
25. Comte 1896, 263.
26. Bernstein 1991, 54.
27. Mill 1865, 13.

Chapter 13: Intellectual Dynamite

1. Everyone else translates this as *Book on Games of Chance*. I agree it has a nicer ring to it, but *ludo* is ablative singular.
2. Pascal was so close that Leibniz used one of Pascal's diagrams in his own treatise on calculus. Indeed, Leibniz said in a letter to Bernoulli that sometimes it seemed that Pascal had a bandage over his eyes. Boyer 1989, 153.
3. Bernstein 1996, 3.
4. "Fermat and Pascal on Probability," 2.
5. "Fermat and Pascal on Probability," 7.

Chapter 14: The Kinetic Theory of Gases

1. The son's mathematical skill practically drove the father mad. A year after Daniel published his *Hydrodynamica*, Johann plagiarized many of the results in his own work, *Hydraulica*. The father even lied and backdated publication of *Hydraulica* by seven years to discredit his own son, the true author of the results, as the plagiarist.
2. Fox 1971, 304.
3. Fox 1971, 304.
4. Maxwell 1860, 148.
5. Narasimhan 1999, 119.
6. Sen 2022.
7. Schilpp 1949, 33.

Chapter 15: Parable of Atoms

1. Sen 2022, 97.
2. *Bring' vor, was wahr ist;*
   Schreib' so, daß klar ist
   Und verficht's, bis es mit dir gar ist
3. Cercignani 2006, 185.
4. There are many roads to positivism. The most obvious is to insist that nothing but observable entities are admissible in theory construction. Even Comte did not go that far. I have a friend, a very fine German scholar, who insists that Mach was not a positivist. Was he not, in fact, a constructivist? Those who really want to have an opinion on this should examine the primary sources. At the very least, when I call Mach's thought positivistic, it is the

main line. Planck, Einstein, Heisenberg, and Schrödinger all thought of Mach in this way. Pauli, who was Mach's godson, said that Mach's influence on him was an "anti-metaphysical baptism." Almost unanimously, modern scholarship assumes a close connection and says things like "The Machean positivist principle of the elimination of unobservables," and finally, there are close connections between Comte's primary move, the denial of metaphysics and hidden essence, and Mach's principle of thought economy. The main line is not necessarily correct, of course, only comfortable.

5. Mach 1902, 98.
6. Mach 1986, 113.
7. Cercignani 2006, 184.
8. Cercignani 2006, 13.
9. Einstein 1916, 102.
10. "The energeticists were thoroughly defeated at every point, above all by Boltzmann, who brilliantly expounded the elements of kinetic theory.... Ostwald was quite exhausted when the discussion ended, and Helm spoke of having been lured into an ambush." Lindley 2001, 128.
11. Cercignani 2006, 193.
12. Feynman 2010, vol. 1, ch. 1, sec. 2. (Section 1.1 is basically housekeeping.)
13. Cercignani 2006, 209.
14. Einstein 1998, 451.

Chapter 16: Quantization

1. "Max Planck."
2. Wollaston 1802, 378.
3. This is why Comte's claim that we will never know anything about the chemical composition of stars was such a fatal blunder. We can know a great deal about stellar composition by breaking the light apart and looking for the dark lines. With the same method we can know about the atmospheres of planets, even those in different solar systems.
4. Bohr 2013.
5. Kumar 2008, 83. In fact, physical and mathematical ability ran in the family. Bohr's brother Harald was a world-famous mathematician who played football for Denmark and won silver in the 1908 Olympics, where they trounced France 17–1.
6. Rutherford 1938, 68.
7. Bohr 1928, 580.
8. Wheeler 1986, 304.

Chapter 17: Hydrodynamic Quantization

1. Bush et al. forthcoming.
2. Fort et al. 2010, 17515.
3. Fort et al. 2010, 17515.

Part V: What We Talk About When We Never Know What We Are Talking About

1. Holland 1993, xvii.

Chapter 18: Schrödinger's Equation

1. It is sometimes called "the heat equation in imaginary time," a parlance which I imagine arises from the fact that stationary states are real functions times a complex exponential in time: $exp(-iEt/\hbar)$ for some energy E.
2. Named after Wilhelm Ostwald, who coincidentally argued against Boltzmann's atomic hypothesis, favoring a now obsolete idea called *energeticism*.
3. Rosenblum and Kuttner 2006. The title of Chapter 8 is "One-Third of Our Economy."
4. Faye 2019.

Chapter 19: The Dice of God

1. Technically, the procedure is to take the square of the modulus, so you look at the quantity $\psi^*\psi = |\psi|^2$. Here ψ^* is the complex conjugate.
2. de Broglie 1987, 3.
3. In Bricmont 2016, 33.
4. Nasar 1998, 221.
5. Nasar 1998, 221.
6. Nasar 1998, 221.
7. Popper 1992, 48.
8. Taylor 1909.
9. Feynman 2010, vol. 1, ch. 37, sec. 1.

Chapter 20: The Measurement Problem

1. Jammer 2004, 397.

Chapter 21: Positivism, Again

1. In Kronig 1960, x.
2. Einstein 2012, 104.
3. Heisenberg 1925, 879.
4. Segrè 2007, 115.
5. Heisenberg 1925, 880.
6. Segrè 2007, 130.
7. Carson 2010, 145–146.
8. Heisenberg 1971, 63.
9. Heisenberg 1971, 63.
10. Heisenberg 1983, 84.
11. Comte 1830, 4.
12. Mach 1914, 35.

13. Heilbron 1988, 218.
14. Heilbron 1988, 203.
15. Heilbron 1988, 203.
16. Pais 1991, 319.
17. Bohr 1935, 697.
18. Heilbron 1988, 211.
19. Heilbron 1988, 211.
20. Fine 2007, 35.
21. Beller 1999b, 205.
22. de Ronde 2021.
23. Fine 2007.
24. Schrödinger 1940.
25. Schrödinger 1983, 328.
26. Davies and Brown 1993, 51.
27. Einstein 1953b.
28. Mermin 1989, 9.

Chapter 22: The Choice

1. Popper 1992, 8–9.
2. Schilpp 1949, 212.
3. Schilpp 1949, 672.
4. Rovelli 2021, 62.
5. Wiseman 2014, 468.
6. Bohm 1980, 87.
7. Jammer first made this point and it is very well made.
8. Jammer 1974, 261.

Chapter 23: Parable of Comets

1. Heidarzadeh 2008, 16.
2. In Laplace 1951, 4.
3. Heidarzadeh 2008, 47.
4. Heidarzadeh 2008, 83.
5. Heidarzadeh 2008, 84.
6. *Georgics* 1.487–488.
7. Sagan and Druyan 1997, 29.
8. Sagan and Druyan 1997, 32–33.
9. Laplace 1830, 79.

Chapter 24: Diffracting Droplets

1. Couder and Fort 2006, 154101–154103.
2. Couder and Fort 2006, 154101–154104.
3. Andersen et al. 2015, in the abstract.
4. Ellegaard and Levinsen 2020, in the abstract.

5. Ellegaard and Levinsen 2020, 2.
6. Wolchover 2018.
7. Cox and Forshaw 2011, 24.

Part VII: Wrong Turning

1. Hossenfelder and Palmer 2020, 1.

Chapter 25: The Double Solution

1. Bacciagaluppi and Valentini 2009, 61.
2. de Broglie 1927, 225.
3. de Broglie 1953, 226.
4. de Broglie 1925b, 500.
5. Bush 2015, 286.
6. de Broglie 1953, 225.
7. Dürr, Goldstein, and Zanghì 2013, vii.

Chapter 26: This Subtle Doctrine

1. Schrödinger 1926.
2. Rosenblum and Kuttner 2006, 105.
3. Pais 1991, 310.
4. Beller 1999b, 144.
5. Sen, Basu, and Sengupta 1995, 426.
6. Beller 1999b, 119.
7. Beller 1999b, 119.
8. Scheibe 1973, 12.
9. Murdoch 1987, 60.
10. Bohr 1985, 21.
11. Schilpp 1949, 211.
12. Schilpp 1949, 220.
13. Schilpp 1949, 221.
14. Schilpp 1949, 209.
15. Beller 1999b, 164.
16. Beller 1999b, 228.
17. Beller 1999a, 262.
18. Schilpp 1949, 211.
19. Dresden 1987, 295.
20. Pais 1991, 309–310.
21. Rosenfeld 1963, 54.
22. Bohr 2013.
23. "Niels Bohr" 2023.

Chapter 27: Solvay

1. Davisson and Germer 1927.

2. de Broglie 1953, 227.
3. de Broglie 1927, 241. Emphasis in original.
4. de Broglie 1953, 228.
5. Hegel 1977, 2.
6. See the book of that title by George Gamow.
7. Carroll 2019, 67.
8. In Bacciagaluppi and Valentini 2009, 210.
9. Heilbron 1988, 200.
10. In Bacciagaluppi and Valentini 2009, 489.
11. Born 1949, 109.
12. Becker 2018, 36.
13. James 2019, 108.
14. Mehra 1975, xvii.
15. Nash 2003, 4.
16. Bacciagaluppi and Valentini 2009, 441.
17. Mehra 1975, xvi.
18. de Broglie 1953, 229.
19. de Broglie 1962, 184.
20. de Broglie 1953, 229.
21. Bricmont 2017, 199.

Chapter 28: The Bohr-Einstein Debate

1. Kumar 2008, 274.
2. In Bohr 1985, 37–39.
3. In Pais 1991, 427.
4. The fancy rhetorical term is *chiasmus*.
5. Kumar 2008, 274.
6. Einstein 1952.
7. Einstein 2011, 34.
8. In Fine 1986, 2.
9. Einstein 2011, 44.
10. Pais 1991, 22.
11. Pais 2005, 8.
12. Kumar 2008, 262. Emphasis in original.
13. Pais 1991, 8.
14. Pais 2005, 9.

Chapter 29: The Proof of a Martian Anthropologist

1. "I have known a great many intelligent people in my life. I knew Max Planck, Max von Laue, and Werner Heisenberg. Paul Dirac was my brother-in-law; Leo Szilard and Edward Teller have been among my closest friends; and Albert Einstein was a good friend, too. And I have known many of the brightest younger scientists. But none of them had a mind as quick and acute as Jancsi von Neumann. I have often remarked this in the presence of those men, and no one ever disputed me.

"You saw immediately the quickness and power of von Neumann's mind. He understood mathematical problems not only in their initial aspect, but in their full complexity. Swiftly, effortlessly, he delved deeply into the details of the most complex scientific problem. He retained it all. His mind seemed a perfect instrument, with gears machined to mesh accurately to one thousandth of an inch." Szanton 1992, 58.

2. In Bhattacharya 2021, 15.
3. Polya 1957, xv.
4. Bhattacharya 2021, xii.
5. In Bhattacharya 2021, xii.
6. In Bhattacharya 2021, 15.
7. In Bhattacharya 2021, xi.
8. "Leonardo da Vinci" 2023.
9. von Neumann 2018, xi.
10. von Neumann 2018, 2.
11. von Neumann 2018, 109.
12. Born 1949, 109.
13. Beller 1999b, 279.
14. von Neumann 2018, 213.
15. von Neumann 2018, 213.
16. Beller 1999b, 213.
17. Feyerabend 1995, 78.
18. Beller 1999b, 278.
19. In Bricmont 2016, 281.
20. Born and Einstein 1971, 149.

Chapter 30: Parable of Miasma

1. Johnson 2006, 123.
2. Johnson 2006, 124.
3. Johnson 2006, 118.
4. Johnson 2006, 120.
5. Johnson 2006, 121.
6. Johnson 2006, 2.
7. Johnson 2006, 125.
8. Johnson 2006, 125.
9. Johnson 2006, 69.
10. Johnson 2006, 70.
11. Johnson 2006, 168.
12. Johnson 2006, 154.
13. Johnson 2006, 163.
14. Johnson 2006, 122.
15. Comte 1831, 436.
16. "Cholera" 2023.

Chapter 31: Many Droplets

1. Couchman et al. 2022, 9.

Part IX: The Tranquilizing Philosophy

1. Einstein 2011.
2. Jaynes 1996, 4.

Chapter 32: The Only Mystery

1. Feynman 2010, vol. 3, ch. 1, sec. 5.
2. Feynman 2010, vol. 3, ch. 1, sec. 5.
3. Feynman 2010, vol. 3, ch. 1, sec. 7.
4. Heisenberg 1959, 51.
5. In Bricmont 2017, 24.
6. Heisenberg 1959, 129.
7. In Bricmont 2016, 6.
8. Carroll 2012, 35.
9. Putnam 2005, 624.
10. In Bricmont 2016, 8.
11. Davies and Brown 1993, 67–68.
12. Born 1949, 109.
13. Holland 1993, 1.
14. The writer concludes that, as a consequence, "Free will is preserved."
15. O'Connor and Franklin 2022.
16. Born and Einstein 1971, 91.
17. Born 1949, 109.
18. Feynman 1967, 145.
19. Krauss 2012.
20. "Quantum Tunnelling" 2023.
21. Jammer 1974, 440.
22. Bricmont 2017, 173.
23. Carroll 2019, 20.
24. Rovelli 1996, 1650.
25. Ballentine 1986, 66.
26. Kumar 2008, 376.
27. Suplee 2002, 76.
28. Beller 1999b, 279.
29. Popper 1992, 5–6.
30. "The Mystery of the Multiverse" 2022.
31. In case you thought I was bluffing, here they are:

    Heisenberg said that the Copenhagen interpretation "only states that an addition of parameters [hidden variables] in the sense of classical physics would be useless." Ballentine 1988, 48.

"If quantum mechanics provides a complete description of the electron—as Bohr insisted—this diffuseness is not merely a reflection of our ignorance about where the electron is, it is a characteristic of the electron itself." Maudlin 2018.

"A wave function provides a complete description of an individual quantum system." Khrennikov 2010, 34.

"The acceptance of QM, in accordance with the orthodox point of view, as expressing the ultimate and irreducible random character of physical phenomena." Peña et al. 1972, 177.

"To many the experimental verification of the violation of [Bell's] inequalities is sufficient evidence for the completeness of quantum theory." Geurdes 2010, 1.

"Most physicists, including QBists, believe that quantum mechanics is the full and correct theory of the world." Von Baeyer 2016, 158.

"Yet, we assert that the wave function (in principle) can provide the most complete possible description of the system that is consistent with the actual structure of matter." Bohm 1989, 620.

"Bohr believed that quantum mechanics was a complete fundamental theory of nature." Kumar 2008, 305.

"The usual assumption that the wave function describes a complete description of reality." Bohm 1989, 613.

"In quantum mechanics it is usually assumed that the wave function does contain a complete description of the physical reality of the system in the state to which it corresponds." Einstein et al. 1935, 778.

"Although we believe that after having specified [the wave function] we know the state of the system completely, nevertheless only statistical statements can be made concerning the values of the physical quantities involved." von Neumann 2018, 134.

"This idea is founded on the belief that the Schrödinger equation is fundamental; that nothing underpins it." Hossenfelder and Palmer 2020, 2.

"[What bothered Einstein was] the idea that there was nothing beyond 'dice-playing,' that quantum theory was the complete description of reality." Bernstein 1991, 40.

"The wave function with its probabilistic interpretation offers the most complete possible specification." Jammer 1974, 280.

"L'interprétation de Bohr et Heisenberg, non seulement ramène toute la Physique à la probabilité, mais elle donne à cette notion un sens qui est tout nouveau dans la Science. Tandis que tous les grands maîtres de l'époque classique, depuis Laplace jusqu'à Henri Poincaré, ont toujours proclamé que les phénomènes naturels étaient déterminés et que la probabilité, quand elle s'introduit dans les théories scientifiques, résultait de notre ignorance ou de notre incapacité à suivre un déterminisme trop compliqué, dans l'interprétation actuellement admise de la Physique quantique, nous avons affaire à de la « probabilité pure » qui ne résulterait pas d'un déterminisme caché." de Broglie 1952, 303.

"The usual interpretation of the quantum theory... [assumes] that the most complete possible specification of an individual system is in terms of a wave function that determines only probable results of actual measurement processes." Bohm 1952a, 166.

"The [Copenhagen] interpretation also rejects the possibility that the seemingly probabilistic behavior of quantum systems stems from underlying, deterministic mechanisms." Horgan 2018.

"In quantum mechanics, we are not dealing with an arbitrary renunciation of a more detailed analysis of the atomic phenomena, but with a recognition that such an analysis is in principle excluded." Bohr 1996, 258.

"In fact, one of the assumptions made was that the wave function gives a complete description of reality as far as this could be done. This is what Einstein objected to." Bohm 1986.

"It's not possible to know if our subject of investigation is a particle or wave, just like it's not possible to know its location and momentum. You can know one, but not the other. Neither can be known at the same time. There is an inherent uncertainty, an inherent randomness to our universe. It is ingrained in our very existence, and to deny it is futile." Sweatman 2015.

"But Bohr was insistent that the quantum state of a specific system was a *complete* description of the state of that system." Sklar 1992, 280.

"Bohr took an unexpected approach to this question: instead of asking if the theory was too young to be fully understood, he declared that the theory was complete; you cannot visualize what the electron is doing because the microworld of the electron is not, in principle, visualizable (*anschaulich*)." Maudlin 2018.

"The central point is that quantum theory is fundamentally pragmatic, but nonetheless complete. The principal difficulty in understanding quantum theory lies in the fact that its completeness is incompatible with external existence of the space-time continuum of classical physics." Ballentine 1988, p. 33.

"If the state vector is assumed to *completely* describe the individual system... then one will be forced to the absurd conclusion that the 'pointer' of the apparatus (a *macroscopic* object) has no definite position. The supposed *reduction of the state vector*, and all the difficulties and complications associated with it, are only artifacts of the vain attempt to retain the above assumption. But to what purpose? Under the more modest assumption that a state vector represents an *ensemble* of similarly prepared systems, the measurement process poses no particular problem." Ballentine 1986, 30.

"According to quantum mechanics, this probabilistic prediction is all there is to say. It is probabilistic not because we are missing information. There just isn't any more information. The wave function is the full description of the particle—that's what it means for the theory to be fundamental." Hossenfelder 2022, 22.

"Einstein was wrong with his 1927 Solvay Conference claim that quantum mechanics is incomplete." Nordén 2016, first sentence of the abstract.

"Positivism in quantum philosophy has two central strands—operationalism and instrumentalism.... Both varieties of positivism are aimed at arguing the finality of quantum mechanics." Beller 1999b, 203.

"One can convince oneself that statistical distributions, such as are given by quantum mechanics and verified by experiment, have such a structure that they cannot be reproduced by hidden parameters." London and Bauer in Gouesbet 2013, 41.

"The essential nature of this physical reality which reveals itself both as a particle and as a wave remains obscure and the Copenhagen interpretation implies that no further insight into it is possible." Prosser 1976, 181.

"It is the assumption that the quantum state description is the most, complete possible description of an individual physical system." Ballentine 1970, 358.

"Every attempt, theoretical or observational, to defend such a hypothesis [of hidden variables] has been struck down." Wheeler 2016, 188.

"There is again a huge majority of physicists... who would claim that the debate is over, that ψ is complete, and that there is nothing hidden beyond ψ." Gouesbet 2013, 42.

"No concealed parameters can be introduced with the help of which the indeterministic description could be transformed into a deterministic one." Born 1949, 109.

"John Stewart Bell developed the mathematical inequality that is named after him. This states that if there are hidden variables, the correlation between the results of a large number of measurements will never exceed a certain value. However, quantum mechanics predicts that a certain type of experiment will violate Bell's inequality, thus resulting in a stronger correlation than would otherwise be possible." NobelPrize.org 2022.

"The acceptance of quantum mechanics, in accordance with the orthodox point of view, as expressing the ultimate and irreducible random character of physical phenomena.... The 'doctrinaire assumption' of the Copenhagen interpretation, according to which ψ is the most complete description ever attainable." Jammer 1974, 291.

"The usual interpretation of the quantum theory... involves an assumption... that the most complete possible specification of an individual system is in terms of a wave function that determines only probable results of actual measurement processes." Bohm in Jammer 1974, 283.

"Any standard interpretation then states that no further specification of the quantum mechanical state by additional parameters (hidden variables), which might enable one to predict individual outcomes rather than probabilities over an ensemble of systems, is possible." Home and Whitaker 2007, 60.

"The wave function offers the most complete possible description of an individual system." Jammer 1974, 280.

"Einstein first diagnosed the difficulty as arising from a mistaken assumption that the quantum mechanical wave function, the ψ-function, gives a complete description of the one case, as opposed to an average over the descriptions of many cases. It was the loss of an independent, objective reality...that Einstein found unacceptable. Quantum mechanics, therefore, could not be the fundamental theory of nature that Bohr claimed it to be." Kumar 2008, 273.

"This double nature of radiation (and material corpuscles) is a major property of reality, which has been interpreted by quantum mechanics in an ingenious and amazingly successful fashion. This interpretation, which is looked upon as essentially final by almost all contemporary physicists, appears to me as only a temporary way out." Einstein in Schilpp 1949, 51.

"In order to defend the completeness of quantum theory one is thus forced to assume that it makes no sense whatsoever to talk about localization of an unobserved particle." Selleri 1990, 11.

"Unless, indeed, one clings *a priori* to the thesis that the description of nature by the statistical scheme of quantum mechanics is final." Schilpp 1949, 674.

"Einstein speaking the utterances of a hypothetical person: 'True, I admit that the quantum-theoretical description is an incomplete description of the individual system. I even admit that a complete theoretical description is, in principle, thinkable. But I consider it proven that the search for such a complete description would be aimless.'" Schilpp 1949, 672.

"Born, Pauli, Heitler, Bohr and Margenau...consider it proved that a theoretically complete description of a system can, in essence, involve only statistical assertions." Einstein in Schilpp 1949, 666.

"But occasionally at night, when the full moon is bright, I do what in the physics community is the intellectual equivalent of turning into a werewolf: I question whether quantum mechanics is the complete and ultimate truth about the physical universe." Sir Anthony Leggett in Becker 2018, 270.

"However, this does not indicate an incompleteness of quantum theory within physics, but an incompleteness of physics within the totality of life." Pauli in Bricmont 2017, 237.

"Rosenfeld...suggested that the introduction of 'hidden variables' would be 'empty talk.'" Bricmont 2017, 205.

"This consistent mathematical scheme tells us everything which can be observed. Nothing is in nature which cannot be described by this mathematical scheme." Heisenberg in Pais 1991, 310.

"[Contemporary physicists] somehow believe that the quantum theory provides a description of reality, and even a complete description." Einstein 2011, 43.

"Einstein: 'The attempt to conceive the quantum-theoretical description as the complete description of the individual systems leads to unnatural theoretical interpretations.'" Einstein in Schilpp 1949, 671.

"However, there's no language 'deeper than quantum mechanics' that could be used to interpret quantum mechanics." Motl 2011.

"The view of the status of quantum mechanics which Bohr and Heisenberg defended was, quite simply, that quantum mechanics was the last, the final, the never-to-be-surpassed revolution in physics." Popper 1992, 6.

"Bohr was saying that nature at the quantum level is probabilistic and that there is no deeper, underlying reality." Sen 2022, 161.

"No completely deterministic mechanism that could explain correctly the observed wave-particle duality of the properties of matter is even conceivable." Bohm 1989, 115.

"It seems, therefore, almost certainly of no use to search for hidden variables." Bohm 1951, 115.

"It [the Copenhagen interpretation] said that quantum theory...was a complete description of reality." Deutsch 2011, 308.

Chapter 33: Everything a Great and Good Man Could Be

1. Bernstein 1991, 66.
2. Bohm 1980, 107.
3. Pais 1991, 14.
4. Beller 1999b, 270.
5. Popper 1992, 6.
6. Popper 1992, 9.
7. In Beller 1999b, 265.
8. von Weizsäcker 1988, 3.
9. French 1985, 353.
10. Wheeler 1985, 226.
11. Schrödinger 1926.
12. Gilder 2009, 53.
13. In Kumar 2008, 227.
14. In Kumar 2008, 131.
15. Pais 1991, 29.
16. In Becker 2018, 31.
17. In Becker 2018, 31.
18. Pais 1991, 14.
19. Feynman 1985, 132.
20. Beller 1999a, 252.
21. Pais 1991, 5.
22. Beller 1999a, 254.
23. Beller 1999b, 274–275.
24. Pais 1991, 6.
25. Beller 1999a, 260.
26. In Beller 1999b, 273.

Chapter 34: The Compton Effect

1. Bohr 2013.
2. Kumar 2008, 106.

3. Beller 1999a, 260.
4. Heisenberg 1971, 75.
5. Beller 1999a, 260.
6. Beller 1999a, 260.
7. Heisenberg 1963.
8. Dresden 1987, 97.
9. Dresden 1987, 292.
10. Dresden 1987, 294.
11. Einstein 1990, 395.
12. Millikan 1917, 230.
13. Millikian 1916, 355.
14. Bohr 1922, 14.
15. Dresden 1987, 293.
16. Dresden 1987, 290.
17. Dresden 1987, 296.
18. Dresden 1987, 293.
19. Dresden 1987, 293.
20. Kramers and Holst 1923, 174.
21. Kramers and Holst 1923, 175.
22. Dresden 1987, 194.

Chapter 35: Our Quantum "Culture"

1. I have stolen the title of this chapter from the excellent *Making Sense of Quantum Mechanics* by Jean Bricmont.
2. Schilpp 1949, 671.
3. Motl 2011.
4. "Wigner's Friend" 2023.
5. Wigner 1967, 172.
6. Bell 1987, 117.
7. Chopra 1989, 108.
8. Herbert 1993, 5.
9. Herbert 1993, 179.
10. Today the film is probably unwatchable, which is for the best since the film-makers seemed to have the same scruples as a lowlife guru. For example, they interviewed the philosopher David Albert for four hours, and he told them that quantum physics has nothing to do with consciousness. Since this was not the material they wanted, they just aggressively edited everything so that he seemed to be saying quantum physics had everything to do with consciousness.
11. Patterson 2022.
12. Bell 1987, 170.
13. Rosenblum and Kuttner 2006, 12, 15, 99.
14. Rosenblum and Kuttner 2006, 201.
15. Kriss 2016.
16. Becker 2018, 136.

17. Becker 2018, 123.
18. Becker 2018, 123–124.
19. Steven Strogatz, in an endorsement of Sean Carroll's *Something Deeply Hidden*.
20. Oxford University Press publicizing Saunders et al.'s *Many Worlds?*
21. Ball 2018.
22. Wells 2013.
23. Carroll 2019, title of Chapter 2.
24. Hossenfelder 2018, 101.
25. Hossenfelder 2018, 205.
26. Ball 2018.
27. Everett 1956, 8.
28. Everett 1956, 9.
29. Carroll 2019, 40–41.
30. Carroll 2019, 20.
31. Sean Carroll, in conversation with the author, January 21, 2022.
32. Vaidman 2021.

Chapter 36: The Shadow Physics of Our Time

1. Mendyk et al. 2019, 161.
2. Holland 1993, 1.
3. Heisenberg 1927, 83.
4. Jaynes 2003, 328.
5. Reitman 1984.
6. Mead 2002, 1.
7. Carver Mead, conversation with the author, May 3, 2022.
8. Baggott 2013, x.
9. Hossenfelder 2020.
10. Smolin 2006, 75.
11. Horgan 1996, 89.
12. Horgan 2003, 175.
13. Ferry 2019, 359.
14. Woit 2006, 255.
15. Baggott 2013, 209.
16. Jaynes 2003, 329.
17. Lakatos 1978, 58.
18. Popper 1992, 156.
19. Russell 1967, 374.
20. Hossenfelder 2018, 205.
21. Hossenfelder 2018, 127.
22. Hossenfelder 2018, 127.
23. Mermin 2012, 8.
24. Jaynes 1989, 15.
25. Einstein 1987, 119.

Part X: Interlude

1. Hardesty 2014.

Chapter 37: Parable of d'Alembert's Paradox

1. Bush 2015, 288.
2. "Physics of Gas Flow" 1956, 343.
3. Stewartson 1981, 308.

Chapter 38: Droplet Statistics

1. Shukla et al. 2020, 2.
2. Harris et al. 2013.
3. Harris et al. 2013.
4. Bush and Oza 2021, 24.
5. Sáenz et al. 2018, 315.

Part XI: Return to Clarity

1. Dirac 1963.

Chapter 39: Bene Respondere!

1. Sen, Basu, and Sengupta 1995, 426.
2. Bohm 1989, 160.
3. Beller 1999b, 228.
4. Becker 2018, 40.
5. In Heilbron 1988, 206.
6. Pais 1991, 315.
7. In Beller 1999b, 159.
8. Heilbron 1988, 206.
9. Beller 1999b, 119.
10. Wheeler 1963, 30.
11. Pais 1991, 23.
12. Bohm 1989, 161.
13. Beller 1999b, 200.
14. In Beller 1999b, 279.
15. In Howard 1985, 178.
16. Bohr 1933, 458.
17. Quoted in Beller 1999b, 119.
18. Beller 1999b, 264.
19. Beller 1999b, 243–244.
20. Heisenberg 1927, 83.
21. Beller 1999b, 257.
22. Landsman 2006, 215.

Chapter 40: One Might Say Far-Fetched

1. Becker 2018, 75.
2. Labatut 2021, 104–105.
3. Beller 1999b, 195.
4. Heisenberg 1979, 44.
5. Beller 1999b, 288.
6. Beller 1999b, 199.
7. Heisenberg 1979, 16–17.
8. Heisenberg 1979, 17.
9. Heisenberg 1979, 17–18.
10. Heisenberg 1979, 25.
11. Beller 1999b, 196.
12. Beller 1999b, 191.
13. Beller 1999b, 210.
14. Beller 1999b, the title of Chapter 9.
15. Beller 1999b, 192.
16. Beller 1999b, 192.
17. In Bricmont 2016, 5.
18. Heisenberg 1959, 129.
19. In Bricmont 2016, 8.
20. Davies and Brown 1993, 67.
21. In Bricmont 2016, 10.
22. Beller 1999b, 278.
23. Becker 2018, 92.
24. Becker 2018, 93.
25. Bohm 1986.
26. Beller 1999b, 206.
27. Bohm 1989, 29.
28. Bohm 1989, 623.
29. Bohm 1989, 135.
30. Bohm 1957, 95.
31. Bohm 1957, 82.
32. Bohm 1957, 64.
33. Bohm 1957, 82.

Chapter 41: Discarded Diamonds

1. Gell-Mann 1994, 170.
2. Gell-Mann 1994, 170.
3. Bohm 1986.
4. Bohm 1986.
5. Bohm 1986.
6. Becker 2018, 149.
7. Becker 2018, 90.

8. Becker 2018, 90.
9. Becker 2018, 90.
10. Jacobsen 2007, 3.
11. Becker 2018, 109.
12. Becker 2018, 109.
13. Becker 2018, 107–108.
14. Becker 2018, 107.
15. Becker 2018, 107.
16. de Broglie 1952, 289.
17. de Broglie 1952, 290.
18. de Broglie 1952, 301.
19. de Broglie 1952, 302.
20. de Broglie 1952, 303.
21. de Broglie 1952, 308.
22. de Broglie 1952, 310.
23. de Broglie 1952, 311.
24. *Hâtez-vous lentement, et sans perdre courage,*
    vingt fois sur le métier, remettez votre ouvrage.
    Polissez-le sans cesse, et le repolissez.
    Ajoutez quelquefois, et souvent effacez.
25. Einstein 1953a.

Chapter 42: It Falls Apart in Your Hands

1. In Jammer 1974, 272.
2. Jammer 1974, 277.
3. de Broglie 1953, 217.
4. de Broglie 1952, 306.
5. Crull and Bacciagaluppi 2016, 7.
6. Crull and Bacciagaluppi 2016, 251.
7. Jammer 1974, 273.
8. Soler 2009, 330.
9. Wick 1995, 68.
10. Bohm 1957, 95.
11. Bernstein 1991, 12.
12. Bernstein 1991, 13.
13. Bernstein 2011.
14. Bernstein 1991, 53.
15. Bernstein 1991, 65.
16. Bernstein 1991, 65.
17. Bernstein 1991, 54.
18. Bell 1987, 1.
19. Bell 1987, 5.
20. Bell 1987, 6.

21. Bell 1987, 8.
22. In Bricmont, 2016, 257.

Chapter 43: Unintuitive Droplets

1. Einstein 1972, 425.
2. Wolff and Haselhurst 2004, 1.
3. Griffiths 2017, 171.
4. Ohanian 1986, 500.
5. Hestenes 1990, 1213.
6. Hestenes 1990, 1213.
7. Hestenes 1990, 1214.
8. Bernard-Bernardet et al. 2022.
9. Kumar 2008, 243.
10. Technically, we are also assuming the absence of all external heat sources or sinks.
11. Englert et al. 1992, 1175.
12. Scully 1998, 42.
13. Frumkin et al. 2022, L010203-1.

Chapter 44: Parable of Plate Tectonics

1. Romm 1994, 407.
2. Lewis 2000, 5.
3. Conniff 2012.
4. Conniff 2012.
5. Conniff 2012.
6. Conniff 2012.
7. Conniff 2012.
8. Conniff 2012.
9. Conniff 2012.
10. Conniff 2012.
11. Conniff 2012.
12. Lewis 2000, 157.
13. McKie 2012.
14. Lewis 2000, 238.
15. Lewis 2000, 238.
16. Lewis 2000, 159.
17. Sand-Jensen 2007, 725.

Part XIII: Hidden Variables Today

1. Valentini 2000.

Chapter 45: A Bolt from the Blue

1. Isaacson 2007, 401.

2. "Boris Podolsky" 2023.
3. Becker 2018, 56–57.
4. Bohr 1996, 251.
5. Bohr 1996, 252.
6. Landsman 2006, 217.
7. Bohr 1996, 251.
8. See Becker 2018, 55.
9. Heisenberg 1959, 160.
10. Isaacson 2007, 449.

Chapter 46: Entanglement

1. Bohm 1989, 614.
2. Bell 1987, 30.
3. Einstein et al. 1935, 780.
4. Bell 1987, 83.
5. Chen 1979.
6. Bernstein 1991, 84.
7. Maudlin 2022.
8. Brunner et al. 2014, 471.

Chapter 47: Bell's Theorem

1. Myrvold, Genovese, and Shimony 2019.
2. Quoted in Myrvold, Genovese, and Shimony 2019.

Chapter 48: It's as If Reality...Didn't Exist

1. Becker 2018, 211.
2. Wick 1995, 117.
3. Becker 2018, 194.
4. See, for example, Adam Becker's *What Is Real?* and David Kaiser's *How the Hippies Saved Physics.*
5. NobelPrize.org 2022.
6. NobelPrize.org 2022.
7. In Bricmont 2016, 10.
8. Rovelli 2021, xiii.
9. Motl 2011.
10. Adams 2013.
11. Maudlin 2014, 424010-2.
12. Wiseman 2014, 468.
13. Brunner et al. 2014, 471.
14. Bernstein 1991, 76.

Chapter 49: The Fourth Assumption

1. Khrennikov 2009, 170.

 2. Gill 2014, 2.
 3. Nieuwenhuizen 2009, 129.
 4. Jaynes 2003, xxvii.
 5. Boole 1862, 230.
 6. Khrennikov 2010, 28.
 7. Hess and Philipp 2005.
 8. Vorob'ev 1962.
 9. Pitowsky 1989, 38.
10. Khrennikov 2008b, 1451.
11. Hess and Philipp 2005, 1757.
12. Nieuwenhuizen 2009, 131.
13. Ana María Cetto, phone call with the author, November 29, 2022.
14. Hess and Philipp 2005, 1765–1766.
15. Khrennikov 2008b, 1451.
16. Khrennikov 2010, 26.
17. Khrennikov 2008b, 1449.
18. Kupczynski 2023, 9.
19. Khrennikov 2009, viii.
20. See Cetto 2022 and Kupczynski 2017a.
21. Khrennikov 2008b, 1452.
22. Cetto 2022, 6.
23. Khrennikov 2008a, 22.
24. Nieuwenhuizen 2009, 130.
25. Khrennikov 2008b, 1452.

Chapter 51: Parable of Icicles

 1. Ogawa and Furukawa 1999, 1.
 2. Gorman 2015.
 3. Gorman 2015.

Part XV: The Ancient Strength of Rational Thinking

 1. Comte 1988, 53.
 2. Jeans 1930, 133.

Chapter 52: Positivism, Again?

 1. Ferry 2019, 338.
 2. Motl 2011.
 3. Szymborski 1989, 163.
 4. Jaynes 2003, 327–328.
 5. Nasar 1998, 220–221.
 6. Becker 2018, 212.
 7. Bell 1987, 167.

Chapter 53: The Lion's Share

1. Frankel 2012, 39.
2. Frankel 2012, 41.
3. Hesse 1966, 4–5.
4. Wolchover 2014.
5. Wolchover 2014.
6. Sean Carroll, conversation with the author, January 21, 2022.
7. John Bush, conversation with the author, May 7, 2021.
8. *The Colbert Report* 2014.
9. Cetto and de la Peña 2015.
10. The best theory of particle physics, called the standard model, predicts that neutrinos are massless. However, this prediction is at odds with experiments.
11. Bush et al. forthcoming.
12. John Bush, conversation with the author, July 4, 2023.
13. In Bricmont 2017, 24.
14. Bush and Oza 2021, 25.
15. Jean Bricmont, conversation with the author, February 10, 2022.
16. de Broglie 1953, 228.
17. de Broglie 1987, 22.
18. Garuccio and Laurora 2021.

Chapter 54: Parable of Extramission

1. Winer et al. 2002.
2. Zubairy 2016, 9.
3. Koestler 1969, 14.
4. O'Regan 2010, note 2.

Chapter 55: Wisdom and Folly

1. Einstein 2011, 44.
2. Garuccio and Laurora 2021.
3. Bush and Oza 2021, 017001-34.
4. Kupczynski 2023.
5. Bohr 1985, 13.
6. Pais 1991, 13.
7. In Laplace 1951, 4.
8. Griffiths 2017, 643.

Bibliography

Abragam, Anatole. 1988. "Louis Victor Pierre Raymond De Broglie. 15 August 1892–19 March 1987." In *Biographical Memoirs of Fellows of the Royal Society* 34 (December), 21–41. https://doi.org/10.1098/rsbm.1988.0002.

Adams, Allan. 2013. "Lecture 1: Introduction to Superposition." MIT 8.04 Quantum Physics. https://www.youtube.com/watch?v=lZ3bPUKo5zc.

Al-Khalili, Jim. 2003. *Quantum: A Guide for the Perplexed*. London: Weidenfeld & Nicolson.

Andersen, Anders, Jacob Madsen, Christian Reichelt, Sonja Rosenlund Ahl, Benny Lautrup, Clive Ellegaard, Mogens T. Levinsen, and Tomas Bohr. 2015. "Double-Slit Experiment with Single Wave-Driven Particles and Its Relation to Quantum Mechanics." *Physical Review E* 92, 013006 (July). https://doi.org/10.1103/PhysRevE.92.013006.

Arago, François. 1858. "Fresnel, biographie lue en séance publique de l'Académie des Sciences le 26 juillet 1830." In *Oeuvres complètes de François Arago*, vol 1. Edited by J. A. Barral, 185–187. Paris: Gide.

Aristotle. 1933. *Metaphysics, Volume I: Books 1–9*. Translated by Hugh Tredennick. Loeb Classical Library 271. Cambridge, MA: Harvard University Press.

Bacciagaluppi, Guido, and Antony Valentini. 2009. *Quantum Theory at the Crossroads: Reconsidering the 1927 Solvay Conference*. Cambridge, UK: Cambridge University Press.

Baggott, Jim. 2013. *Farewell to Reality: How Modern Physics Has Betrayed the Search for Scientific Truth*. New York: Pegasus Books.

Baggott, Jim. 2020. *Quantum Reality: The Quest for the Real Meaning of Quantum Mechanics—a Game of Theories*. Oxford, UK: Oxford University Press.

Baker, F. Todd. 2015. *Atoms and Photons and Quanta, Oh My! Ask the Physicist About Atomic, Nuclear, and Quantum Physics*. San Rafael, CA: Morgan & Claypool Publishers.

Ball, Philip. 2018. "Why the Many-Worlds Interpretation Has Many Problems." *Quanta Magazine*, October 18. https://www.quantamagazine.org/why-the-many-worlds-interpretation-has-many-problems-20181018/.

Ballentine, Leslie E. 1970. "The Statistical Interpretation of Quantum Mechanics." *Reviews of Modern Physics* 42, no. 4 (October), 358–381.

Ballentine, Leslie E., ed. 1986. *Foundations of Quantum Mechanics Since the Bell Inequalities: Selected Reprints*. College Park, MD: American Association of Physics Teachers.

Ballentine, Leslie. 1988. *Foundations of Quantum Mechanics Since the Bell Inequalities*. American Association of Physics Teachers.

Becker, Adam. 2018. *What Is Real? The Unfinished Quest for the Meaning of Quantum Physics*. New York: Basic Books.

Bell, John Stewart. 1987. *Speakable and Unspeakable in Quantum Mechanics: Collected Papers on Quantum Philosophy*. Cambridge, UK: Cambridge University Press.

Beller, Mara. 1999a. "Jocular Commemorations: The Copenhagen Spirit." *Osiris* 14, 252–273.

Beller, Mara. 1999b. *Quantum Dialogue: The Making of a Revolution*. Chicago: University of Chicago Press.

Bernard-Bernardet, S., M. Fleury, and E. Fort. 2022. "Spontaneous Emergence of a Spin State for an Emitter in a Time-Varying Medium." *The European Physical Journal Plus* 137, no. 4, 1–8.

Bernstein, Jeremy. 1991. *Quantum Profiles*. New York: Oxford University Press.

Bernstein, Jeremy. 2011. "Von Neumann, Bell and Bohm." arXiv preprint 1102.2222.

Bernstein, Peter L. 1996. *Against the Gods: The Remarkable Story of Risk*. New York: John Wiley & Sons.

Bhattacharya, Ananyo. 2021. *The Man from the Future: The Visionary Life of John von Neumann*. W. W. Norton & Company.

Bloch, Felix. 1976. "Heisenberg and the Early Days of Quantum Mechanics." *Physics Today* 29, no. 12 (December), 23–27.

Bohm, David. 1952a. "A Suggested Interpretation of the Quantum Theory in Terms of Hidden Variables, I." *Physical Review* 85 (January), 166–179.

Bohm, David. 1952b. "A Suggested Interpretation of the Quantum Theory in Terms of Hidden Variables, II." *Physical Review* 85 (January), 180–193.

Bohm, David. 1957. *Causality and Chance in Modern Physics*. London: Routledge.

Bohm, David. 1980. *Wholeness and the Implicate Order*. London: Routledge.

Bohm, David. 1986. "David Bohm—Session IV." Interview of David Bohm by Maurice Wilkins. September 25. Niels Bohr Library & Archives, American Institute of Physics, College Park, MD. www.aip.org/history-programs/niels-bohr-library/oral-histories/32977-4.

Bohm, David. 1989. *Quantum Theory*. New York: Dover Publications.

Bohr, Niels. 1922. "The Structure of the Atom." Nobel lecture, December 11, 1922. https://www.nobelprize.org/uploads/2018/06/bohr-lecture.pdf.

Bohr, Niels. 1928. "The Quantum Postulate and the Recent Development of Atomic Theory." *Nature* 121, 580–590.

Bohr, Niels. 1933. "Light and Life." *Nature* 131, 457–459.

Bohr, Niels. 1935. "Can Quantum-Mechanical Description of Physical Reality Be Considered Complete?" *Physical Review* 48 (October), 696–702.

Bohr, Niels. 1985. *Foundations of Quantum Physics I (1926–1932)*. Edited by Jørgen Kalckar. *Niels Bohr Collected Works Volume 6*. Amsterdam: North Holland.

Bohr, Niels. 1996. *Foundations of Quantum Physics II (1933–1958)*. Edited by Jørgen Kalckar. *Niels Bohr Collected Works Volume 7*. Amsterdam: North Holland.

Bohr, Vilhelm. 2013. "Niels Bohr—Life Behind the Physics." Perimeter Institute for Theoretical Physics. https://www.youtube.com/watch?v=Zt5SdirL-ys%7D.

Boole, George. 1862. "On the Theory of Probabilities." *Philosophical Transactions of the Royal Society of London* 152, no. 152 (December), 225–252.

"Boris Podolsky." 2023. Wikipedia, accessed November 27, 2023. https://en.wikipedia.org/wiki/Boris_Podolsky.

Born, Max. 1949. *Natural Philosophy of Cause and Chance*. Oxford: Clarendon Press. Scanned version available via Internet Archive: https://ia601501.us.archive.org/1/items/in.ernet.dli.2015.204850/2015.204850.Natural-Philosophy_text.pdf.

Born, Max, and Albert Einstein. 1971. *The Born–Einstein Letters: Friendship, Politics and Physics in Uncertain Times*. Translated by Irene Born. London: Macmillan.

Born, Max, and Emil Wolf. 1999. *Principles of Optics: Electromagnetic Theory of Propagation, Interference and Diffraction of Light*. 7th edition. Cambridge, UK: Cambridge University Press.

Bourdeau, Michel. 2023. "Auguste Comte." *Stanford Encyclopedia of Philosophy Archive*. https://plato.stanford.edu/archives/spr2023/entries/comte/.

Boyer, Carl B. 1989. *A History of the Calculus and Its Conceptual Development*. New York: Dover Publications.

Boyle, Robert. 1744–1765. "On the Mechanical Origin of Heat and Cold." In *The Works of the Honourable Robert Boyle*. Vol. 4. London: A. Millar.

Bricmont, Jean. 2016. *Making Sense of Quantum Mechanics*. Cham, Switzerland: Springer.

Bricmont, Jean. 2017. *Quantum Sense and Nonsense*. Cham, Switzerland: Springer.

Briggs, William L., and Van Emden Henson. 1995. *The DFT: An Owner's Manual for the Discrete Fourier Transform*. Philadelphia: SIAM.

Bromley, David G., and J. Reed Braden. 2016. "Religion of Humanity." World Religions and Spirituality Project. https://wrldrels.org/2016/10/08/reliigion-of-humanity/.

Brougham, Henry. 1803. "Bakerian Lecture on Light and Colours." *Edinburgh Review* 1, 450–456.

Brunner, Nicolas, Daniel Cavalcanti, Stefano Pironio, Valerio Scarani, and Stephanie Wehner. 2014. "Bell Nonlocality." *Reviews of Modern Physics* 86, no. 2 (April), 419–478.

Bryson, Gladys. 1936. "Early English Positivists and the Religion of Humanity." *American Sociological Review* 1, no. 3 (June), 343–362.

Buchwald, Jed Z., and Robert Fox, eds. 2013. *The Oxford Handbook of the History of Physics*. Oxford: Oxford University Press.

Burton, Harry Edwin, trans. 1943. "The Optics of Euclid." *Journal of the Optical Society of America* 35, no. 5 (May), 357–372.

Bush, John W. M. 2015. "Pilot-Wave Hydrodynamics." *Annual Review of Fluid Mechanics* 47 (January), 269–292. https://doi.org/10.1146/annurev-fluid-010814-014506.

Bush, John W. M., Valeri Frumkin, and Konstantinos Papatryfonos. Forthcoming. "The State of Play in Hydrodynamic Quantum Analogs." In *Advances in Pilot Wave Theory*, edited by Paulo Castro, John W. M. Bush, and José Croca. Berlin: Springer.

Bush, John W. M., and Anand U. Oza. 2021. "Hydrodynamic Quantum Analogues." *Reports on Progress in Physics* 84, no. 1 (January), 017001.

Byrne, Oliver. 2010. *The First Six Books of the Elements of Euclid.* Cologne: Taschen.

Carroll, Sean. 2012. *The Particle at the End of the Universe: How the Hunt for the Higgs Boson Leads Us to the Edge of a New World.* New York: Dutton.

Carroll, Sean. 2019. *Something Deeply Hidden: Quantum Worlds and the Emergence of Spacetime.* New York: Dutton.

Carson, Cathryn. 2010. *Heisenberg in the Atomic Age: Science and the Public Sphere.* New York: Cambridge University Press.

Cercignani, Carlo. 2006. *Ludwig Boltzmann: The Man Who Trusted Atoms.* Oxford, UK: Oxford University Press.

Cetto, Ana María. 2022. "Electron Spin Correlations: Probabilistic Description and Geometric Representation." *Entropy* 24, no. 10, 1439. https://doi.org/10.3390/e24101439.

Cetto, Ana María, and Louis de la Peña. 2015. *The Emerging Quantum: The Physics Behind Quantum Mechanics.* Berlin: Springer.

Chen, Edwin. 1979. "Twins Reared Apart: A Living Lab." *New York Times*, December 9. https://www.nytimes.com/1979/12/09/archives/twins-reared-apart-a-living-lab.html.

"Cholera." 2023. Wikipedia, accessed October 14, 2023. https://en.wikipedia.org/wiki/Cholera.

Chopra, Deepak. 1989. *Quantum Healing: Exploring the Frontiers of Mind/Body Medicine.* New York: Bantam Books.

Colbert Report. Season 11, "Terence Tao." Aired November 12, 2014, on Comedy Central. https://www.cc.com/video/6wtwlg/the-colbert-report-terence-tao.

Comte, Auguste. 1830. *Cours de Philosophie Positive.* Vol. 1. Paris: Bachelier.

Comte, Auguste. 1831. *Cours de Philosophie Positive.* Vol. 2. Paris: Bachelier.

Comte, Auguste. 1838. *Cours de Philosophie Positive.* Vol. 3. Paris: Bachelier.

Comte, Auguste. 1896. *The Positive Philosophy of Auguste Comte.* Translated by Harriet Martineau. London: George Bell and Sons.

Comte, Auguste. 1988. *Introduction to Positive Philosophy.* Edited by Frederick Ferré. Indianapolis, IN: Hackett Publishing Company.

Comte, Auguste. 2009. *The Catechism of Positive Religion.* Translated by Richard Conureve. Cambridge: Cambridge University Press.

Conniff, Richard. 2012. "When Continental Drift Was Considered Pseudoscience." *Smithsonian Magazine* (June). https://www.smithsonianmag.com/science-nature/when-continental-drift-was-considered-pseudoscience-90353214/.

Couchman, Miles M. P., Davis J. Evans, and John W. M. Bush. 2022. "The Stability of a Hydrodynamic Bravais Lattice." *Symmetry* 14, no. 8, 1524. https://doi.org/10.3390/sym14081524.

Couder, Yves, and Emmanuel Fort. 2006. "Single-Particle Diffraction and Interference at a Macroscopic Scale." *Physical Review Letters* 97, no. 15 (October), 154101.

Cox, Brian, and Jeff Forshaw. 2011. *The Quantum Universe: Everything That Can Happen Does Happen.* London: Allen Lane.

Crull, Elise, and Guido Bacciagaluppi, eds. 2016. *Grete Hermann—Between Physics and Philosophy.* Dordrecht: Springer.

"D'Alembert's Paradox." 2023. Wikipedia, revised February 25, 2023. https://en.wikipedia.org/wiki/D%27Alembert%27s_paradox.

Davies, P. C. W., and J. R. Brown. 1993. *The Ghost in the Atom: A Discussion of the Mysteries of Quantum Physics.* Cambridge, UK: Cambridge University Press.

Davies, Tony. 1997. *Humanism.* The New Critical Idiom. New York: Routledge.

Davisson, C., and L. H. Germer. 1927. "The Scattering of Electrons by a Single Crystal of Nickel." *Nature* 119 (April 16), 558–560. https://doi.org/10.1038/119558a0.

Davy, Sir Humphry. 1840. *The Collected Works of Sir Humphry Davy.* Edited by John Davy. Vol. 8. London: Smith, Elder.

de Broglie, Louis. 1925a. "Recherches sur la théorie des quanta." *Annales de Physique* 10, no. 3 (April), 22–128. https://doi.org/10.1051/anphys/192510030022.

de Broglie, Louis. 1925b. "Sur la fréquence propre de l'électron." *Comptes rendus hebdomadaires des séances de l'Académie des sciences,* 180, 498–500.

de Broglie, Louis. 1927. "La mécanique ondulatoire et la structure atomique de la matière et du rayonnement." *Journal de Physique et le Radium* 8, no. 5 (May), 225–241. https://doi.org/10.1051/jphysrad:0192700805022500.

de Broglie, Louis. 1952. "La physique quantique restera-t-elle indéterministe?" *Revue d'histoire des sciences et de leurs applications* 5, no. 4 (October–December), 289–311.

de Broglie, Louis. 1953. *The Revolution in Physics.* New York: Noonday Press.

de Broglie, Louis. 1962. *New Perspectives in Physics: Where Does Physical Theory Stand Today?* Translated by A. J. Pomerans. New York: Basic Books.

de Broglie, Louis. 1967. "Monsieur De Broglie." French News interviews Louis de Broglie on January 1, 1967. INA. https://www.ina.fr/video/AFE04002106.

de Broglie, Louis. 1987. "Interpretation of Quantum Mechanics by the Double Solution Theory." *Annales de la Fondation Louis de Broglie* 12, no. 4, 1–23.

de Ronde, Christian. 2021. "Mythical Thought in Bohr's Anti-Realist Realism (Or: Lessons on How to Capture and Defeat Smoky Dragons)." arXiv preprint 2101.00255.

Deutsch, David. 1997. *The Fabric of Reality: The Science of Parallel Universes—and Its Implications.* New York: Penguin Books.

Deutsch, David. 2011. *The Beginning of Infinity: Explanations That Transform the World.* New York: Penguin Books.

Dirac, P. A. M. 1963. "P. A. M. Dirac—Session V." Interview of P. A. M. Dirac by Thomas S. Kuhn. May 14. Niels Bohr Library & Archives, American Institute of Physics, College Park, MD. www.aip.org/history-programs/niels-bohr-library/oral-histories/4575-5.

Drake, Ellen Tan. 2006. "Hooke's Ideas of the Terraqueous Globe and a Theory of Evolution." In *Robert Hooke: Tercentennial Studies*, edited by Michael Cooper and Michael Hunter, 135–152. London: Routledge.

Dresden, Max. 1987. *H. A. Kramers: Between Tradition and Revolution*. Berlin: Springer.

Dürr, Detlef, Sheldon Goldstein, and Nino Zanghì. 2013. *Quantum Physics Without Quantum Philosophy*. Berlin: Springer.

Dyson, Freeman J. 2007. "Why Is Maxwell's Theory So Hard to Understand?" Second European Conference on Antennas and Propagation (EuCAP 2007). IEEE Xplore. https://doi.org/10.1049/ic.2007.1146.

Eddi, A., E. Fort, F. Moisy, and Y. Couder. 2009. "Unpredictable Tunneling of a Classical Wave-Particle Association." *Physical Review Letters* 102, no. 24, 240401. https://doi.org/10.1103/PhysRevLett.102.240401.

Eddi, A., J. Moukhtar, S. Perrard, E. Fort, and Y. Couder. 2012. "Level Splitting at Macroscopic Scale." *Physical Review Letters* 108, no. 26, 264503. https://doi.org/10.1103/PhysRevLett.108.264503.

Einstein, Albert. 1916. "Ernst Mach." *Physikalische Zeitschrift* 17, 101–104.

Einstein, Albert. 1931. "Maxwell's Influence on the Development of the Conception of Physical Reality." In *James Clerk Maxwell: A Commemoration Volume, 1831–1931*, edited by Joseph John Thomson, Max Planck, Albert Einstein, and Others, 66–73. Cambridge, UK: Cambridge University Press.

Einstein, Albert. 1949. "Notes for an Autobiography." *The Saturday Review of Literature*, November 26, 9–12. https://archive.org/details/EinsteinAutobiography.

Einstein, Albert. 1951. "The Advent of the Quantum Theory." *Science* 113, no. 2926 (January), 82–84.

Einstein, Albert. 1952. "Typed Letter Signed ('A. Einstein') to Daniel M. Lipkin, Princeton, 5 July, 1952." https://www.christies.com/en/lot/lot-6210437.

Einstein, Albert. 1953a. Letter to David Bohm, February 17, 1953. Available from the Raab Collection. https://www.raabcollection.com/literary-autographs/einstein-bohm.

Einstein, Albert. 1953b. Letter to Erwin Schrödinger, June 17, 1953.

Einstein, Albert. 1954. *Ideas and Opinions*. New York: Crown Publishers.

Einstein, Albert. 1972. *Correspondance avec Michele Besso 1903–1955*. Paris: Hermann.

Einstein, Albert. 1987. *Letters to Solovine*. New York: Philosophical Library.

Einstein, Albert. 1990. "On the Development of Our Views Concerning the Nature and Constitution of Radiation." In *The Collected Papers of Albert Einstein, Volume 2: The Swiss Years: Writings, 1900–1909 (English Translation Supplement)*, edited by John Stachel, David C. Cassidy, Jürgen Renn, and Robert Schulmann. Translated by Anna Beck. Princeton, NJ: Princeton University Press, 379–394.

Einstein, Albert. 1998. *The Collected Papers of Albert Einstein, Volume 8: The Berlin Years: Correspondence, 1914–1918,* edited by József Illy, A. J. Kox, and Michel Janssen. Princeton, NJ: Princeton University Press.

Einstein, Albert. 2011. *Letters on Wave Mechanics: Correspondence with H. A. Lorentz, Max Planck, and Erwin Schrödinger.* Edited by K. Przibram. New York: Philosophical Library.

Einstein, Albert. 2012. *The Collected Papers of Albert Einstein, Volume 13: The Berlin Years: Correspondence, January 1922–March 1923,* edited by Diana Kormos Buchwald, József Illy, Ze'ev Rosenkranz, and Tilman Sauer. Princeton, NJ: Princeton University Press.

Einstein, Albert. 2015. *The Collected Papers of Albert Einstein, Volume 14: The Berlin Years: Writings & Correspondence, April 1923–May 1925 (English Translation Supplement),* edited by Diana Kormos Buchwald, József Illy, Ze'ev Rosenkranz, Tilman Sauer, and Osik Moses. Princeton, NJ: Princeton University Press.

Einstein, Albert, Boris Podolsky, and Nathan Rosen. 1935. "Can Quantum-Mechanical Description of Physical Reality Be Considered Complete?" *Physical Review* 47, no. 10, 777–780.

Ellegaard, Clive, and Mogens T. Levinsen. 2020. "Interaction of Wave-Driven Particles with Slit Structures." *Physical Review E* 102 (August), 023115. https://doi.org/10.1103/PhysRevE.102.023115.

Englert, Berthold-Georg, Marian O. Scully, Georg Süssmann, and Herbert Walther. 1992. "Surrealistic Bohm Trajectories." *Zeitschrift für Naturforschung A* 47, no. 12, 1175–1186.

"Erwin Schrödinger." 2023. Wikipedia. Revised November 6, 2023. https://en.wikipedia.org/wiki/Erwin_Schr%C3%B6dinger.

Everett, Hugh III. 1956. "The Theory of the Universal Wave Function." PhD diss., Princeton University. https://archive.org/details/TheTheoryOfTheUniversalWaveFunction/mode/1up.

Fargue, D. 2017. "Louis de Broglie's 'Double Solution' a Promising but Unfinished Theory." *Annales de la Fondation Louis de Broglie* 42, numéro spécial, 9–18.

Faye, Jan. 2019. "Copenhagen Interpretation of Quantum Mechanics." Stanford Encyclopedia of Philosophy Archive. https://plato.stanford.edu/archives/win2019/entries/qm-copenhagen/.

Fermat and Pascal on Probability. 1654–1660. https://www.york.ac.uk/depts/maths/histstat/pascal.pdf. Accessed Jan 28, 2024.

Ferry, David. 2019. *The Copenhagen Conspiracy.* Dubai, UAE: Jenny Stanford Publishing.

Feyerabend, Paul. 1995. *Killing Time: The Autobiography of Paul Feyerabend.* Chicago: University of Chicago Press.

Feynman, Richard P. 1964. "Probability and Uncertainty: The Quantum Mechanical View of Nature." Messenger Lectures at Cornell University. BBC video. https://www.feynmanlectures.caltech.edu/fml.html#6.

Feynman, Richard P. 1967. *The Character of Physical Law*. Cambridge, MA: MIT Press.

Feynman, Richard P. 1985. *Surely You're Joking, Mr. Feynman!* New York: W. W. Norton & Company.

Feynman, Richard P. 2010. *The Feynman Lectures on Physics: The New Millennium Edition*. Edited by Robert B. Leighton and Matthew Sands. 3 vols. Available online through Caltech, https://www.feynmanlectures.caltech.edu/.

Fine, Arthur. 1986. *The Shaky Game: Einstein, Realism, and the Quantum Theory*. Chicago: University of Chicago Press.

Fine, Arthur. 2007. "Bohr's Response to EPR: Criticism and Defense." *Iyyun: The Jerusalem Philosophical Quarterly* 56 (January), 1–26. http://faculty.washington.edu/afine/Iyyun.pdf.

Fort, Emmanuel, Antonin Eddi, Arezki Boudaoud, Julien Moukhtar, and Yves Couder. 2010. "Path-Memory Induced Quantization of Classical Orbits." *PNAS* 107, no. 41, 17515–17520.

Fourier, Joseph. 1822. *Théorie analytique de la chaleur*. Paris: Chez Firmin Didot, père et fils.

Fox, R. 1971. *The Caloric Theory of Gases: From Lavoisier to Regnault*. Oxford, UK: Clarendon Press.

Frankel, Henry R. 2012. *The Continental Drift Controversy: Volume 1: Wegener and the Early Debate*. Cambridge, UK: Cambridge University Press.

French, A. P. 1985. "Some Closing Reflections." In *Niels Bohr: A Centenary Volume*. Edited by A. P. French and P. J. Kennedy. Cambridge, MA: Harvard University Press, 351–353.

Frumkin, Valeri, David Darrow, John W. M. Bush, and Ward Struyve. 2022. "Real Surreal Trajectories in Pilot-Wave Hydrodynamics." *Physical Review A* 106, no. 1, L010203. https://doi.org/10.1103/PhysRevA.106.L010203.

Garuccio, Augusto, and Angela Laurora. 2021. "Augusto Garuccio & Angela Laurora—John Bell's Unpublished Notes About de Broglie's Pilot Wave." Centro de Filosofia das Ciências UL. September 13, 2021. From the International Conference on Advances in Pilot Wave Theory & HQA-2021, July 26–30, 2021. https://www.youtube.com/watch?v=tPdNIYZXVL4.

Gell-Mann, Murray. 1994. *The Quark and the Jaguar: Adventures in the Simple and the Complex*. New York: Henry Holt.

Geurdes, J. F. 2010. "CHSH and Local Hidden Causality." *Advanced Studies in Theoretical Physics* 4, no. 20, 945–949.

Gilder, L. 2009. *The Age of Entanglement: When Quantum Physics Was Reborn*. New York: Vintage.

Gill, R. D. 2014. "Comment on 'Quantum Correlations Are Weaved by the Spinors of the Euclidean Primitives.'" In *Royal Society Open Science*, 1–6. https://www.math.leidenuniv.nl/~gill/Author_tex-v2.pdf.

Gorman, James. 2015. "Why Icicles Look the Way They Do." *New York Times*, March 16. https://www.nytimes.com/2015/03/17/science/why-icicles-look-the-way-they-do.html.

Gouesbet, Gérard. 2013. *Hidden Worlds in Quantum Physics*. New York: Dover.

Green, James N., and Thomas E. Skidmore. 2021. "Positivism." In *Brazil: Five Centuries of Change*, available online at https://library.brown.edu/create /fivecenturiesofchange/chapters/ chapter-4/positivism/.

Gribbin, John. 2019. *Six Impossible Things: The Mystery of the Quantum World*. Cambridge, MA: MIT Press.

Griffiths, David J. 2017. *Introduction to Quantum Mechanics*. Second edition. Cambridge, UK: Cambridge University Press.

Grimaldi, Francesco Maria. 1665. *Physico-mathesis de lumine, coloribus, et iride, aliisque annexis*. Bologna, Italy: Victorii Benatii.

Gros, Johannès. 1908. "The Religion of Humanity and Its High Priestess (Illustrated)." *Open Court* 1908, no. 1, article 3. Available via Southern Illinois University: https://opensiuc.lib.siu.edu/cgi/viewcontent.cgi?article=2193&context =ocj.

Guillin, Vincent. 2016. "Aspects of Scientific Explanation in Auguste Comte." *European Journal of Social Sciences* 54, no. 2, 17–41.

Haldane, J. B. S. 1928 (1927). *Possible Worlds and Other Papers*. New York: Harper & Brothers.

Hall, A. R. 1948. "Sir Isaac Newton's Note-Book, 1661–65." *The Cambridge Historical Journal* 9, no. 2, 239–250.

Hardesty, Larry. 2014. "Fluid Mechanics Suggests Alternative to Quantum Orthodoxy." *MIT News*, September 12. https://news.mit.edu/2014/fluid-systems -quantum-mechanics-0912.

Harris, Daniel M., Julien Moukhtar, Emmanuel Fort, Yves Couder, and John W. M. Bush. 2013. "Wavelike Statistics from Pilot-Wave Dynamics in a Circular Corral." *Physical Review E* 88, no. 1, 011001.

Hartsfield, Tom. 2013. *Quantum Mechanics Supports Free Will*. Big Think, April 3, https://bigthink.com/articles/quantum-mechanics-supports-free-will/.

Hearnshaw, John. 2010. "Auguste Comte's Blunder: An Account of the First Century of Stellar Spectroscopy and How It Took One Hundred Years to Prove That Comte Was Wrong!" *Journal of Astronomical History and Heritage* 13, no. 2 (July), 90–104.

Hegel, G. W. F. 1977. *Phenomenology of Spirit*. Translated by A. V. Miller. Oxford, UK: Oxford University Press.

Heidarzadeh, Tofigh. 2008. *A History of Physical Theories of Comets, from Aristotle to Whipple*. Archimedes: New Studies in the History and Philosophy of Science and Technology, vol. 19. New York: Springer.

Heilbron, John L. 1988. "The Earliest Missionaries of the Copenhagen Spirit." In *Science in Reflection: The Israel Colloquium: Studies in History, Philosophy, and Sociology of Science: Volume 3*, edited by Edna Ullmann-Margalit, 201–233. Dordrecht: Springer.

Heisenberg, Werner. 1925. "Quantum-Mechanical Reinterpretation of Kinematic and Mechanical Relations." *Zeitschrift für Physik* 33, 879–893.

Heisenberg, Werner. 1927. "The Physical Content of Quantum Kinematics and Mechanics." In *Quantum Theory and Measurement*, edited by J. A. Wheeler and W. H. Zurek, 62–84. Princeton: Princeton University Press.

Heisenberg, Werner. 1958. *The Physicist's Conception of Nature*. Translated by A. J. Pomerans. New York: Harcourt Brace.

Heisenberg, Werner. 1959. *Physics and Philosophy: The Revolution in Modern Science*. London: Allen and Unwin.

Heisenberg, Werner. 1963. "Werner Heisenberg—Session VIII." Interview of Werner Heisenberg by Thomas S. Kuhn. February 25. Max Planck Institute, American Institute of Physics, College Park, MD. https://www.aip.org /history-programs/niels-bohr-library/oral-histories/4661-8.

Heisenberg, Werner. 1971. *Physics and Beyond: Encounters and Conversations*. New York: Harper & Row.

Heisenberg, Werner. 1979. *Philosophical Problems of Quantum Physics*. Woodbridge, CT: Ox Bow Press.

Heisenberg, Werner. 1983. "The Physical Content of Quantum Kinematics and Mechanics." In *Quantum Theory and Measurement*, edited by John Archibald Wheeler and Wojciech Hubert Zurek, 62–84. Princeton, NJ: Princeton University Press.

Herbert, Nick. 1993. *Elemental Mind: Human Consciousness and the New Physics*. New York: Dutton.

Herivel, John. 1975. *Joseph Fourier: The Man and the Physicist*. Oxford, UK: Clarendon Press.

Hess, Karl, and Walter Philipp. 2005. "The Bell Theorem as a Special Case of a Theorem of Bass." *Foundations of Physics* 35, no. 10, 1749–1767.

Hesse, Mary B. 1966. *Models and Analogies in Science*. Notre Dame, IN: University of Notre Dame Press.

Hestenes, David. 1990. "The Zitterbewegung Interpretation of Quantum Mechanics." *Foundations of Physics* 20, no. 10, 1213–1232.

Holland, Peter. R. 1993. *The Quantum Theory of Motion: An Account of the de Broglie–Bohm Causal Interpretation of Quantum Mechanics*. Cambridge, UK: Cambridge University Press.

Home, Dipankar, and Andrew Whitaker. 2007. *Einstein's Struggles with Quantum Theory: A Reappraisal*. New York: Springer.

Horgan, John. 1996. *The End of Science: Facing the Limits of Knowledge in the Twilight of the Scientific Age*. New York: Broadway Books.

Horgan, John. 2003. *Rational Mysticism: Dispatches from the Border Between Science and Spirituality*. New York: Houghton Mifflin.

Horgan, John. 2018. "David Bohm, Quantum Mechanics and Enlightenment." *Scientific American* blog, July 23. https://blogs.scientificamerican.com/cross -check/david-bohm- quantum-mechanics-and-enlightenment/.

Hossenfelder, Sabine. 2018. *Lost in Math: How Beauty Leads Physics Astray*. New York: Basic Books.

Hossenfelder, Sabine. 2019. "The Crisis in Physics Is Not Only About Physics." *BackReAction* blog, October 30. https://backreaction.blogspot.com/2019/10 /the-crisis-in-physics-is-not-only-about.html.

Hossenfelder, Sabine. 2020. "Why the Foundations of Physics Have Not Progressed for 40 Years." *IAI News*. https://iai.tv/articles/why-physics-has-made -no-progress-in-50-years-auid-1292.

Hossenfelder, Sabine. 2022. *Existential Physics: A Scientist's Guide to Life's Biggest Questions.* New York: Viking.

Hossenfelder, Sabine, and Tim Palmer. 2020. "Rethinking Superdeterminism." *Frontiers in Physics* 8 (May). https://doi.org/10.3389/fphy.2020.00139.

Howard, Don. 1985. "Einstein on Locality and Separability." *Studies in History and Philosophy of Science Part A* 16, no. 3, 171–201.

Huerre, Patrick. 2019. "Yves Couder 1941–2019." Obituary available online via EUROMECH News. https://euromech.org/news/Yves%20Couder/Prof.%20Yves%20Couder%20obituary.pdf.

Irion, Robert. 2000. "They've Seen a Ghost." *New Scientist*, July 8. https://www.newscientist.com/article/mg16722464-400-theyve-seen-a-ghost/.

Isaacson, Walter. 2007. *Einstein: His Life and Universe.* New York: Simon & Schuster.

Jacobsen, Anja Skaar. 2007. "Léon Rosenfeld's Marxist Defense of Complementarity." *Historical Studies in the Physical and Biological Sciences* 37, supplement 2007, 3–34. https://doi.org/10.1525/hsps.2007.37.s.3.

James, Tim. 2019. *Elemental: How the Periodic Table Can Now Explain (Nearly) Everything.* New York: Abrams.

Jammer, Max. 1974. *The Philosophy of Quantum Mechanics: The Interpretations of Quantum Mechanics in Historical Perspective.* New York: Wiley.

Jammer, Max. 1999. *Einstein and Religion: Physics and Theology.* Princeton, NJ: Princeton University Press.

Jammer, Max. 2004. "Review of 'Synthetische Quantentheorie' by Ulrich Hoyer." *Journal for General Philosophy of Science* 35, no. 2, 397–402.

Jauch, J. M., and C. Piron. 1963. "Can Hidden Variables Be Excluded in Quantum Mechanics?" *Helvetica Physica Acta* 36, no. 7, 827–837. https://access.archive-ouverte.unige.ch/access/metadata/810b38b2-cabe-4e12-9aea-85791e061bb9/download.

Jaynes, Edwin T. 1989. "Clearing Up Mysteries—the Original Goal." In *Maximum-Entropy and Bayesian Methods: Proceedings of the 8th Maximum Entropy Workshop*, edited by J. Skilling, 1–28. Dordrecht: Kluwer.

Jaynes, Edwin T. 1996 (1990). "Probability in Quantum Theory." *Proceedings Volume, Complexity, Entropy and the Physics of Information*, ed. W. H. Zurek. Reading, MA: Wesley.

Jaynes, Edwin T. 2003. *Probability Theory: The Logic of Science.* New York: Cambridge University Press.

Jeans, James. 1930. *The Universe Around Us.* London: Cambridge University Press.

Johnson, Steven. 2006. *The Ghost Map: The Story of London's Most Terrifying Epidemic—and How It Changed Science, Cities, and the Modern World.* New York: Riverhead Books.

Kaiser, David. 2012. *How the Hippies Saved Physics: Science, Counterculture, and the Quantum Revival.* New York: W. W. Norton & Company.

Khrennikov, Andrei. 2008a. "Bell-Boole Inequality: Nonlocality or Probabilistic Incompatibility of Random Variables?" *Entropy* 10, 19–32.

Khrennikov, Andrei. 2008b. "EPR-Bohm Experiment and Bell's Inequality: Quantum Physics Meets Probability Theory." *Theoretical and Mathematical Physics* 157, no. 1, 1448–1460.

Khrennikov, Andrei. 2009. *Contextual Approach to Quantum Formalism.* Fundamental Theories of Physics, vol. 160. Dordrecht: Springer.

Khrennikov, Andrei. 2010. *Ubiquitous Quantum Structure: From Psychology to Finance.* Heidelberg: Springer Berlin.

Koestler, Arthur. 1969. *The Sleepwalkers.* New York: Macmillan.

Kramers, Hendrik Anthony, and Helge Holst. 1923. *The Atom and the Bohr Theory of Its Structure.* London: Gyldendal.

Krauss, Lawrence M. 2012. "The Consolation of Philosophy." *Scientific American,* April 27. https://www.scientificamerican.com/article/the-consolation-of-philos/.

Kriss, Sam. 2016. "The Multiverse Idea Is Rotting Culture." *The Atlantic,* August 29. https://www.theatlantic.com/science/archive/2016/08/the-multiverse-as-imagination-killer/497417/.

Kronig, R. 1960. "The Turning Point." In *Theoretical Physics in the Twentieth Century: A Memorial Volume to Wolfgang Pauli,* edited by M. Fierz and V. F. Weisskopf. New York: Interscience Publishers.

Kumar, Krishna. 1996. "Linear Theory of Faraday Instability in Viscous Liquids." *Proceedings of the Royal Society A* 452, no. 1948, 1113–1126.

Kumar, Manjit. 2008. *Quantum: Einstein, Bohr, and the Great Debate About the Nature of Reality.* London: Icon Books.

Kupczynski, Marian. 2017a. "Can We Close the Bohr-Einstein Quantum Debate?" *Philosophical Transactions of the Royal Society A* 375, no. 2106 (November), 20160392.

Kupczynski, Marian. 2017b. "Is Einsteinian No-Signalling Violated in Bell Tests?" *Open Physics* 15, no. 1, 87–101.

Kupczynski, Marian. 2023. "Contextuality or Nonlocality: What Would John Bell Choose Today?" *Entropy* 25, no. 2, 280.

Labatut, Benjamín. 2021. *When We Cease to Understand the World.* New York: New York Review of Books.

Lakatos, Imre. 1978. *The Methodology of Scientific Research Programmes: Philosophical Papers Volume 1.* Edited by John Worrall and Gregory Currie. Cambridge, UK: Cambridge University Press.

Landsman, N. P. 2006. "When Champions Meet: Rethinking the Bohr-Einstein Debate." *Studies in History and Philosophy of Science Part B: Studies in History and Philosophy of Modern Physics* 37, no. 1 (March), 212–242.

Laplace, Pierre-Simon. 1830. *The System of the World.* Translated by Henry H. Harte. Vol. 1. Dublin: Longman, Rees, Orme.

Laplace, Pierre-Simon. 1951. *A Philosophical Essay on Probabilities.* Translated by F. W. Truscott and F. L. Emory. New York: Dover Publications.

Lavoisier, Antoine. (1790) 1965. *Elements of Chemistry.* Translated by Robert Kerr. Facsimile of the first edition. Edinburgh, Scotland: Dover.

"Leonardo da Vinci." 2023. Wikipedia. Revised November 4, 2023. https://en.wikipedia.org/wiki/Leonardo_da_Vinci.

Lewis, Cherry. 2000. *The Dating Game: One Man's Search for the Age of the Earth.* Cambridge, UK: Cambridge University Press.

Lindley, David. 2001. *Boltzmann's Atom: The Great Debate That Launched a Revolution in Physics.* New York: Free Press.

Mach, Ernst. 1902. "Address of Mr. Ernst Mach, Professor in the Faculty of Philosophy of the University of Vienna (trans. by Mr. Imans, read by Mr. Laporte)." *Revue Occidentale* 26, no. 2, 98.

Mach, Ernst. 1914. *The Analysis of Sensations, and the Relation of the Physical to the Psychical.* Translated by C. M. Williams. Chicago: Open Court Publishing Company.

Mach, Ernst. 1986. *Principles of the Theory of Heat: Historically and Critically Elucidated.* Edited by Brian McGuinness. Vienna Circle Collection, vol. 17. Dordrecht: Springer.

Maudlin, Tim. 2014. "What Bell Did." *Journal of Physics A: Mathematical and Theoretical* 47, no. 42 (October).

Maudlin, Tim. 2018. "The Defeat of Reason." *Boston Review*, June 1. https://www.bostonreview.net/articles/grand-delusion/.

Maudlin, Tim. 2022. "Cracking the Quantum Code: Physicist Exposes Reality." Interview of Tim Maudlin by Curt Jaimungal. October 21. Theories of Everything with Curt Jaimungal. https://www.youtube.com/watch\?v=fU1bs5o3nss.

"Max Planck." Wikipedia. Updated January 23, 2024. https://en.wikipedia.org/wiki/Max_Planck.

Maxwell, James Clerk. 1860. "Illustrations of the Dynamical Theory of Gases." *Philosophical Magazine Series 1* 20, 148–171.

McKie, Robin. 2012. "David Attenborough: Force of Nature." *The Guardian*, October 28. https://www.theguardian.com/tv-and-radio/2012/oct/26/richard-attenborough-climate-global-arctic-environment.

Mead, Carver. 2002. *Collective Electrodynamics: Quantum Foundations of Electromagnetism.* Cambridge, MA: MIT Press.

Mehra, Jagdish. 1975. *The Solvay Conferences on Physics: Aspects of the Development of Physics Since 1911.* Dordrecht: Springer.

Mendyk, Robert W., Adam Weisse, and Will Fullerton. 2019. "A Wake-Up Call for Sleepy Lizards: The Olfactory-Driven Response of Tiliqua rugosa (Reptilia: Squamata: Sauria) to Smoke and Its Implications for Fire Avoidance Behavior." *Journal of Ethology* 38, no. 2 (December), 161–166.

Mermin, N. David. 1981. "Quantum Mysteries for Anyone." *Journal of Philosophy* 78, no. 7 (July), 397–408.

Mermin, N. David. 1985. "Is the Moon There When Nobody Looks? Reality and the Quantum Theory." *Physics Today* 38, no. 4 (April), 38–47.

Mermin, N. David. 1989. "What's Wrong with This Pillow?" *Physics Today* 42, no. 4 (April), 9–11.

Mermin, N. David. 2012. "Commentary: Quantum Mechanics: Fixing the Shifty Split." *Physics Today* 65, no. 7 (July), 8–10.

Mill, John Stuart. 1865. *Auguste Comte and Positivism.* London: Trübner & Co.

Millikan, R. A. 1916. "A Direct Photoelectric Determination of Planck's '*h*.'" *Physical Review* 7, no. 3, 355–388.

Millikan R. A. 1917. *The Electron. Chicago*: University of Chicago Press.

Mills, A. A. 1981. "Newton's Prisms and His Experiments on the Spectrum." *Notes and Records of the Royal Society of London* 36, no. 1, 13–36.

Motl, Lubos. 2011. "How Classical Fields, Particles Emerge from Quantum Theory." Accessed April 11, 2019. https://motls.blogspot.com/2011/11/how-classical-fields-particles-emerge.html.

Murdoch, Dugald. 1987. *Niels Bohr's Philosophy of Physics*. Cambridge, UK: Cambridge University Press.

Myrvold, Wayne, Marco Genovese, and Abner Shimony. 2019. "Bell's Theorem." *Stanford Encyclopedia of Philosophy* Archive. https://plato.stanford.edu/archives/fall2021/entries/bell-theorem/.

"The Mystery of the Multiverse." 2022. Institute of Art and Ideas, uploaded October 6, 2022. https://iai.tv/video/the-mystery-of-the-multiverse.

Narasimhan, T. N. 1999. "Fourier's Heat Conduction Equation: History, Influence and Connections." *Reviews of Geophysics* 37, no. 1 (February), 151–172.

Nasar, Sylvia. 1998. *A Beautiful Mind: A Biography of John Forbes Nash, Jr., Winner of the Nobel Prize in Economics*. New York: Simon & Schuster.

Nash, John. 1994. Nobel Prize biographical essay. NobelPrize.org. https://www.nobelprize.org/prizes/economic-sciences/1994/nash/biographical/.

Nash, John. 2015. *Letter to J. Robert Oppenheimer*. Institute for Advanced Study. https://www.ias.edu/ideas/2015/john-forbes-nash-jr.

Nash, John, Jr. 2003. "An Interesting Equation." John F. Chemerda Lecture in Science. https://web.math.princeton.edu/jfnj/texts_and_graphics/Main.Content/An_Interesting_Equation_and_An_Interesting_Possibility/An_Interesting_Equation.general.vac/From.PennState/intereq.r.pdf.

Newton, Isaac. 1958. *Isaac Newton's Papers and Letters on Natural Philosophy*. Edited by I. Bernard Cohen. Cambridge, MA: Harvard University Press.

Nichols, E. L. 1933. "Sidelights on the Era of Young and Fresnel." *Journal of the Optical Society of America* 23, 1–6.

Niels Bohr. 2023. Wikipedia. Revised March 8, 2010. https://en.wikipedia.org/wiki/Niels_Bohr.

Nietzsche, Friedrich. (1888) 1990. *Twilight of the Idols*. Translated by R. J. Hollingdale. London: Penguin Classics.

Nieuwenhuizen, Th. M. 2009. "Where Bell Went Wrong." AIP Conference Proceedings 1101, no. 1 (March), 127–133.

NobelPrize.org. 2022. *The Nobel Prize in Physics 2022 Press Release*. October 4. https://www.nobelprize.org/prizes/physics/2022/press-release/.

Nordén, Bengt. 2016. "Quantum Entanglement: Facts and Fiction—How Wrong Was Einstein After All?" *Quarterly Reviews of Biophysics* 49, e17.

Norton, John D. 2022. "The Measurement Problem." Part of the Spring 2022 course materials for Einstein for Everyone. Last updated April 19, 2022. https://sites.pitt.edu/~jdnorton/teaching/HPS_0410/chapters/quantum_theory_measurement/index.html.

O'Connor, J. J., and E. F. Robertson. 1997. "Jean Baptiste Joseph Fourier." Biography on MacTutor, accessed October 14, 2023. https://mathshistory.st -andrews.ac.uk/Biographies/Fourier/.

O'Connor, Timothy, and Christopher Franklin. 2022. "Free Will." *Stanford Encyclopedia of Philosophy Archive*. https://plato.stanford.edu/archives/win2022 /entries/free-will/.

Ogawa, Naohisa, and Yoshinori Furukawa. 1999. "Waves on Icicles." arXiv preprint cond-mat/9907381.

Ohanian, Hans C. 1986. "What Is Spin?" *American Journal of Physics* 54, no. 6 (June), 500–505.

O'Regan, Kevin. 2010. "Ancient Visions." April 22. http://nivea.psycho.univ-paris5 .fr/FeelingSupplements/AncientVisions.htm.

Pais, Abraham. 1991. *Niels Bohr's Times, in Physics, Philosophy, and Polity.* Oxford, UK: Oxford University Press.

Pais, Abraham. 2005. *Subtle Is the Lord: The Science and the Life of Albert Einstein.* Oxford, UK: Oxford University Press.

Patterson, Nancy. 2022. *Quantum Physics and the Power of the Mind.* Morrisville, NC: Lulu.com.

Peacock, George. 1855. *Life of Thomas Young.* London: John Murray.

Peña, Luis de la, Ana María Cetto, and T. A. Brody. 1972. "On Hidden-Variable Theories and Bell's Inequality." *Lettere al Nuovo Cimento* 5, no. 2, 177–181.

Perrard, Stéphane, Matthieu Labousse, Marc Miskin, Emmanuel Fort, and Yves Couder. 2014. "Self-Organization into Quantized Eigenstates of a Classical Wave-Driven Particle." *Nature Communications* 5, 3219 (January).

Philippidis, Chris, Chris Dewdney, and Basil J. Hiley. 1979. "Quantum Interference and the Quantum Potential." *Il Nuovo Cimento* B (1971–1996) 52, 15–28. https://doi.org/10.1007/BF02743566.

"Physics of Gas Flow at Very High Speeds." 1956. *Nature* 178, no. 4529 (August), 343–345.

Pitowsky, Itamar. 1989. "From George Boole to John Bell—the Origins of Bell's Inequality." In *Bell's Theorem, Quantum Theory and Conceptions of the Universe*, edited by Menas Kafatos, 37–49. Fundamental Theories of Physics, vol. 37. Dordrecht: Springer.

Plato. 1929. *Timaeus. Critias. Cleitophon. Menexenus. Epistles.* Translated by R. G. Bury. Loeb Classical Library 234. Cambridge, MA: Harvard University Press.

Poinier, Porter. 1877. "History of the Dynamical Theory of Heat." *Popular Science Monthly* 12 (December). Available online at https://en.wikisource.org /wiki/Popular_Science_Monthly/Volume_12/December_1877/History _of_the_Dynamical_Theory_of_Heat_I.

Pólya, George. 1954. *Mathematics and Plausible Reasoning: Vol. I: Induction and Analogy in Mathematics.* Princeton, NJ: Princeton University Press.

Pólya, George. 1957. *How to Solve It: A New Aspect of Mathematical Method.* Second edition. Princeton, NJ: Princeton University Press.

Popper, Karl R. 1992. *Quantum Theory and the Schism in Physics*. New York: Routledge.

Priestley, Joseph. 1796. *Considerations on the Doctrine of Phlogiston and the Decomposition of Water*. Philadelphia, PA: Thomas Dobson.

Prosser, R. D. 1976. "Quantum Theory and the Nature of Interference." *International Journal of Theoretical Physics* 15, no. 3, 181–193.

Putnam, Hilary. 2005. "A Philosopher Looks at Quantum Mechanics (Again)." *British Journal for the Philosophy of Science* 56, no. 4 (December), 615–634.

"Quantum Tunnelling." 2023. Wikipedia. Last updated November 8, 2023. https://en.wikipedia.org/wiki/Quantum_tunnelling.

Redhead, M. L. G. 1977. "Wave-Particle Duality." *British Journal for the Philosophy of Science* 28, 65–80.

Reitman, Ivan. 1984. *Ghostbusters*. Columbia Pictures.

"The Remarkable Story Behind the Most Important Algorithm of All Time." 2022. Veritasium, November 3, 2022. https://www.youtube.com/watch?v=nmgFG7PUHfo.

Rigden, John S. 2005. "Einstein's Revolutionary Paper." Physics World, April 1. https://physicsworld.com/a/einsteins-revolutionary-paper/.

Robinson, Andrew. (2006) 2023. *The Last Man Who Knew Everything: Thomas Young*. Cambridge, UK: Open Book Publishers.

Roller, Duane. 1961. *The Early Development of the Concepts of Temperature and Heat*. Cambridge, MA: Harvard University Press.

Romero, Simon. 2016. "Nearly in Ruins: The Church Where Sages Dreamed of a Modern Brazil." *New York Times*, December 25. https://www.nytimes.com/2016/12/25/world/americas/nearly-in-ruins-the-church-where-sages-dreamed-of-a-modern-brazil.html.

Romm, James. 1994. "A New Forerunner for Continental Drift." *Nature* 367, 407–408.

Rosenblum, Bruce, and Fred Kuttner. 2006. *Quantum Enigma: Physics Encounters Consciousness*. First edition. Oxford, UK: Oxford University Press.

Rosenfeld, Léon. 1963. "Niels Bohr's Contribution to Epistemology." *Physics Today* 16, no. 10, 47–54.

Rovelli, Carlo. 1996. "Relational Quantum Mechanics." *International Journal of Theoretical Physics* 35, 1637–1678.

Rovelli, Carlo. 2021. *Helgoland*. New York: Riverhead Books.

Russell, Bertrand. 1967. *The Autobiography of Bertrand Russell*. New York: Routledge.

Rutherford, Ernest. 1938. "Forty Years of Physics." In *Background to Modern Science,* edited by Joseph Needham and Walter Pagel, 49–76. Cambridge, UK: Cambridge University Press.

Sáenz, Pedro J., Tudor Cristea-Platon, and John W. M. Bush. 2018. "Statistical Projection Effects in a Hydrodynamic Pilot-Wave System." *Nature Physics* 14, no. 3, 315–319.

Sagan, Carl, and Ann Druyan. 1997. *Comet*. New York: Random House.

Sand-Jensen, Kaj. 2007. "How to Write Consistently Boring Scientific Literature." *Oikos* 116, no. 5, 723–727.

Saunders, Simon, Jonathan Barrett, Adrian Kent, and David Wallace, eds. 2012. *Many Worlds? Everett, Quantum Theory, and Reality.* New York: Oxford University Press.

Scheibe, Erhard. 1973. *The Logical Analysis of Quantum Mechanics.* Translated by D. B. Sykes. Oxford, UK: Pergamon Press.

Schilpp, Paul Arthur, ed. 1949. *Albert Einstein: Philosopher-Scientist.* Evanston, IL: The Library of Living Philosophers.

Schrödinger, Erwin. 1926. "Typed Letter Signed ('E. Schrödinger') to [Wilhelm Wien: 'Hochverehrter Herr Geheimrat!'], Zurich, 21 October, 1926." In German. https://onlineonly.christies.com/s/shoulders-giants-brief-history-big-ideas /disagreeing-niels-bohr-20/77152.

Schrödinger, Erwin. 1940. Letter to A. S. Eddington, March 22, 1940. *Österreichische Zentralbibliothek für Physik.* Scanned version available online at https:// www.iqoqi-vienna.at/blogs/blog/schroedingers-letter-to-edington-reconciling -machs-and-boltzmanns-philosophy.

Schrödinger, Erwin. 1983. "The Present Situation in Quantum Mechanics: A Translation of Schrödinger's 'Cat Paradox' Paper." In *Quantum Theory and Measurement,* edited by John Archibald Wheeler and Wojciech Hubert Zurek, 152–167. Princeton, NJ: Princeton University Press.

Scully, Marlan O. 1998. "Do Bohm Trajectories Always Provide a Trustworthy Physical Picture of Particle Motion?" *Physica Scripta* T76, 41–46.

Segrè, Gino. 2007. *Faust in Copenhagen: A Struggle for the Soul of Physics.* New York: Viking.

Selleri, Franco. 1990. *Quantum Paradoxes and Physical Reality.* Edited by Alwyn Merwe. Fundamental Theories of Physics 35. Dordrecht: Springer.

Sen, D., A. N. Basu, and S. Sengupta. 1995. "Wave-Particle Duality and Bohr's Complementarity Principle in Quantum Mechanics." *Current Science* 69, no. 5, 426–433.

Sen, Paul. 2022. *Einstein's Fridge: The Science of Fire, Ice and the Universe.* Glasgow, Scotland: William Collins.

Seneca, Lucius Annaeus. 1910. *Physical Science in the Time of Nero: Being a Translation of the Quaestiones naturales of Seneca.* Translated by John Clarke and Archibald Geikie. London: Macmillan.

Shukla, Khemendra, Po-Sung Chen, Jun-Ren Chen, Yu-Hsuan Change, and Yi-Wei Liu. 2020. "Macroscopic Matter Wave Quantum Tunnelling." *Communications Physics* 3, 101. https://doi.org/10.1038/s42005-020-0371-x.

Sklar, Lawrence. 1992. *Philosophy of Physics.* New York: Oxford University Press.

Smith, A. Mark. 1996. "Ptolemy's Theory of Visual Perception: An English Translation of the 'Optics' with Introduction and Commentary." *Transactions of the American Philosophical Society* 86, no. 2, iii–300.

Smolin, Lee. 2006. *The Trouble with Physics: The Rise of String Theory, the Fall of a Science, and What Comes Next.* New York: Houghton Mifflin.

Soler, Léna. 2009. "The Convergence of Transcendental Philosophy and Quantum Physics: Grete Henry-Hermann's 1935 Pioneering Proposal." In *Constituting Objectivity: Transcendental Perspectives on Modern Physics*, edited by Michel Bitbol, Pierre Kerszberg, and Jean Petitot, 329–344. Berlin: Springer.

Stapp, Henry Pierce. (1971) 1986. "S-Matrix Interpretation of Quantum Theory." In *Foundations of Quantum Mechanics Since the Bell Inequalities: Selected Reprints*, edited by Leslie Ballentine. College Park, MD: American Association of Physics Teachers.

Stapp, Henry Pierce. (1972) 1986. "The Copenhagen Interpretation." In *Foundations of Quantum Mechanics Since the Bell Inequalities: Selected Reprints*, edited by Leslie Ballentine. College Park, MD: American Association of Physics Teachers.

Stewartson, Keith. 1981. "D'Alembert's Paradox." *SIAM Journal on Applied Mathematics* 23, no. 3, 308–343.

Suplee, Curt. 2002. *Physics in the 20th Century*. New York: Harry Abrams.

Sweatman, Will. 2015. "Quantum Mechanics in Your Processor: Complementarity." Hackaday, July 24, 2015. https://hackaday.com/2015/07/24/quantum -mechanics-in-your-processor-complementarity/.

Szanton, Andrew. 1992. *The Recollections of Eugene P. Wigner*. Berlin: Springer.

Szymborski, Kris. 1989. "Bibliographie critique." *AIHS* 39, 162–163.

Taylor, G. I. 1909. "Interference Fringes with Feeble Light. *Proceedings of the Cambridge Philosophical Society* 15, 114–115.

Through the Wormhole. 2011. Season 2, episode 7, "How Does the Universe Work?" Aired July 20, 2011, on Science Channel.

Truesdell, Clifford A. 1980. *The Tragicomical History of Thermodynamics, 1822–1854*. New York: Springer.

Tzu, Ram. 1990. *No Way: A Guide for the Spiritually "Advanced."* Redondo Beach, CA: Advaita Press.

Vaidman, Lev. 2021. "Many-Worlds Interpretation of Quantum Mechanics." *Stanford Encyclopedia of Philosophy* Archive. https://plato.stanford.edu/archives /fall2021/entries/qm- manyworlds/.

Valentini, Antony. 2019. "Scientific Realism—Lecture by Prof. Antony Valentini." IQOQI Vienna, February 5. https://www.youtube.com/watch?v=t9 -m0ImOFyk.

Valentini, Antony. 2000. "Lectures on Scientific Realism: Does a Reality to Be Described by Science Exist?" IQOQI Vienna. https://www.youtube.com /watch?v=t9-m0ImOFyk.

Vickers, B. 1984. "Analogy Versus Identity: The Rejection of Occult Symbolism, 1580–1680." In *Occult and Scientific Mentalities in the Renaissance*, edited by B. Vickers, 95–163. Cambridge, UK: Cambridge University Press.

Virgil. 1999. *Eclogues. Georgics. Aenid: Books 1–6*. Loeb Classical Library. Cambridge, MA: Harvard University Press.

von Baeyer, Hans Christian. 2016. *Future of Quantum Physics*. Cambridge, MA: Harvard University Press.

von Neumann, John. 2018. *Mathematical Foundations of Quantum Mechanics.* Edited by Nicholas A. Wheeler. Translated by Robert T. Beyer. New edition. Princeton, NJ: Princeton University Press.

von Weizsäcker, Carl Friedrich. 1988. *Ideas on the Philosophy of Science: The Meaning of Quantum Mechanics, the Political and Moral Consequences of Science.* Transcription of lectures held at CERN, Geneva, January 1988. European Organization for Nuclear Research. https://cds.cern.ch/record/190270 /files/SCAN-0012007.pdf.

Vorob'ev, N. N. 1962. "Consistent Families of Measures and Their Extensions." *Theory of Probability and Its Applications* 7, no. 2, 153–169.

Walker, Jearl. 1978. "Drops of Liquid Can Be Made to Float on the Liquid. What Enables Them to Do So?" *Scientific American* 238, no. 6, 151–158.

Walsh, Lynda. 2013. *Scientists as Prophets: A Rhetorical Genealogy.* New York: Oxford University Press.

Weinberg, Steven. 1993. *Dreams of a Final Theory: The Scientist's Search for the Ultimate Laws of Nature.* New York: Vintage.

Weinberg, Steven. 2017. "The Trouble with Quantum Mechanics." *New York Review of Books*, January 19. https://www.nybooks.com/articles/2017/01/19 /trouble-with-quantum-mechanics/.

Wells, P. 2013. "Perimeter Institute and the Crisis in Modern Physics." *MacLean's*, September 5. https://www.math.columbia.edu/~woit/wordpress/?p=6238.

Wheeler, John A. 1963. "No Fugitive and Cloistered Virtue—a Tribute to Niels Bohr." *Physics Today* 16, no. 1, 30–32.

Wheeler, John A. 1985. "Physics in Copenhagen in 1934 and 1935." In *Niels Bohr: A Centenary Volume*, edited by A. P. French and P. J. Kennedy, 221–226. Cambridge, MA: Harvard University Press.

Wheeler, John A. 1986. "How Come the Quantum?" *Annals of the New York Academy of Sciences* 480, no. 1 (December), 304–316.

Wheeler, John A. (1983) 2016. "Law Without Law." In *Quantum Theory and Measurement*, edited by John Archibald Wheeler and Wojciech Hubert Zurek, 182–213. Princeton Series in Physics. Princeton, NJ: Princeton University Press.

Wheeler, John A., and Mirjana Gearhart. 1979. "Forum: John A. Wheeler—from the Big Bang to the Big Crunch." *Cosmic Search* 1, no. 4 (Fall).

Whittaker, E. T. 1989. *A History of the Theories of Aether and Electricity.* New York: Dover Publications.

Wick, David. 1995. *The Infamous Boundary: Seven Decades of Controversy in Quantum Physics.* Boston, MA: Birkhäuser.

Wigner, Eugene. 1967. *Symmetries and Reflections.* Bloomington: Indiana University Press.

"Wigner's Friend." 2023. Wikipedia, revised November 4, 2023. https:// en.wikipedia.org/wiki/Wigner%27s_friend.

Winer, Gerald A., Jane E. Cottrell, Virginia Gregg, Jody S. Fournier, and Lori A. Bica. 2002. "Fundamentally Misunderstanding Visual Perception: Adults' Belief in Visual Emissions." *American Psychologist* 57, no. 6/7, 417–424.

Wiseman, Howard. 2014. "Bell's Theorem Still Reverberates." *Nature* 510, 467–469.

Woit, Peter. 2006. *Not Even Wrong: The Failure of String Theory and the Search for Unity in Physical Law.* New York: Basic Books.

Wolchover, Natalie. 2014. "Fluid Tests Hint at Concrete Quantum Reality." *Quanta Magazine*, June 24. https://www.quantamagazine.org/fluid-experiments -support-deterministic-pilot-wave-quantum-theory-20140624/.

Wolchover, Natalie. 2018. "Famous Experiment Dooms Alternative to Quantum Weirdness." *Quanta Magazine*, October 11. https://www.quantamagazine.org /famous-experiment-dooms-pilot-wave-alternative-to-quantum-weirdness -20181011/.

Wolff, Milo, and Geoff Haselhurst. 2004. "Light and the Electron—Einstein's Last Question." Presentation at Beyond Einstein, Stanford University. https:// citeseerx.ist.psu.edu/document?repid=rep1&type=pdf&doi=07be736be90adf 14f3bcf6c14dfcaa8162f63602.

Wollaston, William Hyde. 1802. "A Method of Examining Refractive and Dispersive Powers, by Prismatic Reflection." *Philosophical Transactions of the Royal Society of London* 92, 365–380.

Young, Thomas. 1855. *Miscellaneous Works of the Late Thomas Young.* Volume 1. Edited by George Peacock. London: John Murray.

Zajonc, Arthur. 1995. *Catching the Light: The Entwined History of Light and Mind.* New York: Oxford University Press.

Zubairy, M. Suhail. 2016. "A *Very* Brief History of Light." In *Optics in Our Time*, edited by Mohammad D. Al-Amri, Mohamed El-Gomati, and M. Suhail Zubairy, 3–24. Cham, Switzerland: Springer.

Index